ADVANCES IN BRYOLOGY 3

Advances in Bryology

A Publication of the International Association of Bryologists

Volume 3

Bryophyte Ultrastructure

Editor

Norton G. Miller, Biological Survey, New York State Museum, Albany, U.S.A.

Editorial Board

J. CRAMER

in der Gebrüder Borntraeger Verlagsbuchhandlung

BERLIN · STUTTGART 1988

Advances in Bryology

A Publication of the International Association of Bryologists

Advances in Bryology contains review articles summarizing progress in all aspects of research involving bryophytes. The reviews are written both as an evaluation of present knowledge and with reference to contemporary advances in other botanical and biological disciplines. Contributions are by invitation, although the editor welcomes suggestions for suitable papers from interested readers. Papers are published in English with German and English abstracts. An effort will be made to publish one volume every two years. Papers beginning with volume 3 are organized around a theme.

Norton G. Miller
Editor, *Advances in Bryology*
Biological Survey
New York State Museum
Albany, NY 12230 U.S.A.

Table of Contents

Preface

The six papers comprising volume 3 of *Advances in Bryology* treat various aspects of bryophyte ultrastructure. Together with the paper, "Desiccation and Ultrastructure in Bryophytes," by M. J. Oliver and J. D. Bewley in volume 2 of the *Advances* (1984), they provide a comprehensive summary of current investigations into the ultrastructure of mosses, liverworts, and hornworts. Developmental, functional, and descriptive aspects of the featured structures and organelles receive emphasis in these papers, which contain a wealth of comparative data that is useful in assessments of the phylogenetic relationships of bryophytes. As a number of my colleagues continue to stress, bryophytes offer some unique advantages for research at the cellular and subcellular levels of organization. Ample documentation of this will be found in the papers that follow.

Volumes 1 and 2 of the *Advances in Bryology* were published under the imprint of J. Cramer, but owing to Herr Cramer's untimely death, Gebr. Borntraeger Verlagsbuchhandlung will publish succeeding volumes. I thank Dr. E. Nägele of the Gebr. Borntraeger publishing house for his assistance and support in this endeavor. Volumes 1 and 2 appeared under the editorship of Wolfram Schultze-Motel who ably guided the *Advances* through its birth and establishment. Assisting him was an Editorial Board, the members of which are thanked for their service. Upon my selection as Editor, a new Editorial Board was appointed by Hiroshi Inoue, President of the International Association of Bryologists. The new Board has had an important role in setting policy and in the review process. To all those individuals who served as reviewers of papers in volume 3, a mixture of Editorial Board members and others, I offer my deep appreciation. A special acknowledgment is also extended to Patricia Geissler and Irene Bisang for translating the English abstracts into German.

Norton G. Miller

Albany, New York
4 September 1987

Advances in Bryology 3: 1 – 31 (1988).

The Bryophyte Cytoskeleton: Experimental and Immunofluorescence Studies of Morphogenesis

by

John H. Doonan

Pharmacology Department, Robert Wood Johnson Medical School, Busch Campus, Piscataway, New Jersey, U.S.A.

and

Jeffrey G. Duckett

School of Biological Sciences, Queen Mary College, Mile End Road, London E1 4NS, U.K.

With 32 Figures

Abstract: This article reviews the impact of recent immunological investigations and earlier electron microscopy studies of the role of the cytoskeleton in bryophyte morphogenesis. Particular attention is focused on experimental studies of the protonemata of Funariales. Two-dimensional gel electrophoresis of α and β tubulins from the moss *Physcomitrella* reveal these to have similar biochemical properties to those of flowering plants. Antibody labelling of microtubules and microfilaments provide new insights into protonemal morphogenesis. Experiments with unidirectional illumination, calcium gradients, and colchicine indicate that highly ordered microtubule turnover is essential for the establishment of polarity during spore germination and protoplast regeneration. Whereas conventional electron microscopy has failed to reveal microtubules and microfilaments in the apical domes of elongating protonemata, these are clearly demonstrated by fluorescence microscopy. Both networks persist throughout the cell cycle. Studies using antimicrotubule drugs and cytochalasin D suggest that microfilaments are essential to protonemal development, with microtubules imparting directionality to the process. It would seem likely that further studies will reveal microfilaments to be widespread in bryophytes and that these probably constitute the force-generating elements responsible for many of the organelle movements and shaping processes characteristic of sporogenesis and spermatogenesis. Mitosis in protonemata is highly asymmetrical. Dynamic arrays of microtubules are involved in pre- and post-mitotic nuclear movements and determine the sites of side branch formation. Formation of leaflets by buds involves profound changes in microtubule organization. Endoplasmic microtubule arrays, characteristic of protonemata, largely disappear, and symmetrical spindles lack associated cytoplasmic microtubules. Prior to spindle formation, cortical microtubules are replaced by pre-prophase bands, which are never found in protonemata. In leafy shoots of mosses the microtubule cycle is very similar to that in vascular plants. To date, correlative immunofluorescence and electron microscope data are largely limited to the Funariales. Observations should be extended to the hepatics and hornworts. Similarly, there is a pressing need to adopt new fixation protocols in order to preserve the more fragile domains of bryophyte cytoskeletons that are destroyed by conventional preparative procedures. An increasing range of immunological probes can now provide information on microtubule nucleating sites, the distribution of microtubule-associated proteins, including force-generating molecules, calcium regulation of cytoskeletal assembly, and permit discrimination between dynamic and stable microtubule arrays within the same cell. The future prospects for a comprehensive understanding of interrelationships between the cytoskeleton and bryophyte morphogenesis have never been more exciting.

Zusammenfassung: Dieser Beitrag bespricht die Bedeutung neuerer immunologischer und früherer elektronenmikroskopischer Untersuchungen über die Rolle des Cytoskeletts in der Morphogenese der Bryophyten. Besondere Aufmerksamkeit wird den experimentellen Arbeiten an Protonemata der Funariales geschenkt. Zweidimensionale Gelektrophorese von α und β Tubulin des Laubmooses *Physcomitrella* zeigt, daß dessen biochemische Eigenschaften jenen der Blütenpflanzen gleichen. Antikörpermarkierung der Microtubuli und Microfilamente ermöglichen neue Einsichten in die Entstehung des Protonemas. Experimente mit Beleuchtung aus einer Richtung, Calciumgradienten und Colchicin weisen darauf hin, daß eine hochgeordnete Umsetzung der Microtubuli wesentlich für die Entstehung der Polarität während der Sporenkeimung und Protoplastenregeneration ist. Mit indirekter Immunofluoreszens konnte deutlich gezeigt werden, daß Microtubuli und Microfilamente sich in den Scheiteln wachsender Protonemata befinden, was mit konventioneller Elektronenmikroskopie nicht möglich gewesen war. Beide Systeme sind während des ganzen Zellzyklus vorhanden. Untersuchungen, die Antimicrotubulimittel und Cytochalasin D verwenden, deuten darauf hin, daß die Microfilamente für die Protonemaentwicklung unerläßlich sind und daß die Microtubuli in diesem Prozeß für die Richtungsgebung eine Rolle spielen. Es könnte sich in Zukunft herausstellen, daß Microfilamente bei Moosen weitverbreitet sind und daß sie wahrscheinlich die treibende Kraft für viele Organellenbewegungen und formbildende Prozesse der Sporogenese und der Spermatogenese darstellen. Die Mitose in Protonemata ist deutlich asymmetrisch. Die dynamische Anordnung der Microtubuli in prä- und postmitotischen Kernbewegungen bestimmt die Lage der Seitenastbildung. Die Blattbildung in Knospen zieht tiefgreifende Veränderungen der Microtubuliorganisation nach sich. Endoplasmatische Anordnungen, wie sie für Protonemata kennzeichnend sind, verschwinden weitgehend und dymmetrische Spindeln fehlen den zugehörigen cytoplasmatischen Microtubuli. Vor Beginn der Spindelbildung werden die Rindenmicrotubuli durch Präprophasebänder ersetzt, welche in Protonemata nie gefunden wurden. In beblätterten Sprossen von Moosen ähnelt der Microtubulizyklus stark jenem der Gefäßpflanzen. Ergebnisse aus Untersuchungen mit

Immunofluoreszenz korreliert mit Elektronenmikroskopie liegen bis heute erst für die Funariales vor. Die Beobachtungen sollten auf die Leber- und Hornmoose ausgeweitet werden.
Ebenso ist es dringend notwendig, schonendere Fixierungsmethoden als die konventionellen
zu finden, um die empfindlicheren Teile des Mooscytoskeletts zu erhalten. Ein ständig
wachsendes Angebot an immunologischen Tests verspricht neue Informationen über die
Entstehungsart der Microtubuli, die Verteilung der damit assoziierten Proteine unter Einschluß der energieerzeugenden Moleküle, der Calciumregulation des gesamten Cytoskeletts
und erlaubt somit eine Unterscheidung zwischen dynamischer und stabiler Microtubulinanordnung innerhalb derselben Zelle. Die Aussichten auf ein besseres Verständnis von Zusammenhängen zwischen Cytoskelett und Morphogenese bei Moosen waren noch nie so groß.

Keywords: antimicrotubule drugs, bryophytes, cytoskeleton, immunofluorescence, microfilaments, microtubules, mitosis, morphogenesis, phragmoplast, pre-prophase bands, protonemata, protoplasts, spore germination, taxol, tubulins.

Contents

Introduction

The cytoplasm of eukaryotic cells contains a highly structured network of fibrous proteins collectively known as the cytoskeleton (Alberts et al. 1983). The organization and dynamics of the cytoskeleton are fundamental to the control of spatial aspects of morphogenesis. A full understanding of the involvement of the cytoskeleton in determining the planes of cell division and directed cell expansion is of special importance in plants (Lloyd 1982, 1987) because development is restricted by the constraints imposed by a semi-rigid cell wall.

Electron microscope, biochemical, and, more recently, immunological studies have shown that the cytoskeleton of vertebrate cells comprises three distinct elements: microtubules, actin microfilaments, and intermediate filaments. As far as the plant kingdom is concerned, considerations of relationships between the cytoskeleton and morphogenesis have centered almost exclusively on microtubule (MT) dynamics, MTs being the only cytoskeletal elements that, until about five years ago, have been readily and widely recognized in plant cells. However there is now growing evidence from immunofluorescence studies (e.g., Clayton & Lloyd 1985, Williamson et al. 1986) and electron microscopy (Lancelle et al. 1986, Tiwari et al. 1984) using improved fixation protocols, that actin is not only widespread in plants but is also intimately involved in force-generating processes either in conjunction with microtubules (e.g., the spindle apparatus (Trass et al. 1987), spermatogenesis (Marc & Gunning 1986)) or as a separate system (e.g., cytoplasmic streaming (Shimmen & Yano 1986)). Intermediate filament antigens are also known from plants (Dawson et al. 1985, Parke et al. 1987), but their distribution and possible functions have yet to be explored.

Bryophyte materials have and will continue to make a major contribution to understanding the role of the cytoskeleton in plant morphogenesis. The reasons, which hardly need reiteration, are the special combination of morphological simplicity coupled with highly organized development, which can be accurately predicted and modified experimentally (Knoop 1984, Cove & Ashton 1984, Hartmann & Jenkins 1984). Furthermore, bryophytes contain all of the four principal arrays of microtubules found in vascular plants: cortical networks at interphase, preprophase bands, spindles, and phragmoplasts. To date, the well defined stages in the life cycle of members of the Funariales are an almost unrivalled system for the investigation of spatial and temporal regulation of cell division and cell elongation. Spermatogenesis on the other hand provides the rare opportunity to study the development of a microtubule system (the spline) in which the total number and length of each individual microtubule are precisely defined (Duckett et al. 1984, Duckett & Renzaglia, in preparation).

As with vascular plants, the bulk of our present knowledge of bryophyte cytoskeletons derives from transmission electron microscopy of sectioned material (reviewed by Duckett 1986). However, for the future immunofluorescence techniques (whereby the principal elements of the cytoskeleton, not to mention their associated components, can be investigated in whole cells in 3-dimensions) hold considerable advantages over conventional ultrastructural procedures. This is particularly true for studies on highly dynamic morphogenetic processes. Rather than reiterate conclusions from electron microscope studies, many of which are already the subject of recent reviews (e.g. Steer 1985 on the mitotic apparatus; Schnepf 1982 on protonemal morphogenesis; Duckett et al. 1982, 1984 on spermatogenesis; Duckett 1986 on bryophyte ultrastructure), our aim here is to focus attention on bryophyte cytoskeletons as revealed by immunocytochemical probes. Wherever appropriate, immunofluorescence data will be compared with earlier information from electron microscopy. We will highlight the most promising areas for future research and the special advantages bryophyte materials offer over other plant and animal systems.

Immunofluorescence studies of plants have lagged far behind those on animals, principally because of technical difficulties inherent in dealing with walled cells. However, a range of procedures is now available for making plant materials permeable to antibodies, while at the same time preserving the integrity of the cell contents (Doonan & Clayton 1987, Wick & Duniec 1986). These involve the use of non-ionic detergents, specially buffered fixatives (the majority adapted from recipes widely used on animal materials), and a variety of wall degrading enzymes. Although these procedures are effective on some bryophyte tissues, successful antibody labelling of others has not yet been achieved without an unacceptable level of tissue disruption. Fortuitously, the major events of protonemal morphogenesis and all stages in spermatogenesis fall into the former category. Mature leaf tissues and the early stages in spore germination are in the latter. Permeability to antibodies is greatly affected by culture conditions. For example, *Physcomitrella* protonemata grown on solid media are readily stained, whereas those grown in liquid cultures are not (Doonan, unpublished data). The wall chemistry responsible for these differences is not understood. In this context it is noteworthy that aquatic bryophytes and mosses grown in liquid cultures are among the most difficult of all biological materials to infiltrate with resins for electron microscopy.

A further problem that the advent of immunofluorescence studies and the concomitant search for improved fixation protocols have brought to light (for plant electron microscopy these have hardly changed since the early 1960s, Duckett & Renzaglia 1987b) is the realization that some parts of the cytoskeleton are much more sensitive to preparative procedures than others (Maupin & Pollard 1983). As we shall illustrate in this account, the cytoskeleton of bryophytes is far more extensive than that seen in the vast majority of electron micrographs published to date. We may thus anticipate, in the near future, the discovery and characterization of cytoskeletal elements in locations where they have previously been overlooked.

The Structure and Composition of Bryophyte Microtubules

Both in size (outside diameter 24 nm, inside 15 nm) and substructure (13 subunits following contrast enhancement with tannic acid or negative staining of isolated preparations) microtubules in bryophytes appear indistinguishable from those in other plants (Duckett et al. 1983).

The major microtubule protein tubulin, a heterodimer comprising one α and one β polypeptide, can be isolated easily from cells rich in tubulin by cycles of assembly into microtubules and disassembly into tubulin subunits (Weisenberg 1972, Soifer 1986). Isolation from plant tissues by *in vitro* assembly has proven to be much more difficult partly due to the lower concentrations of the protein. However, use of the microtubule-stabilizing compound taxol (Bajer et al. 1982, Hauser 1986, Horwitz et al. 1982) has led to the successful *in vitro* assembly of microtubules from protein extracts in a variety of flowering plants (Morejohn & Fosket 1982, Dawson & Lloyd 1985). Although capable of forming morphologically normal microtubules, plant tubulins display differences from animal tubulins, including drug sensitivity (Morejohn et al. 1984) and the relative electrophoretic mobility of the α and β polypeptides (Dawson & Lloyd 1985). A low-affinity interaction between colchicine and plant tubulin (Morejohn et al. 1987) could explain why colchicine concentrations in excess of 0.1 % are required in order to induce morphogenetic changes in *Funaria* protonemata (Schnepf 1982). Development of *Sphagnum* protonemata is unaffected by concentrations of colchicine of less than 0.25 % (Duckett et al. unpublished data).

When brain tubulin is separated on one-dimensional SDS-polyacrilamide gels, the α polypeptide migrates slightly behind the β. The reverse is true for carrot cell tubulin (Dawson & Lloyd 1985). *Physcomitrella* tubulins can be isolated by exploiting the resistance of cytoskeletal elements to dissolution by non-ionic detergents (Powell et al. 1980; Doonan & Lloyd, in preparation) as demonstrated in Fig. 1. Two-dimensional gel electrophoresis of protoplasts extracted with the non-ionic detergent Nonidet NP40, in the presence of taxol, reveals two clusters of polypeptides (Fig. 3). Western blotting (Towbin et al. 1979) of these gels (Fig. 3), or of total protein extracts (Fig. 2), shows that the tubulins of *Physcomitrella* possess similar electrophoretic properties to those of flowering plants rather than animal or fungal tubulins, viz., the α subunit migrates faster than the β. (Fig. 4).

Although some differences exist between tubulin proteins, particularly the α polypeptide (Dráber et al. 1986, Geuens et al. 1986, Gundersen & Bulinski 1986, Sasse et al. 1987) from various sources, antigenic determinants seem to be largely conserved. These common antigenic sites have proven extremely useful, permitting the

Figures 1–4. Immunological characterization of the cytoskeleton of *Physcomitrella patens*. **1,** The residual microtubule cytoskeleton of a protoplast, following extraction with the non-ionic detergent Nonidet NP40 in microtubule-stabilizing buffer, as revealed by indirect immunofluorescence using monoclonal anti-tubulin antibody YL1 1/2; note the randomly arranged cagework of microtubules and the diffusely staining nucleus (N); scale bar = 10 μm. **2, 3,** Protein profiles from protoplasts as separated by two-dimensional gel electrophoresis. **2,** Total protein; **3,** Detergent insoluble cytoskeletons prepared by Nonidet NP40 extraction of taxol-treated protoplasts; arrowhead denotes the RUBISCO subunit, α and β tubulin subunits are labelled accordingly. **4,** Western blot of total protoplast protein probed with monoclonal antibodies to α and β tubulin.

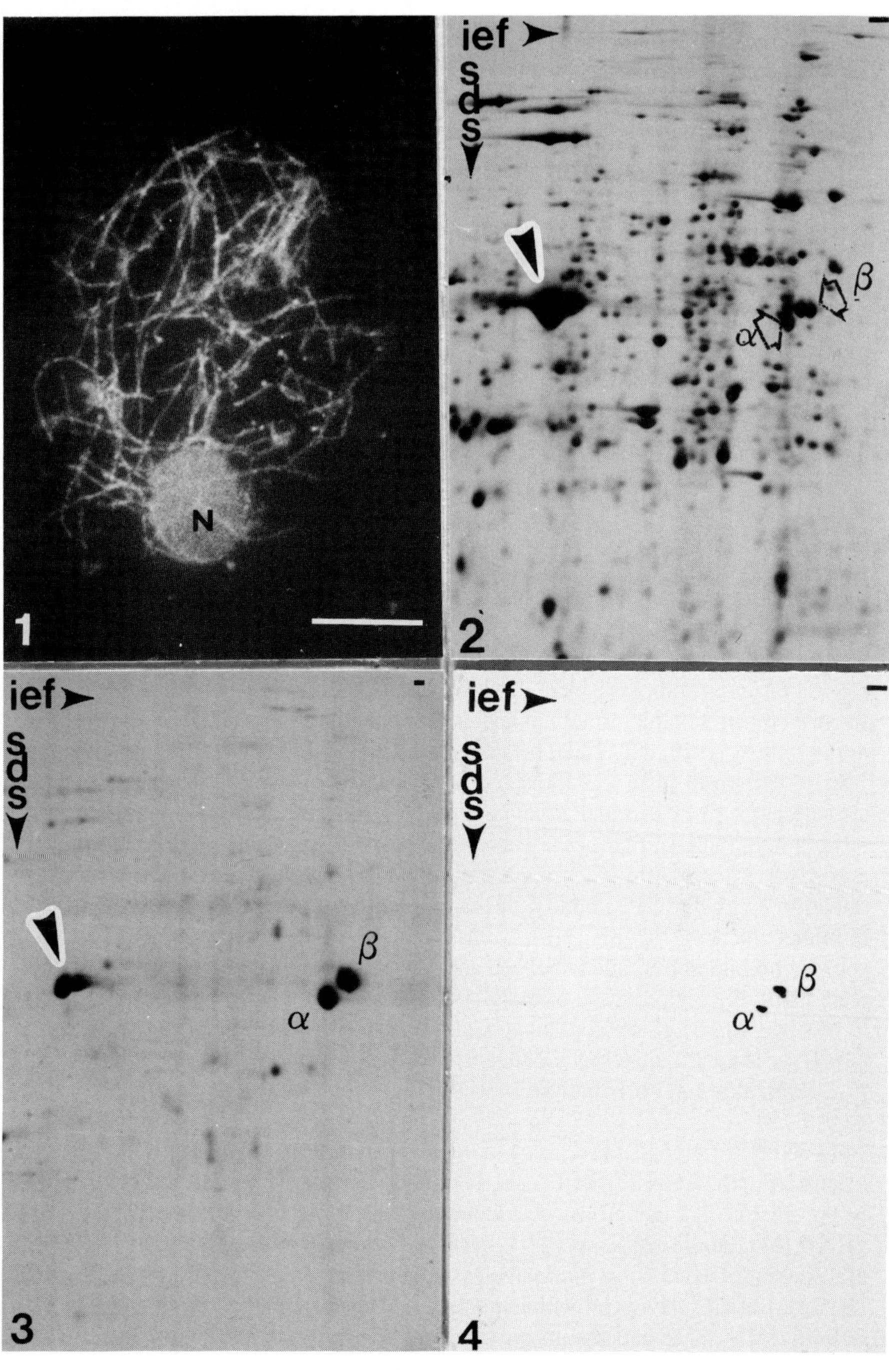

use of heterologous antisera and monoclonal antibodies for immunofluorescent staining in mosses. They may also be used to identify tubulin in biochemical preparations (Fig. 4.) Antisera raised against brain tubulin was used in the production of the first indirect immunofluorescent images of microtubules within intact protoplasts of *Physcomitrella* (Powell et al. 1980) and subsequently to examine spermatogenesis in *Sphagnum* and *Polytrichum* (Miller et al. 1983). Monoclonal anti-yeast α-tubulin antibodies (Kilmartin et al. 1981) have since been used in several studies tracing microtubular distribution during protonemal morphogenesis (Doonan et al. 1985, 1986, 1987). Most recently Mizuno et al. (1985) have produced a monoclonal antibody specific to mung bean α tubulins. This does not cross-react with brain tubulins but has allowed the visualization of all the microtubule arrays in a variety of plant cells including the sporocyte mother cells of *Funaria* (Mizuno et al. 1985). The antibody probably recognizes an epitope shared by most plant α tubulins, but one that is altered or inaccessible in mammalian tubulins.

Spore Germination: Establishment of Cellular Asymmetry

Moss protoplast regeneration is similar to spore germination in that a symmetrical cell is transformed into a polarized system capable of directed expansion. The orientation of a developmental axis both in *Funaria* spores (Schnepf et al. 1982) and in *Physcomitrella* protoplasts (Jenkins & Cove 1983, Burgess & Linstead 1981) is influenced by environmental stimuli and in particular a uni-directional light source. Stimuli produced artificially either by a concentration gradient of the calcium ionophore A23187 (Chen & Jaffe 1979) or an electric field (Burgess & Linstead 1982) may also affect the plane of spore germination and the axis of protoplast differentiation.

In light of these similarities it would seem reasonable to suppose that the cytoskeletal changes accompanying protoplast differentiation might mirror those during germination. Lacking walls initially, the former, unlike spores, are readily amenable to immunocytochemistry (Fig. 5–8). Ultrastructural examination of very early stages in spore germination has failed to reveal any obvious changes in the microtubules associated with the region of the primary outgrowth. Dormant bryophyte spores appear to be singularly bereft of microtubules, and these elements are first detected when the germ tube is approximately 5 µm in length (Schnepf et al. 1982). Microtubules, with predominantly axial orientations, become more numerous as the germ tube elongates.

In contrast to spores protoplasts of mosses contain extensive arrays of randomly oriented microtubules (Powell et al. 1980) (Fig. 5). But, as in spores, there is initially no obvious changes in the cytoskeleton of protoplasts associated with the outgrowth site (Burgess & Linstead 1981; Doonan, unpublished data). During the establishment of asymmetry microtubules become closely associated with the nucleus (Fig. 6). Initially the outgrowth contains randomly arranged microtubules (Fig. 7). Subsequently, these become progressively aligned parallel to the long axis, and the number of microtubules associated with the nucleus vastly increases as it migrates

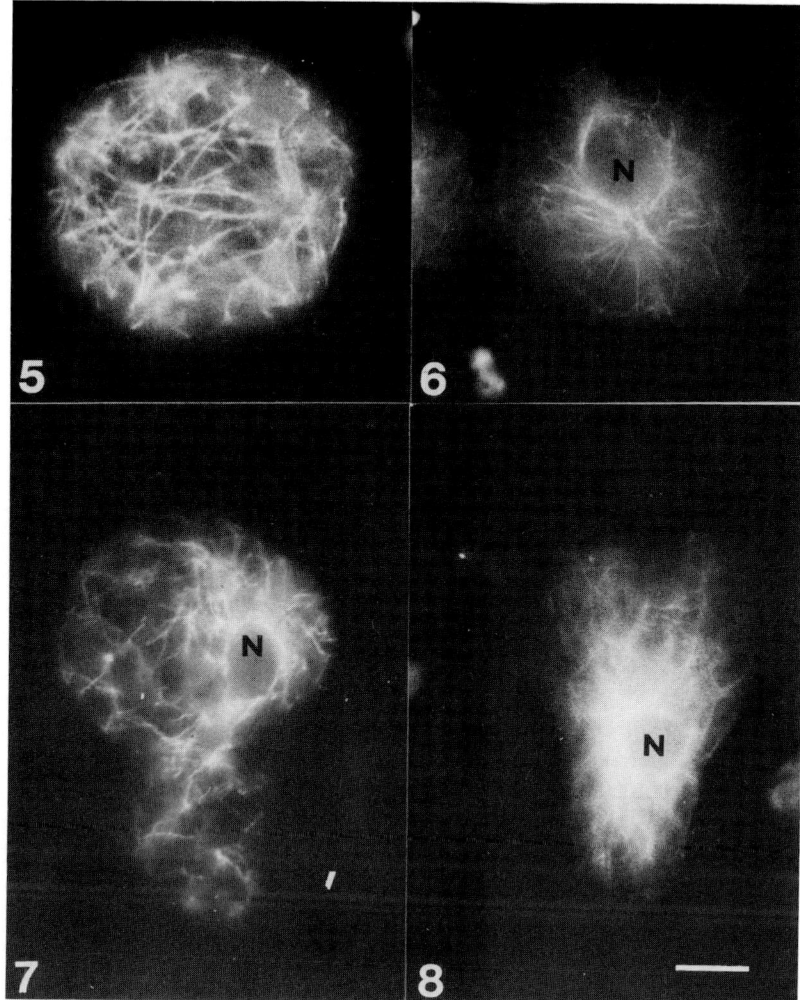

Figures 5–8. Changes in the microtubule arrays in protoplasts of *Physcomitrella* during differentiation, visualized with the anti-tubulin antibody YOL 1/34. Scale bar = 10 μm. **5,** Freshly isolated protoplasts contain a complex random array of cytoplasmic microtubules; **6,** As the protoplasts become asymmetrical, microtubules become associated with the nuclear surface; **7,** After 24 hours outgrowths are well developed; **8,** The nuclei begin to migrate into the outgrowths and the number of nuclear-associated microtubules is vastly increased.

into the outgrowth (Fig. 8). Burgess and Linstead (1981) record that in unidirectionally illuminated protoplasts the longitudinal arrays of microtubules frequently lie in close proximity to chloroplasts and mitochondria, as well as to the nucleus. The organelles themselves are elongated with their long axes parallel to that of the germ tube. Such polarized organization of protoplasts was not observed under conditions of uniform lighting, although the protoplasts did regenerate to form normal protonemata. Underlining the parallels between protoplast regeneration and spore germination, closely associated microtubules, migrating nuclei, and chloroplasts have been observed in electron micrographs of germlings of *Funaria* (Schnepf 1982, Schnepf et al. 1982).

Ultrastructural studies on fern gametophytes reveal microtubule distribution similar to that during bryophyte spore germination (Cran 1979, Kotenko et al. 1987). Microtubules are notably uncommon in fern gametophytes but have been observed around the nuclear envelope during migration phases and in parallel alignment to the long axis of elongating cells. More pertinent to the considerations of protonemal differentiation in mosses is the demonstration in ferns (Kotenko et al. 1987) of localized accumulations of Ca^{2+} at sites where asymmetric cytokinesis will occur. In the light of the well documented effects of Ca^{2+} on microtubule assembly (Margolis 1983) an investigation of relationships between the cytoskeleton and calcium gradients during moss spore germination would appear to be overdue. Bud formation in *Funaria,* which also involves nuclear migration and asymmetric cell division, can be induced by raising Ca^{2+} levels in the medium and adding ionophore A23187 (Saunders & Hepler 1982, 1983). Future studies should include investigation of the distribution of the ubiquitous calcium regulating protein calmodulin (Roberts et al. 1986), which immunofluorescence studies have shown to be closely associated with the cytoskeleton of vascular plants (Lin et al. 1986, Wick 1985, Wick et al. 1985).

Although various workers have failed to find distinctive microtubule arrangements at the site of protuberance formation in either spores or protoplasts (we are not aware, however, of any studies to date that have used fixation protocols specifically designed to preserve labile cytoskeletal elements), both processes are extremely sensitive to antimicrotubule drugs. When grown in the presence of 0.1 % colchicine, protoplasts of *Physcomitrella* no longer respond to unidirectional light, and the polarized organization of their cytoplasmic organelles is lost (Burgess & Linstead 1981). These authors were also unable to find microtubules in electron micrographs of colchicine-treated cells, although these continued to grow and underwent mitosis. Treatment of *Funaria* spores with colchicine causes the formation of spherical cells or abnormal irregular outgrowths (Schnepf 1981, 1982; Schnepf et al. 1982). Cell walls are irregular, indicating that wall material continues to be produced and secreted but that its delivery and deposition are perturbed. Agents that stabilize microtubules (e.g., heavy water) also produce irregular deposition of wall materials, and in *Funaria* cause the rhizoid cell to grow at variable angles with respect to the germ tube. These effects involving both stabilization and destabilization of microtubules clearly indicate that highly ordered microtubule turnover is essential to the establishment of polarity and maintenance of a growing point in spore germination.

Protonemal Elongation: Maintaining a Polar Axis

The hemispherical dome of the *Funaria* caulonema cell is the primary site of cell expansion (Schmiedel & Schnepf 1980). Tip- or apical-growth is found in a variety of plant cells including root hairs, pollen tubes, fungal hyphae, and fern protonemata, and all have many cytological features in common.

The ultrastructure of protonemal apical cells has been described for *Funaria* (Schmiedel & Schnepf 1980), *Physcomitrium* (Jensen & Jensen 1984), and *Dawsonia* (De Maggio & Stetler 1977). The apical dome is devoid of large organelles but rich in Golgi-derived vesicles. This region appears as clear cytoplasm (Fig. 9) when viewed by optical microscopy (Jensen 1981). Behind the apical dome is a region rich in Golgi bodies, mitochondria, and endoplasmic reticulum with characteristic distribution profiles (Jensen & Jensen 1984, Schmiedel & Schnepf 1980, Schnepf 1982). Time-lapse video microscopy of *Physcomitrium* (Jensen 1981) has revealed that organelle distribution is extensively modified during the cell cycle. Elongation of the protonemal cells (just as in fungal hyphae, root hairs, and pollen tubes) is thought to result from preferential discharge of Golgi-derived vesicles at the surface of the apical dome. Together, the polar distribution of vesicles, Golgi bodies, and cell expansion are indicative of a directional transport system. Such a characteristic distribution of the organelles points to the existence of scaffold and force-generating systems that are capable of integrating cellular organization.

It is surprising, therefore, that electron microscopy, albeit using conventional fixatives, has failed to reveal a cytoskeletal system within the apical domes of a variety of tip-growing cells (Doonan & Clayton 1987), including *Funaria* protonemata. Thus the possible role of microtubules in tip-growth has remained a vexed question, although it is well documented that microtubule perturbing agents cause dramatic growth abnormalities in protonemata just as they do in spores (Schnepf 1981, 1982).

However, the recent use of microtubule stabilizing buffers during fixation combined with indirect immunofluorescence (Doonan et al. 1985) clearly demonstrates the presence of microtubules within the apical dome of *Physcomitrella* caulonemata and a variety of other mosses including *Sphagnum* (Doonan & Duckett, unpublished data) where they form characteristic foci at the tip (Fig. 10). The same technique has also revealed microtubules extending to the tips of root hairs (Lloyd 1984, Lloyd et al. 1985), pollen tubes (Pierson et al. 1986), and fungal hyphae (Hoch & Staples 1985, Runeberg et al. 1986). In protonemal cells these microtubules remain in the apical dome throughout mitosis (Fig. 12, 13). As might be anticipated, staining patterns with antitubulin are modified greatly by anti-microtubule drugs (Doonan et al. 1985). The herbicide CIPC (isopropyl-N(3-chlorophenyl)-carbamate) disrupts the microtubule focus in the apical dome. This disruption is associated with abnormalities in growth, including bending at the tip. At higher concentrations of CIPC microtubules disappear completely from the apical dome, and sub-apical sites of cell expansion are initiated. These often form "aborted apices" (Doonan et al. 1985).

Not only microtubules, but also an extensive microfilament system is present in the apical cells of moss protonemata (Doonan et al. 1988). The MF system is extremely sensitive to aldehyde fixation, even using stabilizing buffers, but can be visualized by very gentle permeabilization, which avoids osmotic shock, in conjunction with rhodamine-conjugated phalloidin (Fig. 11). Similar actin networks are also known from pollen tubes (Pierson et al. 1986) and have been particularly well characterized in fungal hyphal tips by use of phalloidin or phallacidin (Hoch & Staples 1985, Runeberg et al. 1986, Tucker et al. 1986).

The partial removal or elimination of the microfilament system in protonemal cells with cytochalasin D slows down or stops growth but does not induce abnormal growth. By contrast the total removal of microtubules using cremart, which leaves the microfilaments intact, causes swelling of the protonemal tips, i.e., undirected cell expansion. These results imply that microfilaments are essential for protonemal growth, but microtubules impart direction to this process. It is interesting that Runeberg et al. (1986) reach the same conclusion about these cytoskeletal elements in fungal hyphae "microtubules appear to be involved in the regulation of the location and movement of various organelles while actin-containing structures are necessary at places where new cell wall material is deposited." At present the nature of these microtubule-microfilament interactions is unknown, but for future studies on cytoskeletal dynamics in bryophytes the implications are clear. These must take full account of the latest works on fungal morphogenesis and cytoskeletal interactions in animal systems.

The recent demonstration of actin by immunofluorescence microscopy of protonemata is perhaps not unexpected in the light of the increasing documentation of the widespread nature of this cytoskeletal element in plants (Jackson 1982, Tiwari et al. 1984, Parthasarathy et al. 1985, Clayton & Lloyd 1985, Williamson et al. 1986). It

Figures 9–13. Cytoskeletal organization in apical protonemal cells of *Physcomitrella*. Scale bar = 15 µm. **9**, Nomarski interference micrograph of a living caulonemal tip cell, note the large basal vacuole (V), cytoplasmic strands posterior to the nucleus (N), and numerous plastids behind the apical dome (A); **10**, Axially aligned arrays of microtubules visualized by indirect immunofluorescence with the anti-tubulin antibody, YOL 1/34; **11**, Microfilaments with a similar alignment to the microtubules, visualized with Rhodomine-conjugated Phalloidin, both microtubules and microfilaments are concentrated immediately to the rear of the apical dome, both extend into the dome where they appear to terminate; **12**, Entry into mitosis is preceded by an increase in the nuclear-associated microtubules particularly anterior to the nucleus; **13**, The mitotic spindle (S) has broad poles, during anaphase the pole to cortex microtubules become more numerous and may be associated with reorientation of the spindle. (**12** and **13** reproduced with permission from Doonan et al. 1985, Fig. 11 from Doonan et al. 1988).

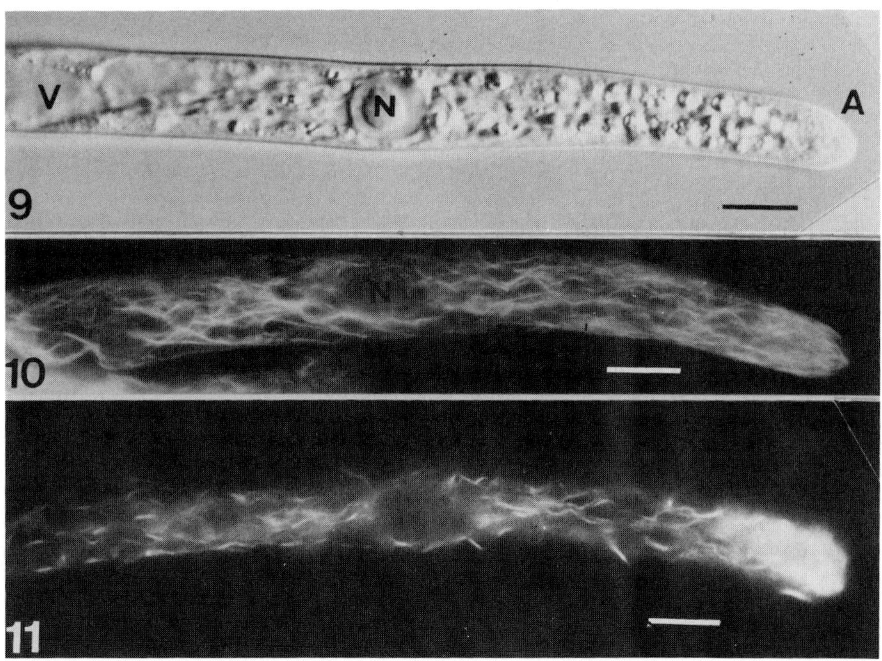
V
N
A
9
10
N
11

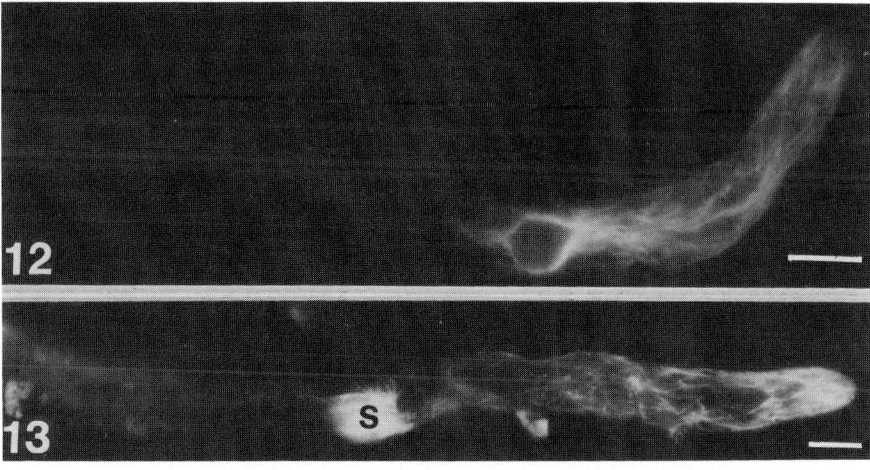
12
13
S

would seem likely that the further applications of immunofluorescence microscopy and new fixation protocols (e.g., Maupin & Pollard 1983, Doonan & Clayton 1986) will soon reveal its real extent in bryophytes. Indeed, promising candidates for actin are already visible in published micrographs. These include filament bundles in differentiating buds and protonemata of *Ceratodon* (Valanne 1971, Idzikowska & Szweykowska 1978) and filament bundles adjacent to the spline in differentiating spermatids of *Petalophyllum* (Duckett et al. 1984). The latter could well be the elusive force-generating elements responsible for spermatid shaping. Abundant fine filaments are also visible in negatively stained preparations of detergent-extracted moss and liverwort spermatids (C. C. J. Miller, unpublished data). Immunofluorescence, using anti-actin, has recently revealed that actin forms a strip along the multilayered structure and is associated with the basal bodies in *Pteridium* spermatozoids (Marc & Gunning 1986). The abundant fibrillar elements associated with microtubules and organelles during sporogenesis (Brown & Lemmon 1982) invite probing for actin, as do the finely fibrillar elements surrounding the centrioles in young spermatids (Duckett et al. 1984). Actin is known to be associated with sperm centrioles in animals (Kleve & Clark 1980). Thus its involvement in the centriolar rearrangements that characterize the early stages in spermatid differentiation in bryophytes would seem likely.

A more speculative hypothesis, but one which nevertheless merits further investigation, is the possibility that the ribonucleoprotein bodies observed by Ligrone (1985) in vegetative gametophyte cells of *Timmiella* and the so called "extruded nucleoli" found during bud morphogenesis in *Funaria* (Conrad et al. 1986) contain unpolymerized cytoskeletal proteins and in particular actin. Our current studies using tannic acid in conjunction with detergents (Maupin & Pollard 1981) reveal these bodies to be extremely widespread in bryophytes. They are particularly conspicuous in cells undergoing differentiation (e.g., derivatives of the apical cell in *Polytrichum*). In animal cells similar dense bodies in the cytoplasm are continuous with actin filaments (Bond & Somlyo 1982), and RNAs and ribonucleoproteins are closely associated with the cytoskeletal framework (Bird & Sells 1986, Pramanik et al. 1986).

Mitosis in Protonemata

Cell division in moss protonemata is a highly asymmetrical process (Knoop 1984). The products of every division receive an unequal portion of the mother cell's cytoplasm. But, regardless of whether the mother cell is apical or a sub-apical cell forming a side branch, the daughter cell at the apex receives the bulk of the cytoplasmic components and re-enters the cell cycle immediately. The sub-apical daughter has only a thin peripheral layer of cytoplasm surrounding a large central vacuole and remains quiescent for a period before becoming competent to divide again. The polarized organization of the interphase cytoplasm, combined with asymmetry of the mitotic spindle itself (Fig. 12, 13), appears to be intimately associated with the unequal partitioning (Doonan et al. 1985, 1986).

A certain minimum length must be achieved by caulonema cells before they can enter mitosis (approx. 200 μm in *Physcomitrium* (Jensen 1981), 400 μm in *Funaria* (Schnepf 1982)), although cell elongation may continue after they achieve competency. It seems, however, that the spatial architecture of the next mitosis is irreversibly determined when the minimal length is reached. First, elongation and enlargement of the basal vacuole (Fig. 9) ensures that the bulk of the cytoplasm lies anterior to the nucleus and will thus end up on the more apical daughter cell. Second, the orientation of the cell plate is fixed with respect to external stimuli (e.g., unidirectional illumination) at the point when the cell achieves mitotic competency. Although this cell may continue to elongate, changing the light source no longer affects the orientation of the cell plate (Jensen 1981). The molecular basis for this fixation of cross-wall orientation is completely unknown, but it has important implications for later development.

During interphase the nucleus is conspicuously associated with numerous microtubules when examined ultrastructurally (Schnepf 1982) or by indirect immunofluorescence (Doonan et al. 1985). In *Funaria* (Schnepf 1982) a constant distance is maintained between the nucleus and apex as the cell elongates. Colchicine treatment causes the nucleus to move backwards (relative to the apex), whereas in the presence of cytochalasin it moves forward, implying that microtubules and microfilaments exert forces on the nucleus in opposing directions (Schmiedel & Schnepf 1980). As the cell approaches mitosis (Fig. 12), the number of nuclear-associated microtubules increases, particularly over the anterior hemisphere (Doonan et al. 1985). Other cytoplasmic microtubules decrease in frequency but never disassemble completely. The decrease in microtubules is most marked rearward of the nucleus, and in many mitotic cells only isolated bundles are detectable near the basal cross-wall. Unlike the mitotic cells of flowering plant meristems (see Gorst et al. 1986 and literature therein cited) a pre-prophase band has never been detected in filamentous protonemal cells of mosses (Doonan et al. 1985).

In suspected pre-prophase apical cells Schmiedel and Schnepf (1980) found that the nucleus could be displaced irreversibly by centrifugation. At all other stages of interphase the nucleus recovered its normal position. The authors suggest that fragmentation of the cortical cytoskeleton during pre-prophase causes nuclear position to become more labile. They also found that colchicine treatment slightly reduced the minimum force needed to displace the nucleus at all stages, presumably due to microtubule fragmentation.

The pre-prophase nucleus in both *Funaria* and *Physcomitrella* is arrow shaped with the point towards the apical dome. The point and flanks of the nucleus are clothed in abundant microtubules (Fig. 12). Examination of such stages by immunofluorescence gives the impression of a uni-polar spindle but with the nuclear envelope still intact (nuclear shaping at prophase is a characteristic feature of mitosis in hepatics, see Steer 1985). When the envelope fragments the spindle poles are broad with numerous smaller microtubular foci (Fig. 13). The spindle-fibers associate laterally to form subgroups. In *Funaria* (Schmiedel et al. 1981) each chromosome is connected by its highly structured kinetochore to a bundle of 5-10 microtubules, which extend

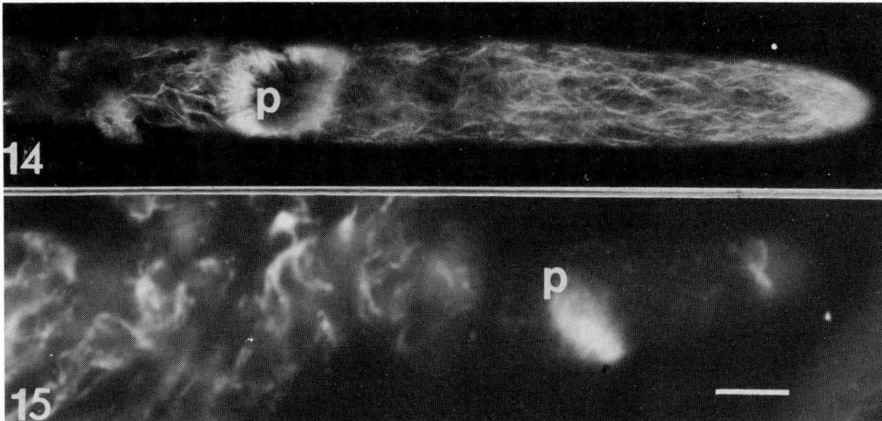

Figures 14, 15. Phragmoplasts in apical protonemal cells of *Physcomitrella*. Scale bar = 14 μm. **14**, Caulonemal cell stained for tubulin; **15**, Chloronemal cell stained with Rhodamine-conjugated Phalloidin. The pharagmoplast (P) contains both microtubule and microfilament components.

to the spindle pole. These bundles also contain microtubules that run past the kinetochores along the surface of the chromosome. In addition the spindle contains bundles of pole-to-pole microtubules that are often associated with ER-like elements.

In mosses spindle elongation is not pronounced, and chromosome separation is presumably achieved by shortening of the kinetochore microtubules. Spindle microtubule dynamics and the mechanism of chromosome separation are currently the subject of intense research activity (see Schulze & Kirschner 1986, Soifer 1986 and literature therein cited).

Cytokineses in plants is brought about by the deposition of a cell plate. The plate is initiated as a raft of coalescing vesicles, usually at what was previously the midzone of the spindle, perpendicular to the spindle axis. The edge of the plate spreads centrifugally eventually fusing with the parental wall. Two interdigitating sets of microtubules, known collectively as the phragmoplast, appear to be associated with the region of active vesicle fusion (Fig. 15). The presumed function of the microtubules is to guide the vesicles to sites of deposition of wall materials at the plate edge. However, Schmiedel et al. (1981) report that microtubules are absent from the periphery of the cell plate in *Funaria* and that colchicine does not inhibit plate development. There is thus some doubt over the precise role of microtubules in this situation. Phalloidin staining demonstrates the presence of microfilaments in the moss phragmoplast at this stage (Fig. 16).

The phragmoplast in caulonema is initiated during late anaphase or early telophase. At this stage the spindle-phragmoplast complex is orientated with its long axis parallel to that of the cell. The complex then begins to tilt so that the daughter nuclei become positioned against the limits of the cell. Pole-to-cortex microtubules become more numerous as reorientation progresses. Anti-microtubule drugs are extremely effective at preventing this process in both *Physcomitrella* and *Funaria*. The use of CIPC, which selectively removes the pole-to-cortex microtubules, causes random orientation of both spindle and phragmoplast (Doonan et al. 1985).

Besides forming the oblique cross-wall characteristic of caulonema, the reorientation of the spindle-phragmoplast complex has important consequences for the subsequent development of the sub-apical daughter cell (Jensen 1981). As mentioned above, the direction of tilt can be predetermined prior to mitosis by the prevailing unidirectional light source and is such that the subapical daughter nucleus is reconstituted on the lit flank of the cell. In *Physcomitrium* (Jensen 1981) and *Physcomitrella* (Doonan, unpublished data) the sub-apical nucleus returns to this position to divide when a side branch is formed. The side branch is formed on what was the lit flank of the filament, thus ensuring that, under conditions of directional illumination, side branches are always formed facing the light.

After mitosis, cytoplasmic microtubules increase in number, and bundles often appear to arise from the nuclear surface (cf. the similar origin of the interphase microtubules from the nuclear surface or perinuclear region in higher plant cells (Clayton et al. 1985)). The nuclei migrate away from the cross-wall, and the sub-apical daughter eventually taking up a position along the cell flank about midway down the cell. The role of microtubules in the post-mitotic migration process is unknown.

Side Branch Formation

After a variable period of quiescence, sub-apical caulonemal cells become competent to divide again. Division is initiated when the competent cells perceive a suitable stimulus, the nature of which varies between species and the prevailing environment (Dyer & Duckett 1984). For *Physcomitrella* caulonema grown in the dark, where side branch formation is a rare event, brief illumination will induce side branches on up to 10 or more of the cells immediately behind the filament tip (Doonan et al. 1986, Cove et al. 1978). The transfer of *Funaria* caulonema from nutrient poor to nutrient rich media (Schmiedel & Schnepf 1980) or the application of cytokinin (Saunders & Hepler 1981, 1983; Conrad et al. 1986) leads to branch formation.

The cytological changes that occur during side branch formation appear to be similar throughout the Funariales. A small bulge becomes evident along the cell flank close to the leading edge of the oblique apex-proximal cross-wall. The nucleus then migrates from its resting position in the peripheral cytoplasm about midway down the cell to the base of the enlarging bulge. Cytoplasmic strands associated with the nucleus become more prominent and tend to run from the nucleus to the base of the bulge. Organelles begin to accumulate in the region of the bulge, but the tip just like apical cells of protonemata (Jensen & Jensen 1984) contains numerous vesicles. Microtubules are not evident in electron micrographs from this region (Schnepf 1982). Mitosis and cytokinesis occur at the bulge and result in a cross wall, which separates the densely cytoplasmic branch initial from the original filament cell.

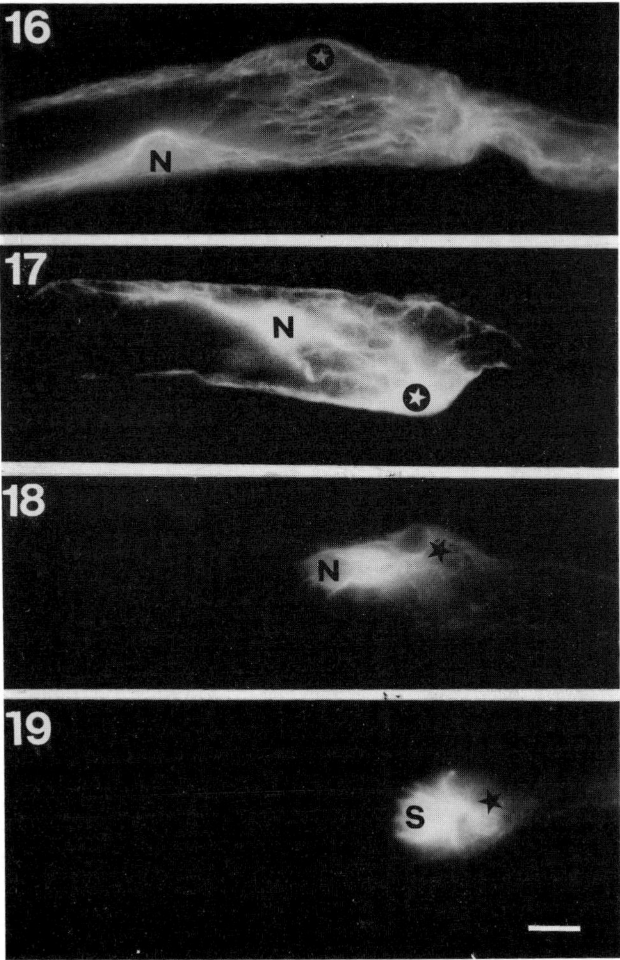

Figures 16–19. Cytoskeletal changes during the formation of side branches from sub-apical caulonemal cells of *Physcomitrella* visualized by indirect immunofluorescence. Scale bar = 10 µm. **16,** The first visible change is nuclear migration to the prospective site of side branching (star); **17,** As the nucleus traverses the cell, microtubules are redistributed, becoming concentrated between the nucleus and the side branch initial; **18,** In pre-mitotic stages the cytoplasm immediately anterior to the nucleus becomes densely stained; **19,** The spindle is highly asymmetrical, the anterior pole is much more highly focused than the posterior. (Reproduced with permission from Doonan et al. 1986.)

The site of cell division at the base of the bulge is approximately the location where the parent nucleus was first constituted (Jensen 1981). This is a particularly striking example of the way in which spatial parameters of one mitotic division apparently determine those of the next. Premitotic movement of the nucleus and other cellular components to specific locations and subsequent asymmetrical cell division is a question of central significance in plant morphogenesis (Lloyd & Barlow 1982, Lloyd 1987). How then does the nucleus find its way back to the original cross-wall in moss protonemata? Ultrastructural studies (e.g., Conrad et al. 1986) indicate that nuclear movement prior to branch or bud formation involves accumulation of microtubules over the nuclear surface in *Funaria*. Examination of branch formation in *Physcomitrella* by anti-tubulin immunofluorescence (see Fig. 16–19) reveals that microtubule bundles form cables, which link the nucleus to the site of side branching, and that these bundles shorten and thicken as migration progresses (Doonan et al. 1986). Extraction of the sub-apical cells with non-ionic detergents prior to fixation leaves a microtubular cytoskeleton that is associated preferentially with the apical cross-wall even prior to the onset of nuclear migration. This suggests a microtubule/cross-wall interaction perhaps provides the basis for the "memory" by which the nucleus returns to the cross-wall. That microtubules alone provide the "memory" seems unlikely, however, in view of the normal nuclear migration that takes place when cremart-treated cells are permitted to recover their microtubule arrays.

Nuclear migration is inhibited in the presence of cremart (Doonan et al. 1986). However, the action of this herbicide on both nuclear migration and microtubule integrity is prevented by the presence of taxol. Taxol alone does not prevent migration, indicating that microtubule disassembly is of minor significance in the process. Schmiedel and Schnepf (1979) found that colchicine could prevent nuclear movement in *Funaria* if applied prior to the onset of migration. If applied later, the nucleus continued to move eventually overshooting its normal division site. These authors suggest that, while microtubule assembly was prevented by the drug, existing microtubules appeared to function normally. This implies that a new set of microtubules is necessary to direct the nucleus into the bulge (i.e. to prevent overshoot).

All the foregoing data point to the necessity of an intact microtubule system for nuclear positioning, but the data do not clearly define the function of the microtubules. One possibility is that the microtubules act as guide ropes, providing directional information while other cytoskeletal systems, actinomyosin for instance (Jackson 1982), provides the motive force. From the insensitivity of the process to cytochalasins (Schmiedel & Schnepf 1979) actin would not appear to be important. The apparent shortening and thickening of the nucleus-bulge microtubule bundles as seen in immunofluorescence images is highly suggestive of either microtubule–microtubule sliding or a population of very highly dynamic elements similar to those found in many animal cells (Schulze & Kirschner 1986).

Side Branch Fate: Alternative Modes of Cell Expansion

The developmental fate of newly formed side branch initials is extremely plastic and largely determined by environmental conditions. Manipulation of their subsequent growth can be achieved by control of environmental variables such as light quality and quantity and by the type and amount of plant growth substances present (reviewed by Cove & Ashton 1984). While the precise reactions may vary between species, most respond to the application of cytokinin by the division of their side branch initials, involving a change from filamentous growth to the formation of buds (Conrad et al. 1986). While protonemata and buds vary in many cytological details, perhaps one of the more obvious is the way in which they expand. Protonemata display highly directed growth as described above, but buds expand to form a globular structure with much less obvious polarity (Fig. 20). Buds have the potential to continue to increase in complexity eventually forming leafy shoots (Fig. 25–28).

In the undetermined branch initial of *Physcomitrella* microtubules form a complex cagework that runs around the nucleus and throughout the cytoplasm (Doonan et al. 1987). In initials committed to filamentous growth, the microtubules gradually become oriented along the axis of expansion, and free microtubule ends are apparent within the apical dome. In caulonemal initials elongation is more rapid, and the free ends are organized into microtubular foci. The cytoplasmic microtubular cytoskeleton in presumptive bud initials shows little polarity (Fig. 21), with the free ends of microtubules being difficult to detect. The entire microtubule system during interphase shows signs of instability and is very difficult to preserve. However, the mitotic microtubules of the spindle and phragmoplast (Fig. 22, 23, 24) preserve well and appear similar to those of filamentous tissues. The application of very high, non-physiological levels of phytohormones to protonemata can fragment the microtubules. The reason for the sensitivity of branch-initial microtubules to cytokinin is unknown, but the relatively unpolarized growth of young buds may be due, at least in part, to a highly labile microtubule system.

In tip-growing cells a polarized microtubular system is thought to direct the transport of vesicles containing cell wall materials to the apex. The vesicles in presumptive buds of *Funaria* are dispersed around the cell cortex (Conrad et al. 1986), and this presumably leads to more uniform deposition of new wall material over a wider area.

As the bud increases in size, the apical cell continues to divide obliquely. The subapical daughter cells, unlike those in protonemata, are capable of further division and contribute to the continued growth of the bud. The interphase microtubules continue to be poorly preserved and, as in protonemata, occupy both cortical and endoplasmic locations within the cytoplasm. Spindles are also similar to those in protonemata in that they are associated with extensive cytoplasmic microtubules. However, microtubule organization undergoes a dramatic change at about the stage when leaflets are first produced (Fig. 25–32). The early leaflets are first seen as finger-like projections from just behind the apex. When these structures are stained with anti-tubulin antibodies, transversely oriented microtubules are observed. Endoplasmic microtu-

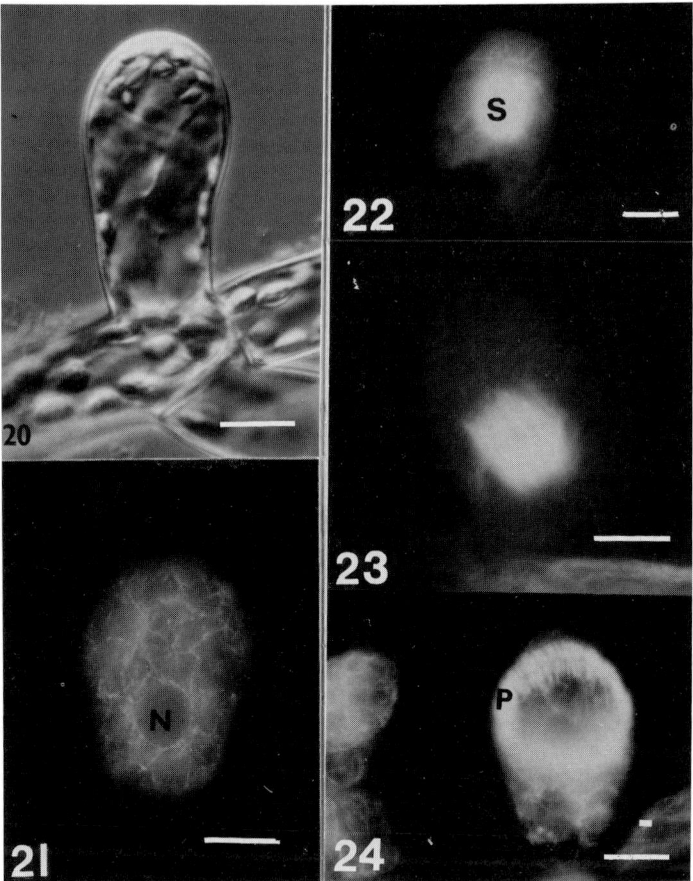

Figures 20–24. Bud formation in *Physcomitrella*. Scale bar = 10 μm. **20,** Nomarski interference micrograph of a presumptive bud during its primary cytokinesis. **21–24,** Immunofluorescence micrographs showing changes in microtubule distribution during the first cell cycle; **21,** Early stages in bud formation are associated with the loss of microtubular foci from the apical region and their random distribution at interphase; **22,** The spindle (S) and associated astral tubules; **23,** Obliquely positioned early phragmoplast (cf. the oblique cross-wall in Fig. **20**); **24,** Late phragmoplast (P). (Reproduced with permission from Doonan et al. 1987.)

bules become less numerous, and when cell squashes from leafy shoots are examined, most cells contain highly organized arrays of transverse cortical microtubules (Fig. 29). The spindles found in the squashes of these tissues lack any associated cytoplasm microtubules (Fig. 32). Prior to spindle formation cortical microtubules are replaced by a pre-prophase band, which encircles the equator of the cell (Fig. 30).

The pre-prophase band of microtubules (PPB) was originally thought to be unique to meristematic regions of vascular plants and thus was suggested as a prerequisite for the series of precise divisions necessary to maintain organized development (Gunning & Hardham 1982, Gorst et al. 1986). The absence of PPBs from the filamentous stages but their presence in the meristems of several species of moss (Schnepf 1973, 1984) does indicate an association with meristematic growth. However, the evidence that the cells of meristematic tissues fix their division sites more accurately than the cells of free-living filaments is far from compelling. Another explanation for the association of the PPB with meristematic tissues is that the band is a "decay" product of the cortical interphase arrays present prior to their disassembly in prophase. During leafy shoot formation in *Physcomitrella* pre-prophase bands are recognizable only after regular cortical interphase arrays are well developed.

Microtubule Organization in Leafy Shoots of Mosses

For the majority of cells within the meristems of mosses the microtubule cycle (see Fig. 29–32) is very similar to that described for flowering plants (Clayton 1985). The interphase cortical microtubules (Fig. 29) are replaced at or just prior to prophase by the pre-prophase band (Fig. 30). This band decays as the mitotic spindle develops (Fig. 31). After the chromosomes have separated the phragmoplast is assembled (Fig. 32), and a cell plate is deposited. Interphase microtubules are initiated from the nuclear surface and fan out towards the cell periphery where they become organized into the cortical interphase array. Similar microtubule patterns have also been found in sporophytic tissues (Sack & Paolillo 1983, 1985), but very few electron microscopic and immunofluorescence studies have so far considered this generation of the life cycle. Cytoskeletal organization during sporogenesis is considered elsewhere in this volume by Brown and Lemmon.

The pre-prophase bands of hepatics are much broader and contain far fewer microtubules than those in mosses (Fowke & Pickett-Heaps 1978; Galatis & Apostolakos 1977, Apostolakos & Galatis 1985a, b, c; reviewed in Duckett 1986) and appear to be absent in hornworts (Brown & Lemmon 1985).

Figures 25–32. Bud development in *Physcomitrella*. **25–28,** Scanning electron micrographs, scale bar = 30 μm. **25,** Pre-leaflet formation; **26–28,** Successive stages in leaflet formation; **29–32,** Cell squashes after leaflets have been formed to reveal a new type of microtubule organization not found in protonemal cells, scale bar = 10 μm; **29,** Interphase cell with transverse arrays of cortical microtubules; **30,** A pre-prophase cell with an equatorial band of microtubules; **31,** The spindle, note the absence of associated cytoplasmic microtubules; **32,** A phragmoplast. (Reproduced with permission from Doonan et al. 1987.)

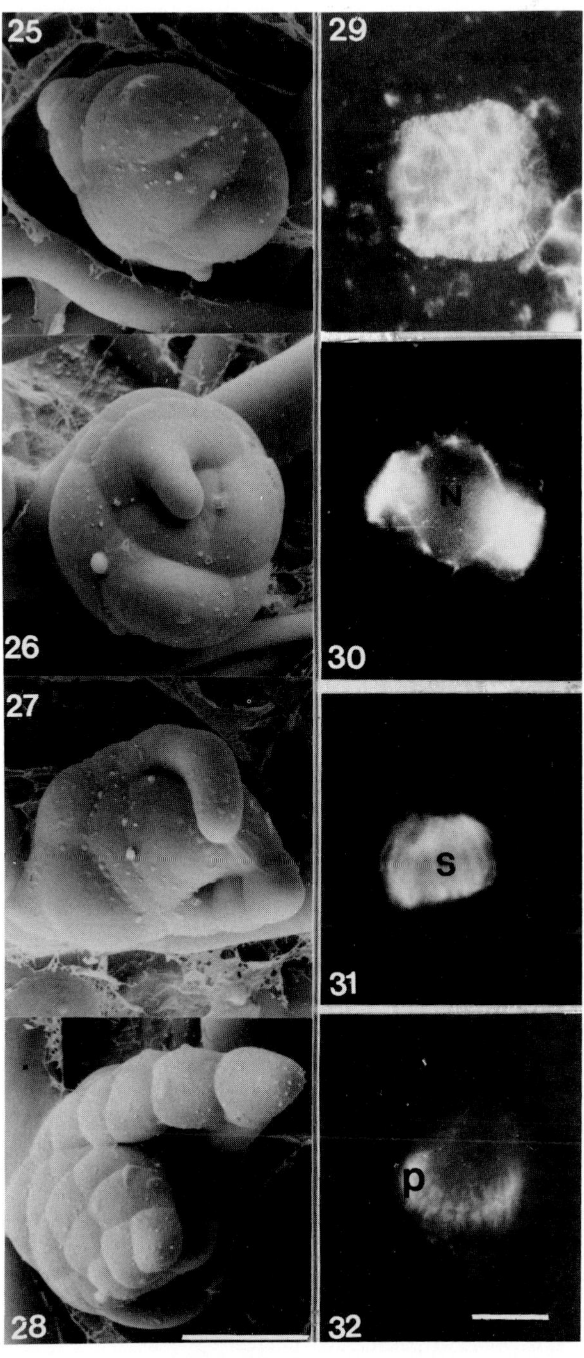
25
26
27
28
29
N
30
S
31
p
32

Intact leaflets of *Physcomitrella* can be stained for microtubules, and examination of mature cells reveals a cortical array organized as a "multi-start helix." The pitch of the helix varies between cells within the same leaf (Doonan et al. 1987). The significance of multi-start helical arrangement of cortical microtubules for cell wall texture is reviewed by Lloyd (1984).

Cytoskeletal organization in *Sphagnum* leaves appears to differ slightly from that in other mosses (Schnepf 1984) in that there is a novel site of microtubule initiation after mitosis. Instead of from the nuclear surface, microtubules are reinitiated from a region between the nucleus and the newly formed cell plate. A further cytoskeletal variation has been described by Wiencke and Schulz (1983) in leaf leptoids of *Funaria*. These contain an extensive system of longitudinally oriented endoplasmic microtubules, which form bundles in close contact with various organelles. We have recently discovered similar microtubule bundles to be numerous in apices of *Polytrichum* (Duckett et al., unpublished data).

Future Prospects

The immunofluorescence studies so far performed on microtubules and microfilaments in moss protonemata mark the beginning of what promises to be a most exciting era of research on bryophyte morphogenesis. However, although the images may be very striking, the shortcomings of immunofluorescence techniques should not be overlooked. Preparative procedures for visualizing the cytoskeleton (e.g., detergent extraction) are extremely destructive and may thus provide misleading information. This problem is particularly acute when mapping the distribution of more soluble proteins such as calmodulin. Ideally, the immunological probes, which allow visualization of specific cytoskeletal components by optical microscopy, should also be used in conjunction with gold labelling to extend the observations to the electron microscope level. The latter option is particularly inviting since fixation protocols are now available that preserve both antigenicity and the more labile cytoskeletal domains not previously seen in electron micrographs. In addition Western blotting enables the isolation and biochemical characterization of these elements from the same specimens.

Antibodies that visualize all the microtubule and microfilament arrays when used in conjunction with appropriate drugs represent only the initial stages in the dissection of cytoskeletal dynamics and function. Microinjection of biotin-labelled tubulin, which has led to dramatic advances in understanding microtubule dynamics in animal cells (Soifer 1986, Schutze & Kirschner 1986), may ultimately not prove feasible with materials as delicate as bryophyte protonemata. However, the application of antibodies against pericentriolar material (Clayton et al. 1985) [a region of animal cells well documented as a microtubule organizing center (Karsenti & Maro 1986, Brinkley 1985)] to bryophytes could provide important new information on microtubule nucleating sites. It would be particularly interesting to see if such probes label the extranuclear polar bodies in hepatics (Steer 1985), nascent blepharoplasts in young spermatids (Duckett et al. 1983), and the putative microtubule organizing centers seen during sporogenesis (Brown & Lemmon 1987). In the light of the identification of actin in moss protonemata investigation of the localization of microtubule-associated proteins (Olmsted 1986, Wiche 1985) may provide clues on possible interactions between microfilaments and microtubules. The exploration for force-generating molecules such as dynein (Gibbons 1981) and kinesin (Vale et al. 1985) should permit discrimination between structural versus force-generating domains of the microtubule network. The localization of calmodulin and Ca^{2+} (Wick & Duniec 1986, Wick et al. 1985) is just one aspect of the regulation of microtubule assembly. Other factors such as the recently identified STOP protein (Margolis et al. 1986) and the acetylation (Sasse et al. 1987) and tyrosination (Geuens et al. 1986, Gundersen & Bulinski 1986) of α-tubulin confer differences in stability between microtubules, even within the same cell. The differing sensitivities of bryophyte microtubules to fixation and drugs suggest the coexistence of different domains within the same cell. Similarly, blepharoplast ontogeny involves differing rates of elongation of specifically identifiable microtubules. Clearly, both protonemata and spermatids are ripe for studies on the possible occurrence and functions of different forms of α tubulin in plants.

Acknowledgments

We thank the Science and Engineering Research Council of Great Britain and the Leverhulme Trust for financial support.

Literature Cited

Alberts, B., Bray, D., Lewis, J., Raff, M., Roberts, K. & Watson, J. D. 1983. Molecular Biology of the Cell. 1146 pp. New York & London. Garland Publishing Inc.

Apostolakos, P. & Galatis, B. 1985. Studies on the development of air pores and air chambers of *Marchantia paleacea*. II. Ultrastructure of the initial microtubule organising centres. Canadian Journal of Botany 63: 744–756.

— — 1985b. Studies on the development of air pores and air chambers of *Marchantia paleacea*. III. Microtubule organization in preprophase-prophase initial aperture cells – formation of incomplete preprophase microtubule bands. Protoplasma 128: 120–135.

— — 1985c. Studies on the development of air pores and air chambers of *Marchantia paleacea*. IV. Cell plate arrangement in initial aperture cells. Protoplasma 128: 136–146.

Bajer, A. S., Cypher, C., Molè-Bajer, J. & Howard, H. M. 1982. Taxol-induced anaphase reversal: evidence that elongating microtubules can exert a pushing force in living cells. Proceedings of the National Academy of Sciences, USA 79: 6569–6573.

Bird, R. C. & Sells, B. H. 1986. Cytoskeleton involvement in the distribution of mRNP complexes and small cytoplasmic RNAs. Biochemica et Biophysica Acta 868:

215–225.

Bond, M. & Somlyo, A. V. 1982. Dense bodies and actin polarity in vertebrate smooth muscle. Journal of Cell Biology 95: 403–413.

Brinkley, B. R. 1985. Microtubule organizing centers. Annual Review of Cell Biology 1: 145–172.

Brown, R. C. & Lemmon, B. E. 1982. Ultrastructural aspects of moss meiosis: cytokinesis and organelle apportionment in *Rhynchostegium serrulatum*. Journal of the Hattori Botanical Laboratory 53: 41–50.

— — 1985. Preprophasic establishment of division polarity in monoplastidic mitosis of hornworts. Protoplasma 124: 175–183.

— — 1988. Sporogenesis in bryophytes. Advances in Bryology 3: this volume.

Burgess, J. & Linstead, P. J. 1981. Studies on the growth and development of protoplasts of the moss *Physcomitrella patens* and its control by light. Planta 151: 331–338.

— — 1982. Cell wall differentiation during growth of electrically polarized protoplasts of *Physcomitrella*. Planta 156: 241–248.

Chen, T.-H. & Jaffe, L. F. 1979. Forced calcium entry and polarized growth of *Funaria* spores. Planta 144: 401–406.

Clayton, L. 1985. The cytoskeleton and the plant cell cycle, pp. 131–141. In: Bryant, J. A. & Francis, D. (Eds.), The Cell Division Cycle in Plants. Society for Experimental Biology Seminar Series 26. Cambridge University Press.

Clayton, L., Black, C. M. & Lloyd, C. W. 1985. Microtubule nucleating sites in higher plant cells identified by an auto-antibody against pericentriolar material. Journal of Cell Biology 101: 319–324.

Clayton, L. & Lloyd, C. W. 1985. Actin organization during the cell cycle in meristematic plant cells. Experimental Cell Research 156: 231–238.

Conrad, P. A., Steucek, G. L. & Hepler, P. K. 1986. Bud formation in *Funaria*: organelle redistribution following cytokinin treatment. Protoplasma 131: 211–223.

Cove, D. J. & Ashton, N. W. 1984. The hormonal regulation of gametophytic development in bryophytes, pp. 177–201. In: Dyer, A. F. & Duckett, J. G. (Eds.), The Experimental Biology of Bryophytes. New York & London. Academic Press.

Cove, D. J., Schild, A., Ashton, N. W. & Hartmann, E. 1978. Genetic and physiological studies of the effect of light on the development of the moss *Physcomitrella patens*. Photochemistry and Photobiology 27: 249–254.

Cran, D. G. 1979. The ultrastructure of fern gametophyte cells, pp. 171–212. In: Dyer, A. F. (Ed.), The Experimental Biology of Ferns. New York & London. Academic Press.

Dawson, P. J., Hulme, J. S. & Lloyd, C. W. 1985. Monoclonal antibody to intermediate filament antigen cross-reacts with higher plant cells. Journal of Cell Biology 100: 1793–1798.

Dawson, P. J. & Lloyd, C. W. 1985. Identification of multiple tubulins in taxol microtubules purified from carrot suspension cells. EMBO Journal 4: 2451–2455.

DeMaggio, A. E. & Stetler, D. A. 1977. Protonemal organization and growth in the moss *Dawsonia superba*: ultrastructural characteristics. American Journal of Botany 64: 449–454.

Doonan, J. H. & Clayton, L. 1986. Immunofluorescence studies on the plant cytoskeleton, pp. 111–136. In: Wang, T. L. (Ed.), Immunology in Plant Science. Society for Experimental Biology Seminar Series 29. Cambridge University Press.

Doonan, J. H., Cove, D. J., Corke, F. M. K. & Lloyd, C. W. 1987. The pre-prophase band of microtubules, absent from tip-growing moss filaments, arises in leafy shoots during transition to intercalary growth. Cell Motility and the Cytoskeleton 7: 138–153.

Doonan, J. H., Cove, D. J. & Lloyd, C. W. 1985. Immunofluorescence microscopy of microtubules in intact cell lineages of the moss, *Physcomitrella patens*. I. Normal and CIPC-treated tip cells. Journal of Cell Science 75: 1–17.

Doonan, J. H., Jenkins, G. I., Cove, D. J. & Lloyd, C. W. 1986. Microtubules connect the migrating nucleus to the prospective division site during side branch formation in the moss *Physcomitrella patens*. European Journal of Cell Biology 41: 157–164.

Doonan, J. H., Cove, D. J. & Lloyd, C. W. 1988. Microtubules and microfilaments in tip growth. Evidence that microtubules impose polarity on protonemal growth in *Physcomitrella patens*. Journal of Cell Science 89: 533–540.

Dráber, P., Dráberová, E., Zicconi, D., Sellito, C., Viklicky, V. & Cappuccinelli, P. 1986. Heterogeneity of microtubules recognised by monoclonal antibodies to alpha-tubulin. European Journal of Cell Biology 41: 82–88.

Duckett, J. G. 1986. Utrastructure in bryophyte systematics and evolution: an evaluation. Journal of Bryology 14: 25–42.

Duckett, J. G., Carothers, Z. B. & Miller, C. C. J. 1982. Comparative spermatology and bryophyte phylogeny. Journal of the Hattori Botanical Laboratory 53: 107–125.

— — — 1983. Gametogenesis, pp. 232–275. In: Schuster, R. M. (Ed.), New Manual of Bryology, Vol. I. Nichinan, Japan. Hattori Botanical Laboratory.

Duckett, J. G. & Renzaglia, K. S. 1988. The ultrastructure of bryophyte plastids. Advances in Bryology 3: this volume.

— — 1987. Transmission electron microscopy of bryophytes. Proceedings of the XIV Botanical Congress, Berlin (In Press).

— — Polarised bidirectional microtubule assembly establishes blepharoplast architecture in mosses (in preparation).

Dyer, A. F. & Duckett, J. G. (Eds.) 1984. The Experimental Biology of Bryophytes. 292 pp. New York & London. Academic Press.

Fowke, L. C. & Pickett-Heaps, J. D. 1978. Electron microscope study of vegetative cell division in two species of *Marchantia*. Canadian Journal of Botany 56: 467–475.

Galatis, B. & Apostolakos, P. 1977. On the fine structure of differentiating mucilage papillae of *Marchantia*. Canadian Journal of Botany 55: 772–795.

Geuens, G., Gunderson, G. G., Nuydens, R., Cornelisson, F., Bulinski, J. C. & DeBrabander, M. 1986. Ultrastructural colocalization of tyrosinated and detyrosinated α-tubulin in interphase and mitotic cells. Journal of Cell Biology 103: 1883–1893.

Gibbons, I. R. 1981. Cilia and flagella of eukaryotes. Journal of Cell Biology 91: 107–124.

Gorst, J., Wernicke, W. & Gunning, B. E. S. 1986. Is the preprophase band of microtubules a marker of organization in suspension cultures? Protoplasma 134: 130–140.

Gundersen, G. G. & Bulinski, J. C. 1986. Microtubule arrays in differentiated cells contain elevated levels of a post-translationally modified form of tubulin. European Journal of Cell Biology 42: 288–294.

Gunning, B. E. S. & Hardham, A. R. 1982. Microtubules. Annual Review of Plant Physiology 33: 651–698.

Hartmann, E. & Jenkins, G. I. 1984. Photomorphogenesis of mosses and liverworts, pp. 203–228. In: Dyer, A. F. & Duckett, J. G. (Eds.), The Experimental Biology of Bryophytes. New York & London. Academic Press.

Hauser, M. 1986. Taxol affects both the microtubular arrays of heliozoan axonemes and their microtubule-organising center. European Journal of Cell Biology 42: 295–304.

Hoch, H. C. & Staples, R. C. 1985. The microtubule cytoskeleton in hyphae of *Uromyces phaseoli* germlings: its relationship to the region of nucleation and to the F-actin cytoskeleton. Protoplasma 124: 112–122.

Horwitz, S. B., Parness, J., Schiff, P. B. & Manfredi, J. J. 1982. Taxol: a new probe for studying the structure and function of microtubules. Cold Spring Harbor Symposia on Quantitative Biology 46: 219–226.

Idzikowska, K. & Szweykowska, A. 1978. The ultrastructural aspects of the cytokinin-induced bud formation in *Ceratodon pupureus*. Protoplasma 94: 41–52.

Jackson, W. T. 1982. Actomyosin, pp. 3–30. In: Lloyd, C. W. (Ed.), The Cytoskeleton in Plant Growth and Development. New York & London. Academic Press.

Jenkins, G. I. & Cove, D. J. 1983. Light requirement for the regeneration of protoplasts of the moss *Physcomitrella patens*. Planta 157: 39–45.

Jensen, L. C. W. 1981. Division, growth and branch formation in protonema of the moss *Physcomitrium turbinatum*: studies of sequential cytological changes in living cells. Protoplasma 107: 301–317.

Jensen, L. C. W. & Jensen, C. G. 1984. Fine structure of protonemal apical cells of the moss *Physcomitrium turbinatum*. Protoplasma 122: 1–10.

Karsenti, E. & Maro, B. 1986. Centrosomes and the spatial distribution of microtubules in animal cells. Trends in Biochemical Sciences 11: 460–463.

Kilmartin, J. V., Wright, B. & Milstein, C. 1981. Rat monoclonal antitubulin antibodies derived using a new nonsecreting rat cell line. Journal of Cell Biology 93: 576–582.

Kleve, M. G. & Clark, W. H., Jr. 1980. Association of actin with sperm centrioles: isolation of centriolar complexes and immunofluorescent localization of actin. Journal of Cell Biology 86: 87–95.

Knoop, B. 1984. Development in bryophytes, pp. 143–176. In: Dyer, A. F. & Duckett, J. G. (Eds.), The Experimental Biology of Bryophytes. New York & London. Academic Press.

Kotenko, K. J., Miller, J. H. & Robinson, A. I. 1987. The role of asymmetic cell division in pteridophyte cell differentiation. I. Localized metal accumulation and differentiation in *Vittaria* gemmae and *Onoclea* prothallia. Protoplasma 136: 81–95.

Lacelle, S. A., Callaham, D. A. & Hepler, P. K. 1986. A method for rapid freeze-fixation of plant cells. Protoplasma 131: 153–165.

Ligrone, R. 1985. Massive ribonucleoprotein bodies in gametophyte vegetative cells of the moss *Timmiella barbuloides* (Brid.) Moenk. Protoplasma 127: 204–211.

Lin, C.-T., Sun, D., Song, G.-X. & Wu, J.-Y. 1986. Calmodulin: localization in plant tissues. Journal of Histochemistry and Cytochemistry 34: 561–567.

Lloyd, C. W. (Ed.). 1982. The Cytoskeleton in Plant Growth and Development. 457 pp. New York & London. Academic Press.
— 1984. Towards a dynamic helical model for the influence of microtubules on wall patterns in plants. International Review of Cytology 86: 1–51.
— 1987. Microtubules and the cellular morphogenesis of plants, pp. 31–57. In: Browder, L. W. (Ed.), Developmental Biology, A Comprehensive Synthesis. Vol. 2. The Cellular Basis of Morphogenesis. New York & London. Academic Press.
Lloyd, C. W. & Barlow, P. W. 1982. The co-ordination of cell division and elongation: the role of the cytoskeleton, pp. 203–228. In: Lloyd, C. W. (Ed.). The Cytoskeleton in Plant Growth and Development. New York & London. Academic Press.
Lloyd, C. W., Clayton, L., Dawson, P. J., Doonan, J. H., Hulme, J. S., Roberts, I. N. & Wells, B. 1985. The cytoskeleton underlying side walls and cross-wall in plants: molecules and macromolecular assemblies. Journal of Cell Science Supplement 2: 143–155.
Marc, J. & Gunning, B. E. S. 1986. Immunofluorescent localization of cytoskeletal tubulin and actin during spermatogenesis in *Pteridium aquilinum* (L.) Kuhn. Protoplasma 134: 163–177.
Margolis, R. L. 1983. Calcium and microtubules, pp. 313–335. In: Cheung, W. Y. (Ed.). Calcium and Cell Function. Vol. 4. New York & London. Academic Press.
Margolis, R. L., Rauch, C. T. & Job, D. 1986. Purification and assay of a 145-KDa protein (STOP145) with microtubule-stabilizing and motility behavior. Proceedings of the National Academy of Sciences, USA 83: 639–643.
Maupin, P. & Pollard, T. D. 1983. Improved preservation and staining of HeLa cell actin filaments, clathrin-coated membranes, and other cytoplasmic structures by tannic acid-glutaraldehyde-saponin fixation. Journal of Cell Biology 96: 51–62.
Miller, C. C. J., Duckett, J. G., Sheterline, P. & Carothers, Z. B. 1983. Immunofluorescence microscopy of the flagella and multilayered structure in two mosses: *Sphagnum palustre* L. and *Polytrichum juniperinum* Hedw. Journal of Cell Science 61: 71–86.
Mizuno, K., Sek, F., Perkin, J., Wick, S., Duniec, J. & Gunning, B. 1985. Monoclonal antibodies specific to plant tubulin. Protoplasma 129: 100–108.
Morejohn, L. C., Bureau, T. E., Tocchi, L. P. & Fosket, D. E. 1984. Tubulins from different higher plant species are immunologically nonidentical and bind colchicine differently. Proceedings of the National Academy of Sciences, USA 81: 1440–1444.
— — — — 1987. Resistance of *Rosa* microtubule polymerization to colchicine results from a low-affinity interaction of colchicine and tubulin. Planta 170: 230–241.
Morejohn, L. C. & Fosket, D. E. 1982. Higher plant tubulin identified by self-assembly into microtubules *in vitro*. Nature 297: 426–428.
Olmstead, J. B. 1986. Microtubule-associated proteins. Annual Review of Cell Biology 2: 421–457.
Parke, J. M., Miller, C. C. J., Cowell, I., Dodson, A., Dowding, A., Downes, M., Duckett, J. D. & Anderton, B. H. 1987. Intermediate filament antigens in higher plant cells. Cell Motility and the Cytoskeleton. In Press.
Parthasarathy, M. V., Perdue, T. D., Witztum, A. & Alvernaz, J. 1985. Actin network as a normal component of the cytoskeleton in many vascular plant cells. American Journal of Botany 72: 1318–1323.

Pierson, E. S., Derksen, J. & Traas, J. A. 1986. Organisation of microfilaments and microtubules in pollen tubes grown *in vitro* or *in vivo* in various angiosperms. European Journal of Cell Biology 41: 14–18.

Powell, A. J., Lloyd, C. W., Slabas, A. R. & Cove, D. J. 1980. Demonstration of the microtubular cytoskeleton of the moss *Physcomitrella patens* using antibodies against brain tubulin. Plant Science Letters 18: 401–404.

Pramanik, S. K., Walsh, R. W. & Bag, J. 1986. Association of messenger RNA with the cytoskeletal framework in rat L6 myogenic cells. European Journal of Biochemistry 160: 221–230.

Roberts, D. M., Lukas, T. J. & Watterson, D. M. 1986. Structure, function and mechanism of action of calmodulin. CRC Critical Reviews on Plant Sciences 4: 311–339.

Runeberg, P., Raudaskoski, M. & Virtanen, I. 1986. Cytoskeletal elements in the hyphae of the homobasidiomycete *Schizophyllum commune* visualised with indirect immunofluorescence and NBD-phallacidin. European Journal of Cell Biology 41: 25–32.

Sack, F. D. & Paolillo, D. J., Jr. 1983. Protoplasmic changes during stomatal development in *Funaria*. Canadian Journal of Botany 61: 2515–2526.

—— 1985. Incomplete cytokinesis in *Funaria* stomata. American Journal of Botany 72: 1325–1333.

Sasse, R., Glyn, M. C. P., Birkett, C. R. & Gull, K. 1987. Acetylated α-tubulin in *Physarum*: immunological characterization of the isotype and its usage in particular microtubular organelles. Journal of Cell Biology 104: 41–49.

Saunders, M. J. & Hepler, P. K. 1981. Localization of membrane-associated calcium following cytokinin treatment in *Funaria* using chlorotetracycline. Planta 152: 272–281.

—— 1982. Calcium ionophore A23187 stimulates cytokinin-like mitosis in *Funaria*. Science 217: 943–945.

—— 1983. Calcium antagonists and calmodulin inhibitors block cytokinin-induced bud formation in *Funaria*. Developmental Biology 99: 41–49.

Schimmen, T. & Yano, M. 1986. Regulation of myosin sliding along *Chara* actin bundles by mature skeletal muscle tropomysin. Protoplasma 132: 129–136.

Schmiedel, G., Reiss, H.-D. & Schnepf, E. 1981. Associations between membranes and microtubules during mitosis and cytokinesis in caulonema tip cells of the moss *Funaria hygrometrica*. Protoplasma 108: 173–190.

Schmiedel, G. & Schnepf, E. 1979. Side branch formation and orientation in the caulonema of the moss, *Funaria hygrometrica*: experiments with inhibitors and with centrifugation. Protoplasma 101: 47–59.

—— 1980 Polarity and growth of caulonema tip cells of the moss *Funaria hygrometrica*. Planta 147: 405–413

Schnepf, E. 1973. Mikrotubulus – Anordnung und Unordnung, Wandbildung und Zellmorphogenese in jungen *Sphagnum* Blättchen. Protoplasma 78: 145–173.

— 1981. Polarity and gradients in tip growing cells, pp. 483–488. In: Schweiger, H.-G. (Ed.). International Cell Biology 1980–1981. Berlin and New York. Springer-Verlag.

— 1982. Morphogenesis in moss protonemata, pp. 321–344. In: Lloyd, C. W. (Ed.). The Cytoskeleton in Plant Growth and Development. New York & London. Academic Press.

— 1984. Pre- and post-mitotic re-orientation of microtubule arrays in young *Sphagnum* leaflets: transitional stages and initation sites. Protoplasma 120: 100–112.

Schnepf, E., Hrdina, B. & Lehne, A. 1982. Spore germination, development of the microtubule system and protonemal cell morphogenesis in the moss *Funaria hygrometrica*. Effects of inhibitors and growth substances. Biochemie Physiologie Pflanzen 177: 461–482.

Schulze, E. & Kirschner, M. 1986. Microtubule dynamics in interphase cells. Journal of Cell Biology 102: 1020–1031.

Soifer, D. (Ed.) 1986. Dynamic Aspects of Microtubule Biology. Annals of the New York Academy of Sciences 466.

Steer, M. W. 1985. Mitosis in bryophytes. Advances in Bryology 2: 1–63.

Tiwari, S. C., Wick, S. M., Williamson, R. E. & Gunning, B. E. S. 1984. Cytoskeleton and integration of cellular function in cells of higher plants. Journal of Cell Biology 99: 63a–69s.

Towbin, H., Staehlin, J. & Gordon, J. 1979. Electrophoretic transfer of proteins from polyacrylamide gels to nitrocellulose sheets: procedure and some applications. Proceedings of the National Academy of Sciences, USA 76: 4350–4354.

Traas, J. A., Doonan, J. H., Rawlins, D. J., Shaw, P. J., Watts, J., K. Lloyd, C. W. 1987. An actin network is present in the cytoplasm throughout the cell cycle of Charrot cells and associated with the dividing nucleus. Journal of Cell Biology 105: 387–395.

Tucker, B. E., Hoch, H. C. & Staples, R. C. 1986. The involvement of F-actin in *Uromyces* cell differentiation: the effects of cytochalasin E and phalloidin. Protoplasma 135: 88–101.

Valanne, N. 1971. The effects of prolonged darkness and light on the fine structure of *Ceratodon purpureus*. Canadian Journal of Botany 49: 547–554.

Vale, R. D., Reese, T. S. & Sheetz, M. P. 1985. Identification of a novel force-generating protein, kinesin, involved in microtubule-based motility. Cell 42: 39–50.

Weisenberg, R. C. 1972. Microtubule formation *in vitro* in solutions containing low calcium concentrations. Science 177: 1104–1105.

Wiche, G. 1985. High-molecular-weight microtubule associated proteins (MAPS): a ubiquitous family of cytoskeletal connecting links. Trends in Biochemical Sciences 10: 67–70.

Wick, S. M. 1985. The higher plant mitotic apparatus: redistribution of microtubules: calmodulin and microtubule initiation material during establishment. Cytobios 43: 285–294.

Wick, S. M. & Duniec, J. 1986. Effects of various fixatives on the reactivity of plant cell tubulin and calmodulin in immunofluorescence microscopy. Protoplasma 133: 1–18.

Wick, S. M., Muto, S. & Duniec, J. 1985. Double immunofluorescence labelling of calmodulin and tubulin in dividing plant cells. Protoplasma 126: 198–206.

Wiencke, C. & Schulz, D. 1983. The fine structure of symplasmic and apoplasmic transport in the ‚nerve‘ of the *Funaria* leaflet. Zeitschrift für Pflanzenphysiologie 112: 337–350.

Williamson, R. E., Perkin, J. L. & McCurdy, D. W. 1986. Production and use of monoclonal antibodies to study the cytoskeleton and other components of the cortical cytoplasm in *Chara*. European Journal of Cell Biology 41: 1–8.

Advances in Bryology 3: 33–93 (1988).

Ultrastructure and Development of Plastids in the Bryophytes

by

Jeffrey G. Duckett

School of Biological Sciences, Queen Mary College,
Mile End Road, London E1 4NS, U.K.

and

Karen S. Renzaglia

Department of Biological Sciences, Box 23590A, East Tennessee
State University, Johnson City, TN 37614, U.S.A.

With 63 Figures

Abstract: This article is a comprehensive analysis of electron micrographs of bryophyte plastids published since the advent of ultrastructural studies on the group over 25 years ago. Our own data on meristematic regions of diverse bryophytes are also included to shed new light on patterns of plastid development. Although bryophyte plastids contain chlorophylls *a* and *b* and possess highly organized granal and intergranal thylakoid systems, their ontogeny is distinctly different from that in vascular plants. In most bryophytes the meristems, including apical cells, contain chloroplasts similar to those in mature photosynthetic tissues. Thylakoid configurations differ greatly, even between closely related taxa. At one extreme are elongate grana comprising few thylakoids and a very limited fret system; at the other are tall stacks and extensive intergranal lamellae. Certain thylakoid arrangements appear to be peculiar to particular groups: vertical intergranal connections in hornworts resemble structures in the charalean alga *Coleochaete*, whereas predominantly peripheral grana in some members of the Metzgeriales are also found in *Chara*. There are no clear-cut differences in plastid ultrastructure between mosses and liverworts nor between gametophytic and sporophytic generations. Unusual thylakoid configurations are apparently specific to cells such as leptoids, spermatids, eggs, mucilage papillae, and transfer cells. Division of fully differentiated plastids is commonplace. Monoplastidic cell lines are established during spermatogenesis in all three groups and in sporogenesis in mosses. Plastid dedifferentiation and redifferentiation accompanies meiosis. In some situations amyloplasts may function in gravity perception. Associations with mitochondria, endoplasmic reticulum, the nucleus, and microtubules are more frequent than in vascular plants. Plastid morphology is far less susceptible to changes in differing light regimes than in vascular plants.

Zusammenfassung: Dieser Beitrag faßt elektronenmikroskopische Studien an Plastiden von Bryophyten zusammen, die seit dem Aufkommen von Ultrastrukturanalysen vor über 25 Jahren erschienen sind. Unsere eigenen Arbeiten über die meristematischen Zonen mancher Bryophyten sind ebenfalls eingeschlossen, um neues Licht auf die Struktur der Plastidenentwicklung zu werfen. Obwohl Moosplastiden Chlorophyll a und b enthalten und ein hochorganisiertes Grana- und Intergrana-Thylakoidsystem besitzen, ist doch deren Ontogenie deutlich anders als bei Gefäßpflanzen. Bei den meisten Bryophyten sind die Chloroplasten der Meristeme einschließlich der Scheitelzelle denen in ausgewachsenen Assimilationsgeweben ähnlich. Die Thylakoidkonfiguration ist äußerst unterschiedlich, sogar bei nah verwandten Gruppen. An einem Extrem finden sich verlängerte Grana mit wenigen Thylakoiden und begrenztem System von Granadurchbrüchen (frets), am andern Ende große Granastapel und ausgedehnte Intergranalamellen. Gewisse Thylakoidanordnungen scheinen nur in bestimmten Gruppen vorzukommen: Senkrechte Intergrana-Verbindungen bei Hornmoosen gleichen Strukturen bei der Grünalge *Coleochaete*, während die vorwiegend peripheren Grana bei einigen Mitgliedern der Metzgeriales auch bei *Chara* gefunden wurden. Es gibt keine scharfen Unterschiede in der Ultrastruktur zwischen Laub- und Lebermoosen, oder zwischen gametophytischer und sporophytischer Generation. Ungewöhnliche Thylakoidkonfigurationen finden sich offensichtlich in speziellen Zellen wie Leptoiden, Spermatiden, Eizellen, Schleimpapillen und Transferzellen. Die Teilung voll differenzierter Plastiden ist häufig. Monoplastidische Zellinien entstehen während der Spermatogenese in allen drei Gruppen und während der Sporogenese bei Laubmoosen. Während der Meiosis dedifferenzieren und redifferenzieren sich die Plastiden. In einigen Fällen mögen Amyloplasten der Schwereperception dienen. Verbindungen zu Mitochondrien, endoplasmatischem Reticulum, Kern und Mikrotubuli sind häufiger als bei Gefäßpflanzen. Wechselnde Lichtbedingungen verändern die Morphologie der Plastiden weit weniger als bei Gefäßpflanzen.

Keywords: Charophytes, chloroplast, differentiation, grana, hornworts, liverworts, meristems, morphogenesis, mosses, photosynthesis, prolamellar bodies, proplastid, thylakoids, ultrastructure.

Contents

Introduction

One of the earliest discoveries attendant on the study of bryophytes by transmission electron microscopy was that the internal organization of the chloroplasts had much in common with green algae and vascular plants (Manton 1957, 1962). Subsequent ultrastructural studies have all confirmed the presence of highly developed granal and intergranal thylakoid systems in the mature photosynthetic tissues of bryophytes, as is the case in vascular plants. However, many green algae (Pickett-Heaps 1975), not to mention *Prochloron* the prokaryote that also contains chlorophylls *a* and *b* (Cox 1986), have irregularly stacked thylakoids. The wealth of electron micrographs published over the last two decades clearly reveals that the plastids of mosses, liverworts, and hornworts are capable of dedifferentiation and redifferentiation along cyclical pathways recalling those in vascular plants (Wellburn 1982, Whatley 1977, 1978).

Perhaps because of their apparent similarity (apart from hornworts) to those in vascular plants, incisive studies of plastid ontogeny in bryophytes are few in number (Duckett 1986). Any general picture of patterns of development and structural diversity must be gleaned from published micrographs in studies with principal aims other than an understanding of plastid morphogenesis. Consequently in most reviews (see Wellburn 1982 for an extensive listing) on chloroplast development, however comprehensive their coverage of the plant kingdom, bryophytes rarely receive more than perfunctory mention. Generalizations are often based on the scrutiny of very few original papers and, as far as liverworts are concerned, on but a single species, *Marchantia polymorpha* L.

Drawing on a comprehensive analysis of published micrographs and our own recent data, this presentation is the first cohesive body of detailed information on the ultrastructure of bryophyte plastids. A recurrent theme will be how plastids in meristematic regions differ from those in fully differentiated tissues. This forms the foundation for detailed comparisons of plastid development between bryophytes, vascular plants, and algae. Where relevant we will highlight features of systematic interest particularly in relation to the plastids of charalean algae. Wherever possible the functional significance of the diverse plastid morphologies will be pointed out. It must be stressed, however, that the systematic and/or functional relevance (if any) of the majority of ultrastructural attributes of bryophyte plastids is presently unknown.

It was not our intention here to consider the photosynthetic apparatus of bryophytes from biochemical and physiological standpoints. Such information is available in other reviews (e.g., Osborn & Raven 1986, Valanne 1984). Similarly we have not reiterated the elegant exposition on plastid cycles by Paolillo (1984) or the purely systematic aspect of spermatid plastids (Duckett et al. 1982, 1984).

The micrographs illustrating this account were of materials fixed by standard aldehyde-osmium protocols, unless noted otherwise in the figure legends.

Anthocerotophyta

Mature Chloroplasts

More ultrastructural studies have been specifically devoted to the plastids of anthocerotes than to those of mosses and liverworts combined; this seems principally due to their algal features, the presence of pyrenoids, and the uniplastidic condition. Burr's comparative study (1970) shows that the chloroplasts in each gametophytic cell of the

Figures 1, 2. *Phaeoceros laevis* (L.) Prosk. 1, Longitudinal section through the thallus apex, note the highly differentiated chloroplast (c) in the apical cell (ac) identical to those in the immediate derivatives (d), n, nucleus, x5500, scale bar = 5 µm; 2, Detail of the pyrenoid (py) in an apical cell plastid (serial section with Fig. 1), s, starch, x26,000, scale bar = 1 µm.

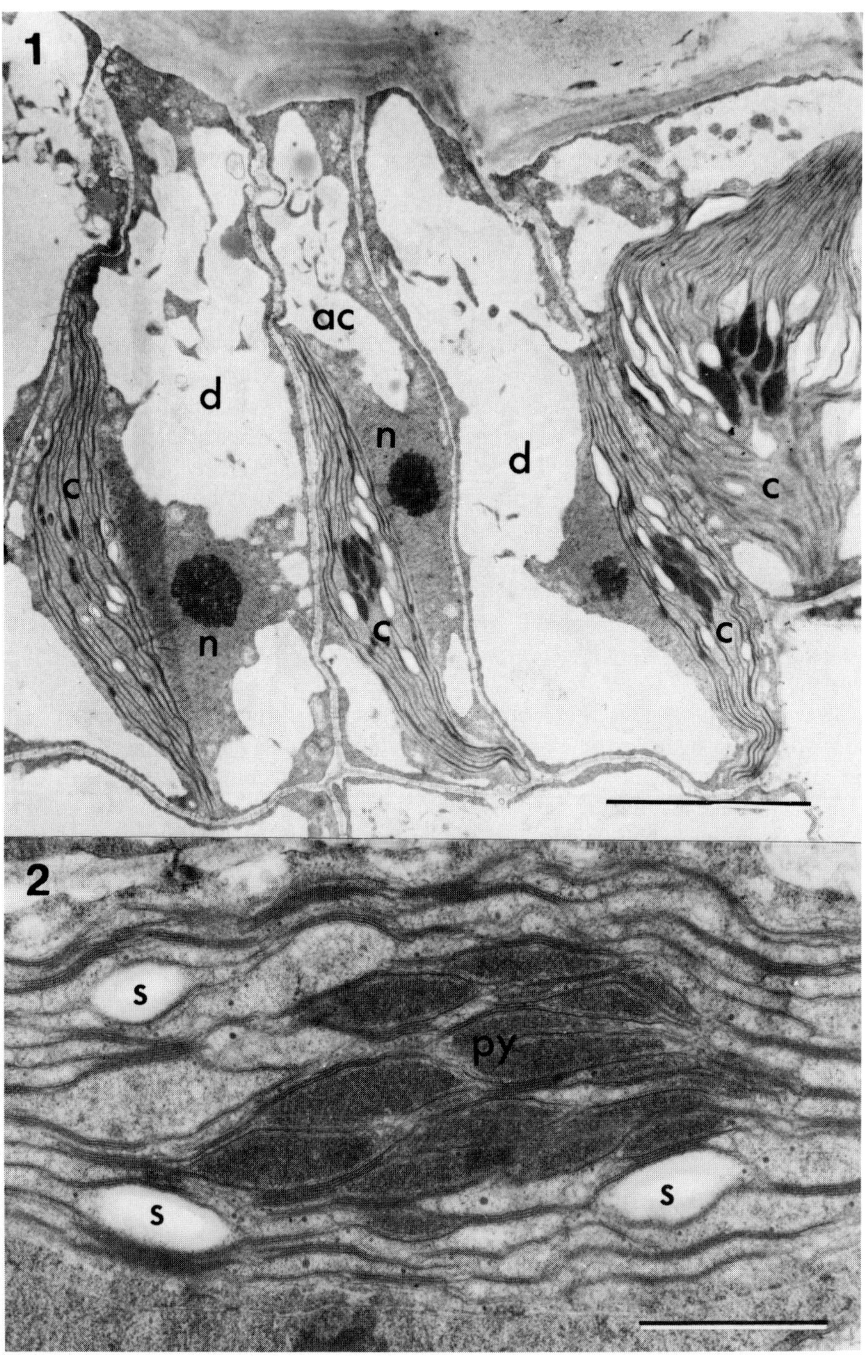
1
ac
d
n
d
c
c
c
n
c
2
s
py
s
s

monoplastidic genera *Phaeoceros* (Fig. 1–3) and *Notothylas* (Fig. 4) possess a centrally-located multiple pyrenoid [i.e., intersected by numerous thylakoids (Griffiths 1970, 1980)] surrounded by abundant starch grains (Fig. 4). Chloroplast organization in *Anthoceros* is identical (Renzaglia & Duckett unpublished data). *Megaceros* displays several kinds of chloroplast morphologies that correlate with the number of plastids per cell. Taxa with one or two plastids per cell possess distinct pyrenoids, while those with four or more lack pyrenoids and contain evenly dispersed starch (Fig. 5). The random distribution of starch and absence of pyrenoids in *Megaceros* have recently been confirmed by Valentine et al. (1986). These authors also found that within *Dendroceros* chloroplast structure varies between species. In *D. giganteus* (Lehm. et Lindenb.) Prosk. starch grains are randomly distributed and pyrenoids absent, whereas *D. granulatus* Mitt. possesses a multiple pyrenoid dissected by highly convoluted thylakoids. As in many algae (Griffiths 1970, 1980), oil droplets are common around the periphery of the pyrenoids and apparently replace starch as the major reserve material in this species. The chloroplasts of *D. validus* Steph. show a range of morphologies from organelles with widely channelled, convoluted pyrenoids surrounded by starch to forms with narrowly channelled thylakoids and no starch.

From these two comparative studies (Burr 1970, Valentine et al. 1986) it is clear that the mature chloroplasts of hornworts vary greatly between species. It would seem likely that other morphologies still await discovery as more species are investigated. To date, the uniformity in *Phaeoceros*, *Notothylas*, and *Anthoceros* contrasts sharply with the situation in *Megaceros* and *Dendroceros*.

Irrespective of the presence or absence of pyrenoids, the mature chloroplasts of all hornworts investigated to date possess clearly defined grana interconnected laterally by numerous thylakoids. The number of thylakoids comprising each granum can vary considerably even within the same organelle. In general the larger the number of thylakoids in the stack the smaller the diameter. Paired or triplet thylakoids may extend for several µm through the stroma, whereas aggregations of four or more rarely exceed 1.5 µm in diameter (Fig. 6–10).

Aside from confirmation that the substructure of the pyrenoid matrix is virtually indistinguishable from that in green algae (possible similarity at the biochemical level has yet to be explored), the most significant discovery from ultrastructural studies of anthocerotalean plastids concerns thylakoid architecture. In addition to the intergranal system linking grana horizontally (i.e., parallel to the long axis of the plastid) other intergranal thylakoids extend outward at right angles from the outer membranes of

Figures 3–5. Mature chloroplasts in the Anthocerotophyta. **3,** Sporophyte chloroplast of *Phaeoceros laevis*, py, pyrenoid, s, starch, x7000 (courtesy of Ligrone & Fioretto 1987); **4,** Gametophyte chloroplast of *Notothylas orbicularis* (Schwein.) Sull. py, pyrenoid, s, starch, x8000; **5,** Gametophyte chloroplast of *Megaceros*, s, starch, x8850. All scale bars = 2 µm.

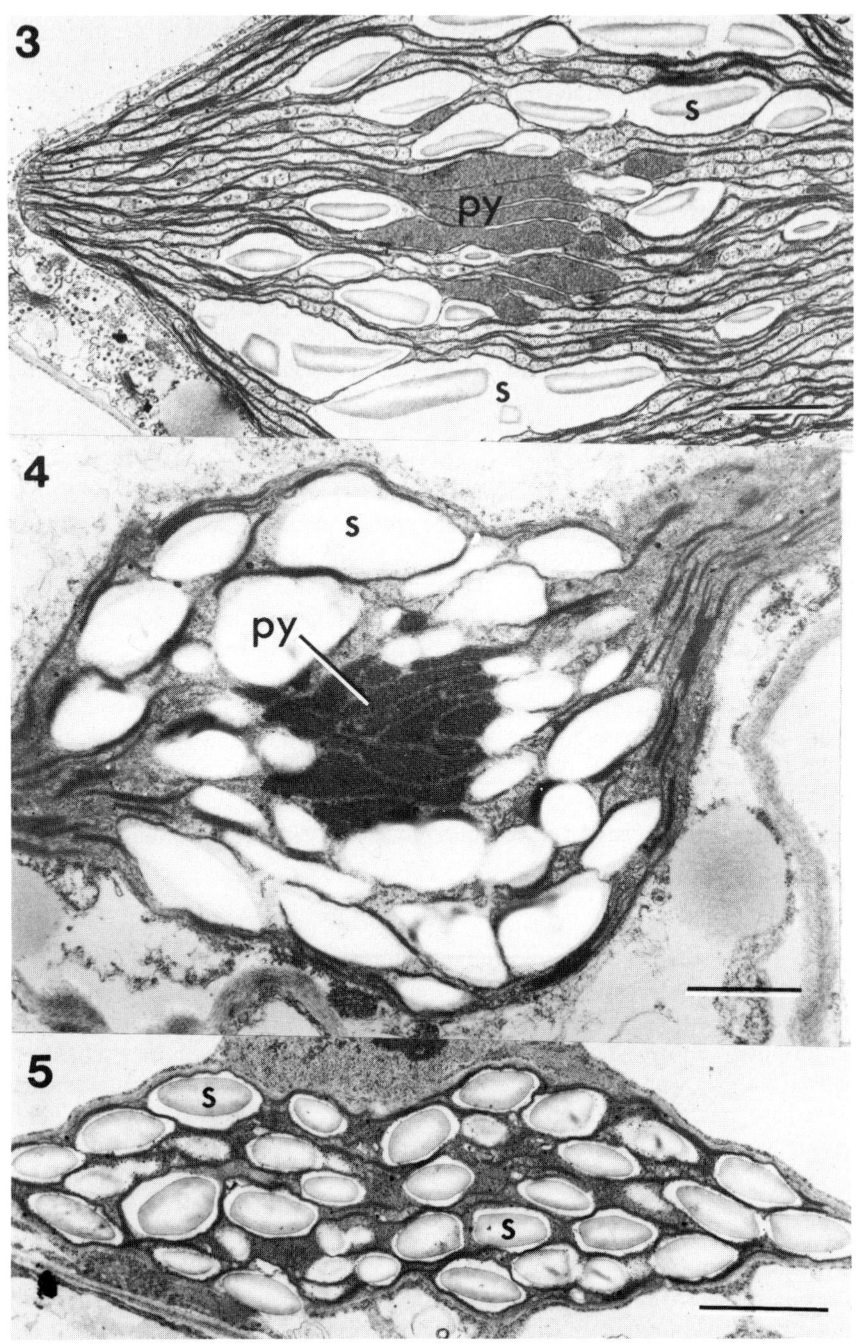

each stack and link adjacent grana vertically (i.e., perpendicular to the long axis). These thylakoids dissect the stroma into a complex series of channels (Fig. 7), which appear in sections as vesicles bounded by the thylakoid membranes (Wilsenach 1963). In addition to elucidating the real nature of the channelled stroma in hornworts following its misinterpretation in earlier studies (Menke 1961, Manton 1962), Wilsenach discovered that the vertical interconnections originate from outfolding and fusion of the outer thylakoids of neighboring grana. It must be underlined that the hornwort "vesicles" between the grana are very different from the networks of vesicles found in other land plants (e.g., the peripheral reticulum of C4 plants). In hornworts the contents of the "vesicles" are plastid stroma (the compartment that also contains the pyrenoid matrix and starch). In other plastids the vesicles lie within the stroma; their contents are intrathylakoidal. The fact that the anthocerotalean "vesicle" system is most well developed in mature chloroplasts, whereas vesicles in other plants are more characteristic of organelles undergoing differentiation (i.e., precursors of a more highly organized membrane system) further underlines this fundamental difference.

In view of the obsession with the systematic and phylogenetic significance of anthocerotalean plastids it is most surprising that comparisons with green algae have hitherto only considered characters visible under the light microscope (viz., number and pyrenoids). Most remarkable is the absence of any critical comparison of plastid ultrastructure in hornworts with that in those green algae [the Characeae of Mattox and Stewart (1984)] now widely accepted as related to the same ancestral stock as archegoniate plants on the basis of motile cell ultrastructure, nuclear and cell division (Mattox & Stewart 1984, Stewart & Mattox 1975, 1978, Pickett-Heaps 1975, Pickett-Heaps & Marchant 1972), and enzymes of the glycolate pathway (Frederick et al. 1973, Gruber & Frederick 1977). Unfortunately, micrographs selected to illustrate features of mitosis, cytokinesis, and the motile apparatus of such algae rarely show the plastids to the best advantage and thus provide only limited opportunity for critical

Figures 6–10. Effects of darkness on the ultrastructure of sporophyte chloroplasts of *Phaeoceros*. **6**, Detail of a chloroplast near the basal meristem differentiating in light, the envelope and associated endoplasmic reticulum (er), invaginations of the inner membrane arrowed, note the absence of vertical interconnections between the grana, x29,000; **7**, Detail of a fully differentiated chloroplast about 3 mm from the sporophyte base, interconnections between adjacent grana at arrows, x22,000; **8**, After 60 days in the dark the plastids in the meristematic cells are irregular in outline and contain massive grana and paracrystalline networks of membranes (arrows), x26,000; **9**, Detail of paracrystalline network from a meristem plastid after 60 days in the dark, x40,000; **10**, Mature chloroplast after 60 days in the dark, well-developed thylakoid system with prominent grana clearly visible but vertical interconnections are absent (cf. Fig. 6), py, pyrenoid, x24,000. All scale bars = 1 μm, except **Fig. 8** = 2 μm (All micrographs courtesy of Ligrone & Fioretto 1987).

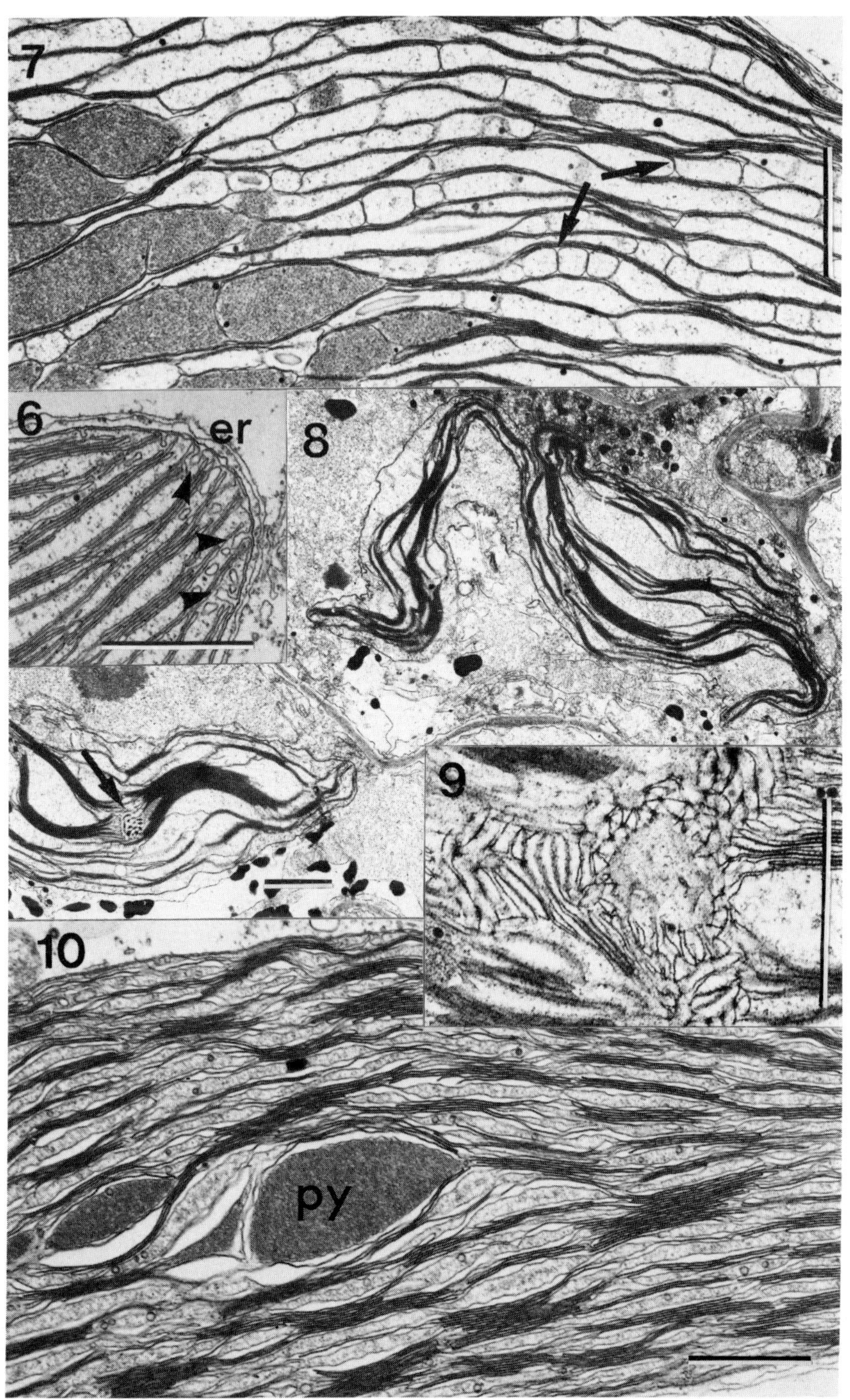
7
6
er
8
9
10
py

comparisons of thylakoid organization. However, careful scrutiny of Pickett-Heaps' encyclopedia (1975) and micrographs in publications on charalean algae reveals features that may have considerable bearing on the affinities of the anthocerotes. As might be expected, compound pyrenoids similar ultrastructurally to those in hornworts are of frequent occurrence throughout the green algae. But the only micrographs that display clear signs of a channelled stroma are of the charalean genus *Coleochaete* (Marchant & Pickett-Heaps 1973: Fig. 12–14). This alga also shares with hornworts the absence of a typical preprophase band of microtubules and exhibits a similar orientation of the plastids relative to the division plane (Brown & Lemmon 1985a). Similar prophasic plastid disposition is also a characteristic of *Selaginella* and *Isoëtes*, but here a discrete preprophase band marks the equatorial plane (Brown & Lemmon 1984a, 1985b, c).

Electron-dense bodies, strongly reminiscent of peroxisomes (microbodies) are associated with the plastid isthmus during the division process in hornworts. Similarly located microbodies are present not only in *Coleochaete* (Marchant & Pickett-Heaps 1973, Pickett-Heaps 1975) but also in a further charalean genus *Klebsormidium* (Pickett-Heaps 1972, Gruber & Frederick 1977). Tantalizingly, it is impossible to ascertain from published micrographs whether or not the plastids of *Klebsormidium* possess a channelled thylakoid system (Pickett-Heaps 1972, Marchant & Pickett-Heaps 1973). What is clear is that the plastids of *Klebsormidium* are similar to those of monoplastidic hornworts with central multiple pyrenoids and numerous starch grains. In overall layout the plastids of *Coleochaete* are strikingly similar to those in some species of *Dendroceros* and *Megaceros*. Some micrographs show no pyrenoids and randomly distributed starch [e.g., Fig. 9 in Marchant and Pickett-Heaps (1973), peripheral cell of the thallus of *C. scutata*], while others reveal numerous starch grains clustered around a centrally located compound pyrenoid [e.g., Fig. 5 in Sluiman (1983) and Fig. 5 in Graham and McBride (1979), the zoospores of *C. pulvinata* and *C. scutata*, respectively; Fig. 7 in Graham (1982), the internal vegetative cell of *C. orbicularis*]. From comparative observations, Burr (1970) suggests that the differences in plastid ultrastructure in the hornworts represent a phylogenetic trend towards an increase in plastid number, a decrease in size, and the loss of the pyrenoid (i.e., from algal-like to a more advanced plant form). The foregoing discussion indicates that such speculation is premature. This trend could equally well be interpreted in the opposite direction. Furthermore, Burr failed to take account of the biplastidic and monoplastidic situations associated with spermatogenesis and sporogenesis in mosses and liverworts (Duckett 1986) and the fact that the plastids of *Chara* (in terms of both number and ontogeny) have far more in common with those of liverworts than with *Klebsormidium*, *Coleochaete*, or hornworts.

As to other green algae with multilayered structures in their motile cells (Mattox & Stewart 1984), the plastids of *Chlorokybus* (Rogers et al. 1980) are quite unlike those of hornworts. They contain two types of pyrenoids, one simple and superficial, the other embedded and complex but not traversed by thylakoids. The chloroplasts of *Trentepohlia* (Graham & McBride 1975) lack pyrenoids and contain long, paired thylakoids that are very different from those of hornworts. Despite the presence of an MLS, Mattox and Stewart (1984) include *Trentepohlia* in the Ulvophyceae and not in the charalean line.

In a study of the zoospores of *Chaetosphaeridium*, a second member of the Coleochaetales, Moestrup (1974) comments that the presence of typical grana and intergranal lamellae (though not itself conclusive) indicates a phyletic affinity between this alga and higher plants. The plastids in question (Moestrup 1974, plate 5E) contain clearly defined grana approximately 0.5 μm in diameter, comprising up to 10 thylakoids and interconnected by numerous intergranal lamellae, i.e., similar to chloroplasts in angiosperm leaves and very different from those in hornworts. Although the precise interrelationships between the various charophycean algae and land plants are still far from clear, these considerations of plastid ultrastructure suggest close affinities between hornworts, *Coleochaete*, and *Klebsormidium*, whereas *Chaetosphaeridium* and the Charales appear to have more in common with mosses and hepatics.

Ontogeny in Gametophytic Tissues

Following the widely quoted assumption (Duckett 1986) based on vascular plants that the meristematic cells of bryophytes contain proplastids (Whatley 1978, Wellburn 1982), we were surprised to discover that the apical cells and their immediate derivatives (Fig. 1, 2) in the thalli of *Phaeoceros* and *Notothylas* contain plastids identical in substructure to those in fully differentiated cells, i.e., with multiple pyrenoids and highly structured thylakoid systems including grana and channelling of the stroma. Uniformity in plastid ultrastructure is characteristic of most cells of *Phaeoceros* and *Notothylas* thalli and presumably other Anthocerotophyta, the only significant departure being variations in amount of starch. The inescapable conclusion from these observations is that in hornworts, just as in the majority of green algae (Pickett-Heaps 1975) and some vascular plants, fully differentiated chloroplasts regularly undergo division.

Towards the ventral side of the thallus, in the cells bordering the *Nostoc* colonies and in the filaments intermingled with the cyanobacterial cells, the plastids have far less complex thylakoid systems (Duckett et al. 1977, Honegger 1980). Grana are still present, but the intergranal system including the vertical anastomoses are much less in evidence.

Further departures from the typical chloroplast ultrastructure have been found in the placental region adjacent to the sporophyte foot in *Phaeoceros* (Gambardella et al. 1981). Passing from typical gametophyte cells to the inner transfer cells, there is a progressive reduction in starch content, simplification of the thylakoid system, dispersion and eventual loss of the pyrenoid, but an increase in the number of plastoglobuli. In the transfer cells the plastids are pleomorphic, with compact grana comprising two to six thylakoids filled with an electron-dense matrix. The stroma occupies most of the plastid interior (cf. normal vegetative cells).

Far more profound changes in the shape, size, and internal organization of the plastids are associated with antheridial development and spermatogenesis (Duckett 1975). In the early generations of spermatogenous cells the pyrenoid disperses and disappears as does the channelling of the stroma between adjacent grana. Later generations (Fig. 49) contain irregular sheet-like plastids with swollen margins. The internal lamellar system is reduced to scattered vesicles. Following the final antheridial mitosis, the plastids rapidly become spherical and are between 0.5 and 1 µm in diameter (cf. over 20 µm in vegetative cells). They are ensheathed by endoplasmic reticulum and intimately associated with the single spermatid mitochondrion lying next to the inner surface of the nucleus. Internally a series of tubular membranes becomes arranged in a cylindrical array in which a single starch grain approximately 1 µm in length and 0.2 µm in diameter (Fig. 50) completely fills the plastid.

In contrast to plastids in spermatogenous cells, jacket-cell plastids remain much the same as those in vegetative cells of the thallus until the onset of spermatid metamorphosis. Thence the pyrenoid disperses and osmiophilic globuli accumulate in sheets between the elongate grana, which now lack vertical interconnections. Accumulation of osmiophilic globuli in sheets is a notable feature of bryophyte antheridia exposed directly to incident radiation and is associated with maturational color changes from green to yellow or orange. It would seem likely that the multilayered screen of globuli, presumably rich in the carotenoid pigments responsible for antheridial coloration, shields the developing gametes from ionizing solar radiation. Layers of osmiophilic globuli characterize the chromoplasts in the petals of many angiosperms (Brett & Somerard 1986) and the mature orange-colored antheridial shield cells in *Chara* (Pickett-Heaps 1975).

Plastid ultrastructure during oogenesis in hornworts has yet to be investigated. In fact the only other monoplastidic eggs to be examined at this level are those of *Isoëtes* and *Selaginella* (Paolillo, personal communication). Here egg maturation is accompanied by the budding of fragments of plastid material, which subsequently degenerate. It would thus be of considerable interest to see if a similar phenomenon occurs in hornworts, not to mention the historical precedent that the cytological continuity of plastids through germ cells was first demonstrated in *Anthoceros* (Schimper 1885).

Ontogeny in Sporophytic Tissues

With fully differentiated chloroplasts the norm throughout the thallus, gametophytic tissues of hornworts provide little opportunity for investigation of the elaboration of the thylakoid system. In light of this it is now clear why an understanding of the differentiation of the thylakoid apparatus in hornworts derives exclusively from observations on the sporophyte. Although mature chloroplasts in fully differentiated sporophytic tissues of hornworts (as yet limited to observations of *Phaeoceros* and *Notothylas*) are apparently indistinguishable from those in the gametophyte (cf. Fig. 1, 2, 3), their patterns of development are somewhat different. While Brown and

Lemmon's study (1985a) confirms the inference from gametophytes that fully differentiated chloroplasts regularly undergo division (Brown & Lemmon 1985a: Fig. 2), it illustrates the replication of forms with less complex thylakoid systems (Brown & Lemmon 1985a: Fig. 1). Originally Wilsenach (1963) and very recently Ligrone and Fioretto (1987) have traced the progressive elaboration of the lamellar network in *Phaeoceros* from the meristematic zone at the base of the sporophyte. Initially the inner membrane of the envelope invaginates to form vesicles. These develop into flattened sacs, which subsequently aggregate and form elongate grana comprising small numbers of thylakoids. Extensive invaginations of the envelope are numerous at this stage, and such plastids are often associated with sheets of endoplasmic reticulum (Fig. 6). At the same time lenses of pyrenoid material begin to accumulate. Finally, outfolding and fusion of the outer-grana thylakoids produce the characteristic channelling of the stroma (Fig. 7). In mature chloroplasts doublet and triplet thylakoid arrays form the greater part of the membrane system with larger grana much more restricted in their distribution (Fig. 3, 7).

The haustorial cells in the sporophyte foot (cf. those in the gametophytic transfer cells described in the previous section) contain plastids different from those found elsewhere in the sporophytes (Gambardella et al. 1981). These plastids are approximately spherical and devoid of grana. They contain several starch grains, each partially ensheathed by single thylakoids, and distinct areas comprising networks of vesicles and plastoglobuli embedded in electron-dense granular material similar to that of the pyrenoid matrix.

In contrast to angiosperms, which in general synthesize chlorophyll only in the light (Bradbeer 1981, Wellburn 1982), gymnosperms, ferns, and mosses are able to synthesize chlorophyll in the dark. Against this background Ligrone and Fioretto (1987) explored the possibility of chlorophyll synthesis in tandem with chloroplast development in sporophytes of *Phaeoceros laevis* (L.) Prosk. kept in the dark for 60 days. After this period, immature plastids contained extremely prominent grana (Fig. 8), with thylakoids converging toward paracrystalline networks of membranes (Fig. 9) reminiscent of prolamellar bodies. In similarly treated mature chloroplasts there was a marked increase in the number of grana per unit area (cf. Fig. 7, 10), an increase in the average number of thylakoids per granum but a dramatic reduction in the channelling of the stroma. The compound pyrenoid was unaffected by this period of darkness. Chlorophyll measurements revealed substantial synthesis in the dark and a significant reduction in the chlorophyll a/b ratio.

Hepatophyta

Apart from comparative data on spermatids (Duckett et al. 1982, 1984), studies of hepatics detailing plastid ultrastructure are limited to a few species. Indeed, only in *Marchantia* and to a lesser extent *Pellia* are there sufficient published micrographs to attempt any interpretation of developmental interrelationships. Particularly acute is the lack of data on the Jungermanniales – almost certainly a reflection of the fact that conspicuous thalloid taxa are the most readily available to electron microscopists lacking a background in hepatic systematics. Drawing on hitherto unpublished data on a variety of metzgerialean, marchantialean, and jungermannialean taxa we will attempt to give an overall picture of plastid diversity in the Hepatophyta. At the same time this will underline that generalizations about hepatic ultrastructure on the basis of one or two genera are highly inappropriate.

Marchantiales

By far the most useful publications to date for understanding plastid ontogeny in *Marchantia* are the works by Galatis and associates on the development of air pores and chambers, mucilage papillae, and oil-body cells (Apostolakos & Galatis 1985a, b, c, Galatis & Apostolakos 1976, 1977, Galatis et al. 1978a, b). One of the most striking features of the apical region of the thallus (Apostolakos & Galatis 1985a, b, c) is the remarkable uniformity in the size, shape, and internal organization of the plastids. Each organelle is spherical or oval and 2–3 μm in diameter. Transversing the stroma are one to three groups of membranes comprising elongate grana and short stroma thylakoids. Each granum is made up of less than six thylakoids, at least 1 μm across and often extending the full diameter of the plastids. Dividing cells (Fowke & Pickett-Heaps 1978, Steer 1985) contain plastids with the same organization thus indicating that, in contrast to angiosperm apices (Whatley 1978, 1980, Bradbeer 1981, Wellburn 1982), proplastids with rudimentary thylakoid systems are not a normal feature of the meristems in *Marchantia*.

In young oil-body cells, well removed from the apical cells of the thallus, the plastids are shortly cylindrical with long grana and short intergranal systems lying parallel to the axis of the organelles (Galatis et al. 1978a, Galatis & Apostolakos 1976). Mature oil-body cells (Galatis et al. 1978b) retain the same thylakoid organization, with the occasional addition of small prolamellar bodies. Umbo-shaped profiles are interpreted as stages in division by constriction. Galatis et al. (1978b) also demonstrate that large osmiophilic globuli associated with the thylakoids are almost certainly a fixation artifact resulting from leakage from the oil bodies during fixation. In the past (Suire 1970) the presence of osmiophilic globuli in the plastids has been interpreted as indicating the origin of lipids from the plastids. There is no evidence from the work of Galatis et al. or our own studies on Aneuraceae (Duckett & Renzaglia 1987) that the plastids of hepatics have any role in oil body elaboration.

In contrast to the oil-body cells, the fully differentiated thallus cells of *Marchantia* contain amylochloroplasts, either discoidal in outline or distended with multiple starch grains. In addition to elongate grana made up of five or fewer thylakoids, these plastids contain more compact grana approximately 0.5 μm in diameter with up to 10 thylakoids (Albertine et al. 1976, Mache & Loiseaux 1973).

During the differentiation of mucilage papillae (Galatis & Apostolakos 1977) the thylakoid system regresses to numerous vesicles scattered through the stroma, and the plastids become highly pleomorphic. At an intermediate stage the plastids become closely associated with mitochondria, and slightly later are completely ensheathed by endoplasmic reticulum. Similar associations of plastids with mitochondria and endoplasmic reticulum are noticeable during oil-body cell differentiation (Galatis et al. 1978a).

Further variations in plastid ultrastructure in *Marchantia* are described by Berrie and Webster (1982) in germinating gemmae. In the most actively meristematic regions of gemmae the plastids possess cytoplasmic caps delimited by a double membrane continuous with the outer membrane of the plastid envelope. Whether or not the caps are derived from endoplasmic reticulum or evagination of the outer membrane of the envelope is unresolved, but their significance may lie in the 20–30 % amplification of the surface area to volume ratio of the plastids in rapidly growing cells.

Berrie and Webster's study (1982) provides good evidence for two distinct modes of plastid division: by partition in meristematic and maturing vegetative cells and by simple fission (constriction) in oil-body cells and rhizoidal initials.

Sphaerocarpales

As far as may be gleaned from the published micrographs in Lehmann and Schulz (1982), Schulz and Lehmann (1969), and Lehmann and Jaster (1981), plastid ontogeny in *Riella* is similar to that in *Marchantia*. The meristem plastids are extremely uniform in size, shape, and internal organization. These ovoid structures, 2–3 μm in diameter, contain five to seven groups of thylakoids extending across the long axis. Each membrane group comprises long grana with mostly two or three but occasionally up to five thylakoids interrupted by shorter unpaired sacs. One to several starch grains lie between the groups of thylakoids. In mature oil-body cells (Lehmann & Jaster 1981) the plastids are shortly cylindrical in shape, but their thylakoid organization is the same as that in meristems. The ultrastructure of the plastids in young sporophytes of *Sphaerocarpos* is similar to that in the gametophytes of *Riella* (Kelley & Doyle 1975), with an internal lamellar system comprising four to six distinct groups of mostly paired thylakoids.

Metzgeriales

In the Metzgeriales published micrographs of plastids in vegetative tissues are limited to *Pellia* (Pihakaski & Pihakaski 1979, 1980). Mature thallus cells contain discoidal to irregularly shaped chloroplasts with thylakoid organization virtually indistinguishable from that in the leaves of many angiosperms. Regular grana, approximately 0.5 μm in diameter and containing up to 12 thylakoids, are interconnected by a regular and extensive intergranal system. Alternate thylakoids in each granum extend laterally to form the cross-links. Up to 75 % of the volume of the plastids is taken up by the thylakoid system compared with rather less than 30 % in *Marchantia*. In the mature thallus tissues of *Conocephalum* (Marchantiales) thylakoid architecture is virtually identical to that in *Pellia* (Gambardella et al. 1979b), which has regular narrow grana, a sharp contrast to the elongate forms found in *Marchantia*.

Our observations on plastid diversity in *Blasia* reveal that the thylakoid organization in the vegetative cells is closely similar to that in *Pellia* but that variations exist that are peculiar to certain cell types (Fig. 11–19). Apical cells and their immediate derivatives contain ovoid to slightly irregularly shaped plastids with two to four lamellar units, which are apparently separate from each other. Each of these comprises a single granum from which single thylakoids extend laterally (Fig. 11). Similar thylakoid organization occurs in the plastids of the apical slime papillae (Fig. 12). Stromal aggregations of phytoferritin (Stead & Duckett 1980) are a regular feature of these plastids, and such aggregations are occasionally encountered in other kinds of gametophytic tissue (Fig. 17). By contrast, the plastids in the cells of the internal slime papilla of the immature *Nostoc* auricle (Fig. 13) contain scattered plastoglobuli and mainly paired thylakoids orientated perpendicularly to the long axis (cf. the moss haustorium described subsequently). Rudimentary thylakoid systems and numerous plastoglobuli are a characteristic feature of the plastids in the transfer cells associated with the mature *Nostoc*-filled auricles in *Blasia* (Duckett et al. 1977).

Passing from the meristem into differentiating thallus cells, there is an increase in the size and number of the grana (Fig. 14) until in the mature tissues each spherical to ovoid plastid contains numerous discrete grana interconnected by an extensive intergranal system (Fig. 15, 16). Most of the grana contain 5–10 thylakoids, but numbers up to 30 are not uncommon. The main structural variation lies in the amount of starch (cf. Fig. 16, 17). Figures 16 and 17 show thylakoid systems divided into two along the plane indicated by the arrows. We suggest that these profiles are of organelles immediately prior to division. Whether this is by partition or constriction has yet to be determined.

The sporophytic tissues of *Blasia* (Fig. 18, 19) contain amylochloroplasts with well-defined granal and intergranal architecture closely similar to that in the gameto-

Figures 11–16. Plastid diversity in the gametophyte of *Blasia pusilla* L. 11, Plastid with compact grana from an apical cell of the thallus, x35,500; **12,** Plastid from a slime papilla at the apex of the thallus, note compact grana and an aggregation of phytoferritin (pf) in the stroma, x35,500; **13,** Plastid from a slime papilla within a *Nostoc* auricle, note the paired thylakoids oriented perpendicularly to the long axis of the organelle, x30,000; **14,** Plastid in slightly older thallus cell, x30,000; **15,** Detail of the granum from a chloroplast in the thallus epidermis, x35,000; **16,** Plastid containing numerous grana and no starch from an epidermal cell of the mature thallus. Plane of division of the thylakoid systems indicated by arrow, x9600. All scale bars = 0.5 μm, except **Fig. 16** = 2 μm.

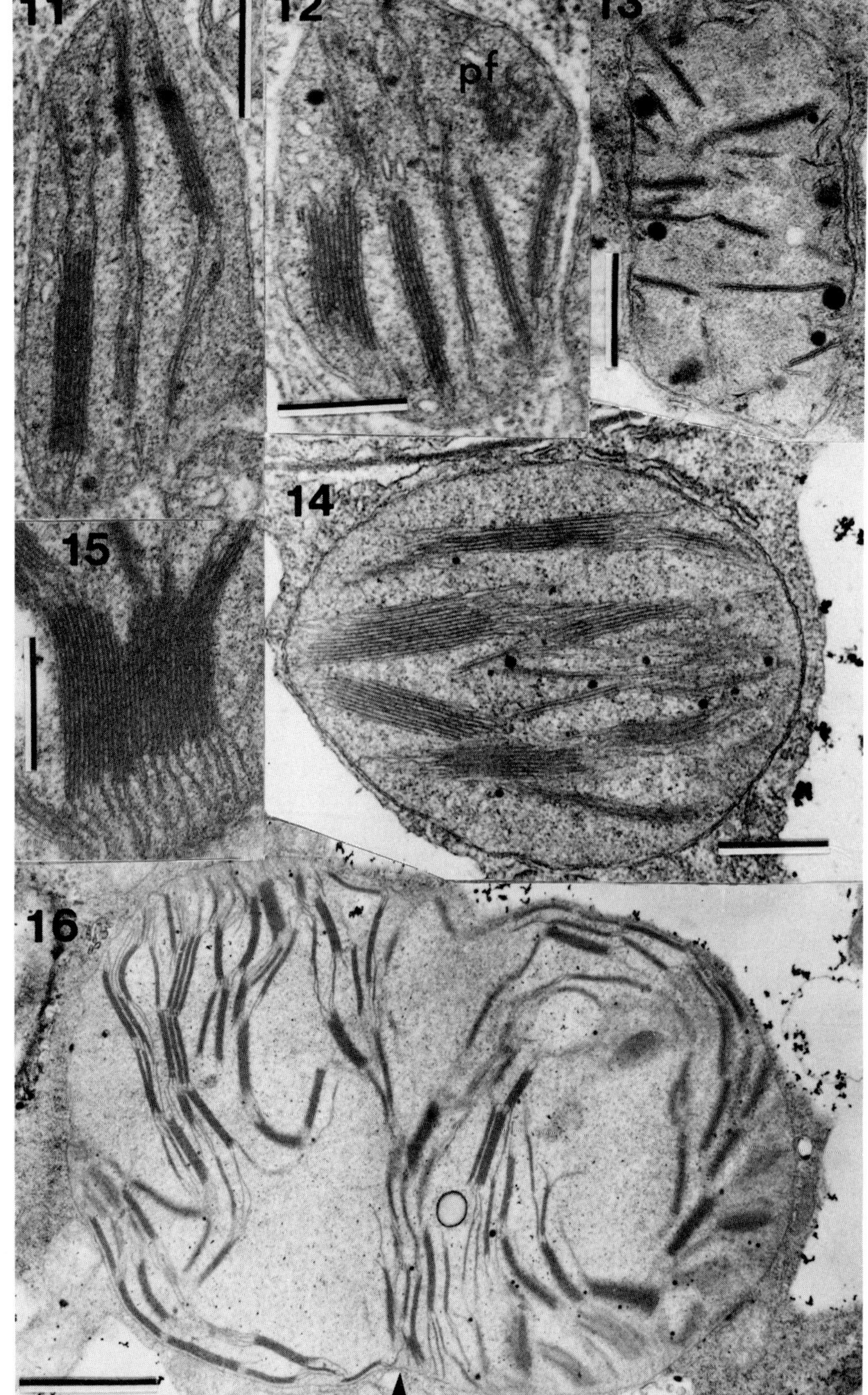
11
12
pf
13
15
14
16

phyte. Sporophytic plastids are, however, readily recognizable by the peripheral location of the starch grains and the consequent aggregation of the lamellar system in the center. In gametophytic tissues the starch grains are invariably separated from the plastid envelope by at least one group of thylakoids (Fig. 17).

Observations on thallus cells of *Pallavicinia* (Fig. 29) reveal plastid ultrastructure similar to that in *Blasia* and *Pellia*, except the grana contain fewer thylakoids. Generalizing from *Pellia*, *Blasia*, and *Pallavicinia* one might conclude that plastid ultrastructure in the Metzgeriales is rather uniform with thylakoid organization closely similar to that in vascular plants. However, this is not the case.

Sections of the thallus of *Metzgeria* (Fig. 20–22) reveal a very different sort of plastid organization and one that changes hardly at all from the apical cell and its immediate derivatives to the mature thallus. The plastids are irregular polyhedra with the thylakoid system largely restricted to the periphery. Parallel to each face lie one or two grana each comprising 3–12 thylakoids. The stroma lamellar-system is poorly developed, and each granum is largely separate from its neighbors.

In the apical cell and its immediate derivatives in *Riccardia* (Fig. 23) the plastids appear triangular or elliptical in section. The majority contain four to six groups of elongate grana each made up of two to four thylakoids. Some apical cell plastids contain single thylakoids traversing the long axis (Fig. 24). In some areas, particularly near the ends, stacks of two to four thylakoids are present. Aggregations of irregular osmiophilic globuli are often present in the stroma of such plastids, whereas mature chloroplasts contain well-defined spherical globuli (Fig. 25, 26).

Irregular polyphedral chloroplasts characterize the cells of the mature thallus (Fig. 25, 26), but their thylakoid network has a higher degree of organization than in *Metzgeria* (Fig. 20–22). Adjacent to each face of the chloroplast and running parallel to it are elongate grana of two or three thylakoids (Fig. 25, 26). Compact grana 0.5 μm in diameter and containing up to 10 thylakoids are convergently grouped near the angles of the polyhedra. These grana are interconnected by elongate pairs and triplets of thylakoids. Occasionally grana occur in the center of the plastids (Fig. 25), but otherwise the central stroma is devoid of membranes. The intergranal system is largely limited to lengths of single thylakoids protruding for short distances (0.3 μm or less) from the compact grana. We interpret the occurrence of small grana at the ends of the thylakoid system in apical cell plastids (Fig. 24) as developmental stages in the elaboration of the grana found near the angles of mature chloroplasts. It is interesting to

Figures 17–19. Amylochloroplasts from fully differentiated cells of *Blasia pusilla*. **17,** Stellate gemma, numerous starch grains between the grana and occasional phytoferritin (pf) aggregations, plane of division of the thylakoid systems indicated by arrow, x16,200; **18,** Seta, large peripheral starch grains alongside large grana, x24,000; **19,** Capsule wall, peripheral starch grains and a central aggregation of grana, x24,000. All scale bars = 1 μm.

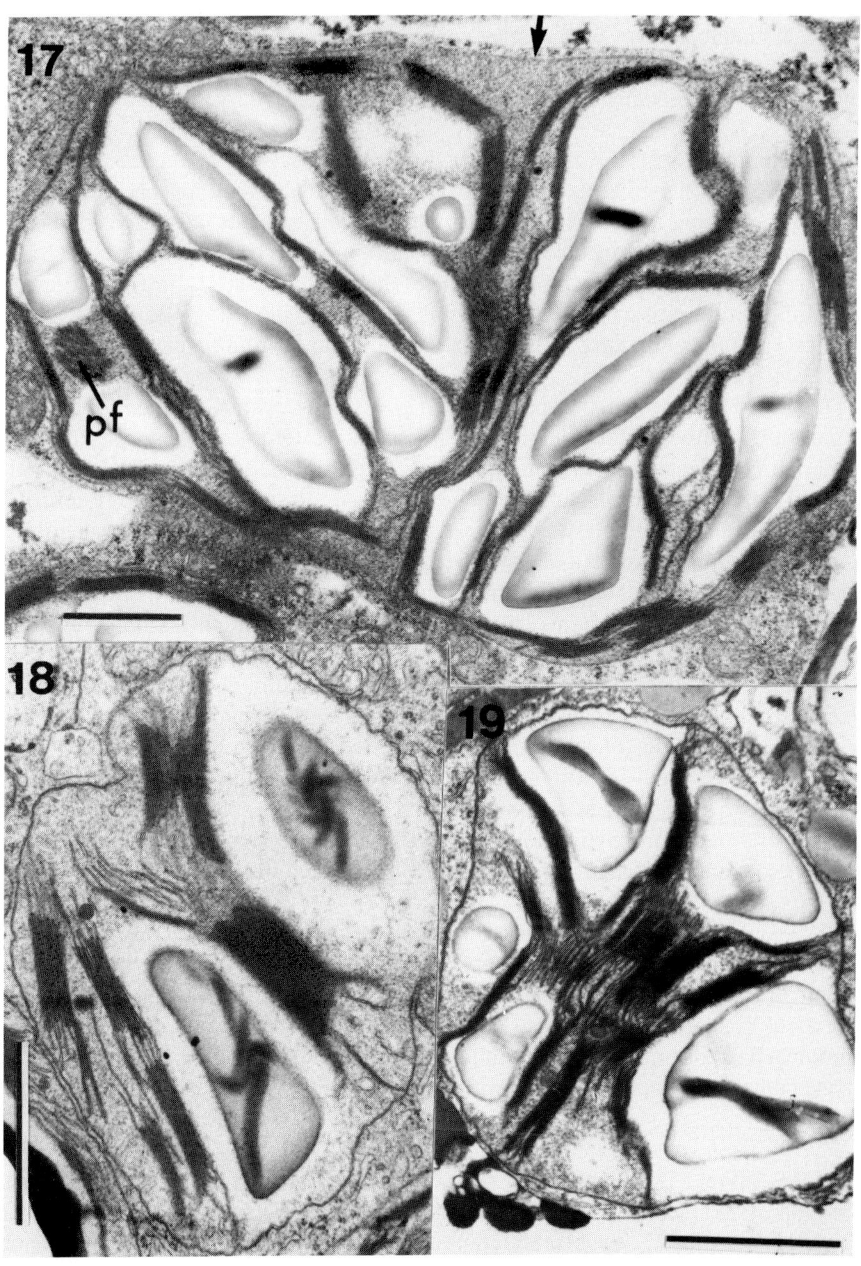
17
pf
18
19

note that mature plastids with angular contours associated with a predominantly peripheral disposition of elongate grana like those found in various metzgerialean taxa are illustrated in internodal and antheridial shield cells of *Chara* by Pickett-Heaps (1975: Fig. 7.12, 7.13, 7.29).

The chloroplasts of *Aneura* are clearly distinct from those of *Riccardia* at the ultrastructural level. In meristems and mature cells of the thallus they are spherical or elliptical in outline and contain compact grana with normally six or fewer thylakoids (occasionally 10) interconnected by paired and single thylakoids (Fig. 28). Granal architecture never impinges on the outline of the chloroplasts. The irregular aggregations of electron-dense material within and adjacent to the chloroplasts (Fig. 28) we interpret as lipids that have leaked from the oil bodies during specimen preparation (cf. Galatis et al. 1978b).

As would be expected, the appearance of plastids in *Cryptothallus mirabilis* Malb., a parasitic member of the Aneuraceae, is distinctly different from those of its sister genera *Riccardia* and *Aneura*. In fact, *Cryptothallus* is unique among all hepatics in clearly possessing proplastids in its gametophytic tissue. Sigee (1969) describes lobed, undifferentiated plastids in the apical cell and its derivatives. Invagination of the inner membrane, often associated with osmiophilic globuli, form tubules or vesicles but never flattened sacs. Even fewer internal membranes occur in the mature regions of the thallus, and in no case do the plastids show any differentiation toward chloroplast structure. Sigee notes that the pigments found in *Cryptothallus* include carotenoids but no chlorophyll. We confirm the absence of chlorophyll in freshly collected specimens of *Cryptothallus*. However our spectrophotometric analysis of material exposed to light for two months (in a sealed polythene bag in the laboratory) showed a chlorophyll content of 0.13 mg g^{-1} fresh weight. The plastids in both freshly collected material and in specimens exposed to light emit red fluorescence characteristic of protochlorophyllide (Whatley 1983). It would seem, therefore, that *Cryptothallus* contains protochlorophyllide, some of which may be converted to chlorophyll on exposure to light.

We confirm that plastid ultrastructure in *Cryptothallus* is similar to that described by Sigee (1969) with multilobed proplastids characterizing the thallus apex (Fig. 57). Some of these plastids (Fig. 58) contain single thylakoids, but such plastids are more frequent in sporophytic tissues (Fig. 60) and in the gametophyte apices following prolonged exposure to light. By contrast, the cells immediately adjacent to the fungal zone contain amyloplasts packed with starch (Fig. 59). A further variation in plastid

Figures 20–22. The thallus apex of *Metzgeria conjugata* Lindb. **20,** Longitudinal section of an apical cell containing numerous chloroplasts (c), n, nucleus, x10,000, scale bar = 2 μm; **21,** Angular chloroplasts (c) in an immediate derivative of an apical cell, note the fibrillar texture in the central stroma and the peripheral location of the grana (g), x24,000, scale bar = 1 μm; **22,** Details of the peripheral grana (g), also illustrating the absence of intergranal thylakoids, x71,000, scale bar = 0.5 μm.

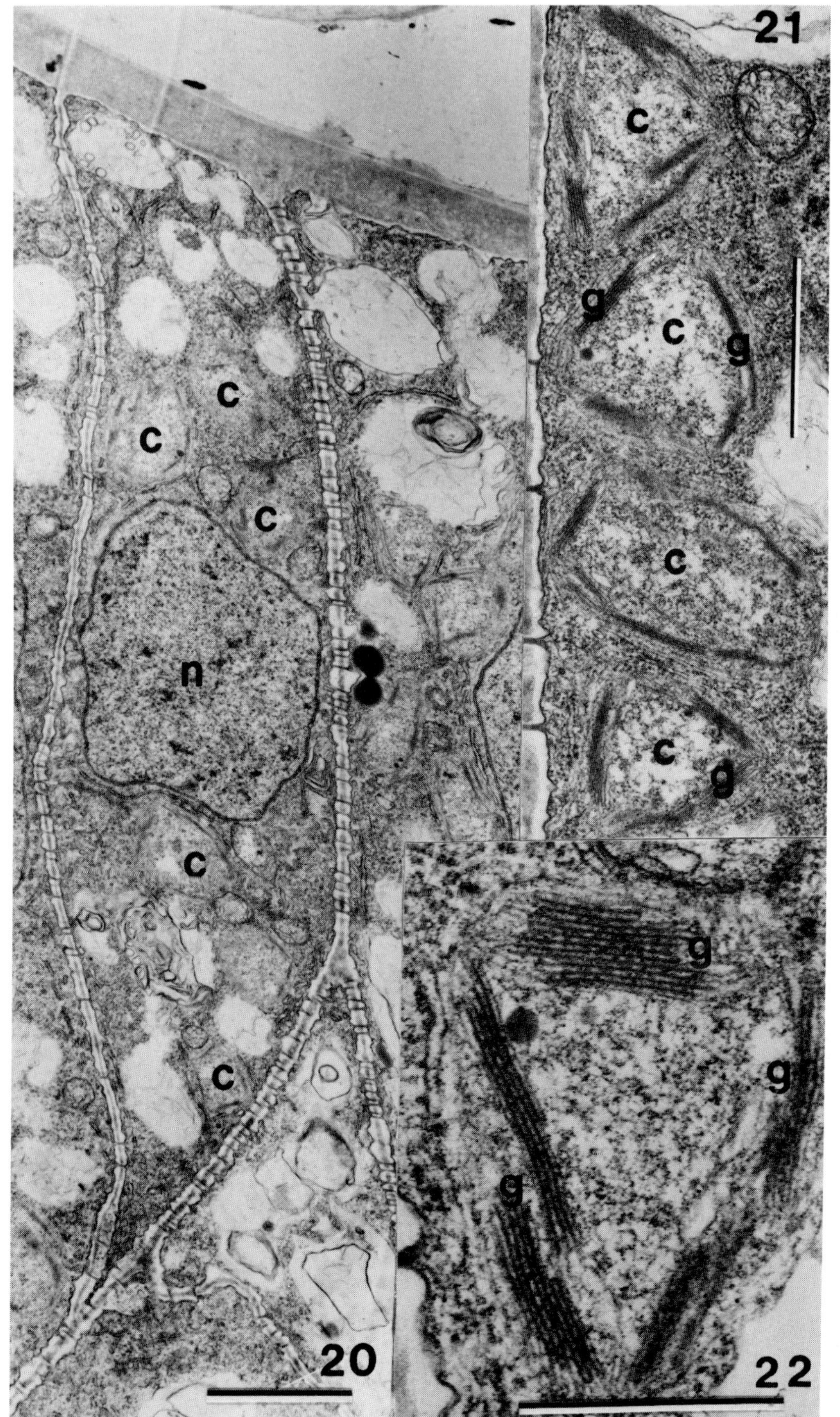

ultrastructure, the occurrence of prolamellar bodies (Fig. 62), is considered in detail later.

Jungermanniales

A huge void in bryological knowledge is cellular organization in the Jungermanniales. Published micrographs that show any details of plastid ultrastructure are few. Hölzl (1969) in a study of oil-body structure in *Mylia* figures cylindrical and irregularly shaped plastids with central aggregations of plastoglobuli and well developed networks of grana and stroma lamellae.

The plastids in dividing cells just below the apex in *Radula obconica* Sull. contain a large, centrally located lipid globule surrounded by large elongate grana mostly in excess of 1 µm in diameter (Crandall-Stotler 1981). By contrast, in mature leaf cells plastoglobuli are less in evidence, and the thylakoid network comprises numerous compact grana less than 1 µm in diameter that are interconnected by numerous stroma lamellae. Suire (1970) figures gemmae cells of *Radula complanata* Dum. with chloroplasts containing extensive grana and intergranal arrays. Many of the thylakoids are swollen. Plastids in the epidermal cells of the stem apex contain scattered grana consisting of five or fewer thylakoids and very few intergranal lamellae.

Our own observations on a range of taxa reveal some with plastids like those in *Radula* and *Mylia* but others with rather different thylakoid architecture (Fig. 30–36).

Perhaps the most striking feature of the plastids of *Nardia scalaris* S. Gray (Fig. 33) is their uniformity within cells from the stem apex to the mature leaf. They are spherical or slightly oval in outline, 1.5–2 µm in diameter, and contain large lipid globules and three to five arrays of thylakoids extending across the diameter of the organelles. Each thylakoid array consists of short grana, approximately 0.5 µm across and containing three to five (occasionally up to eight) thylakoids, interconnected by short lengths of ill-defined stroma lamellae. Starch grains, when present, are always located between the groups of thylakoids never against the envelope. This is also a consistent feature of starch distribution in chloroplasts of *Geocalyx graveolens* (Schrad.) Nees (not illustrated), the thylakoid architecture of which is closely similar to that in *Nardia*. A few chloroplasts in *Nardia* are more elongate in outline (Fig. 32). We interpret these

Figures 23–27. Gametophyte plastids of *Riccardia multifida* (L.) S. Gray. **23,** Apical cell (ac) and derivatives (d), note the angular outline of the plastids and the presence of grana in all the cells, n, nucleus, x6100; **24,** Plastid from an apical cell with small grana (g) and long intergranal thylakoids, x24,500; **25, 26,** Mature chloroplasts in fully differentiated cells of the thallus. Note the polyhedral outline and the predominantly peripheral disposition of the thylakoids, l, lipid droplets, x22,000; **27,** Microbodies (mi) in a mature thallus cell, x20,000. All scale bars = 1 µm, except **Fig. 23** = 5 µm.

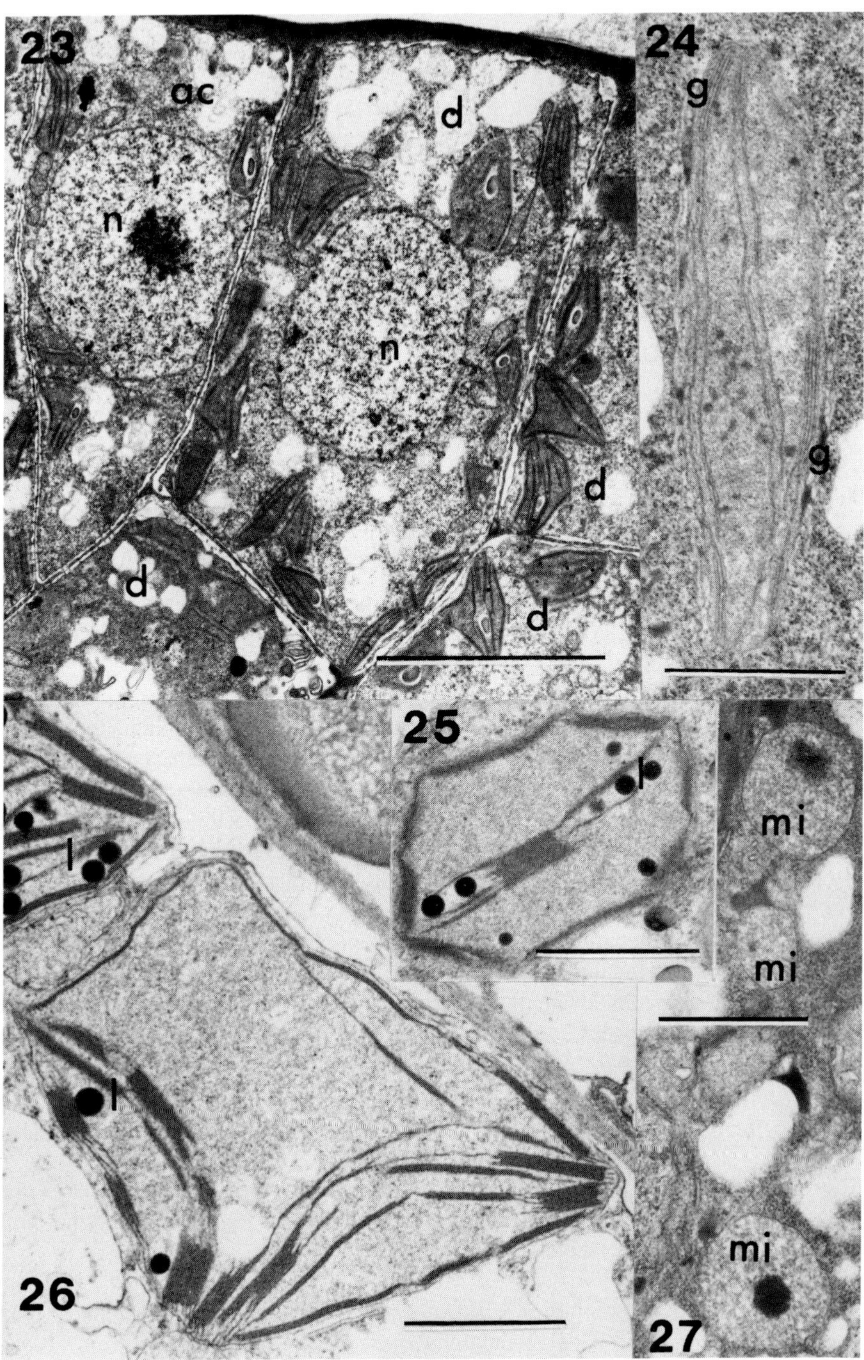

as stages in the division process. They contain distinct grana and poorly developed stroma lamellar systems identical to those in the spherical chloroplasts.

Thylakoid architecture in *Trichocolea tomentella* (Ehrh.) Dum. is similar to that in *Radula*. Plastids in mature leaf cells contain compact grana of 10–20 thylakoids interconnected by a regular and extensive stroma lamellar system (Fig. 31). In young leaf cells and in the meristematic region of the stem the grana contain fewer thylakoids, and long single thylakoids are much more in evidence (Fig. 30).

As noted in the previous section, we have never encountered plastids in the stem apices of hepatics, other than in *Cryptothallus*, that lack a well-defined lamellar system including grana. In an attempt to find proplastids we have examined the stems of *Kurzia pauciflora* (Dicks.) Grolle, *Odontoschisma sphagni* (Dicks.) Dum., and *Cladopodiella fluitans* (Nees) Buch collected from 5–10 cm below the surface of the peaty substratum (Pocock & Duckett 1985). The meristems of these subterranean axes contain plastids like those in the aerial organs (very similar to those of *Nardia*, Fig. 33). However, in fully differentiated cells two clearly distinct kinds of plastids are present. The central cells adjacent to the dividing region contain amyloplasts packed with several grains and lacking any trace of internal lamellae (Fig. 34), whereas the epidermal cells contain pleomorphic proplastids (Fig. 35). Mostly these lack any trace of internal membranes but occasionally ill-defined single thylakoids may be discerned (Fig. 36). Electron-lucent regions of the otherwise highly granular stroma contain fine-fibrils, most likely the plastid DNA.

Although the precise nature of the responses of the underground axes of *Odontoschisma*, *Cladopodiella*, and *Kurzia* has yet to be investigated experimentally, the fact that these grow downwards into the substratum suggests the possibility of gravitropism. The occurrence of amyloplasts in a subapical location in these axes (they are absent from similar cells in creeping aerial axes of the same species) strengthens such a notion because the action of amyloplasts as the sensing organs for gravitropic responses is well documented in the roots of vascular plants (Whatley 1983, Hensel 1986, Sack et al. 1986). Both the statolith-containing root cells and the subapical cells of the subterranean hepatic axes contain abundant sheets of endoplasmic reticulum. Apart from those in the underground stems of *Kurzia*, *Cladopodiella*, and *Odontoschisma* (as far as we are aware the first report of the possible occurrence of statoliths in gametophytes of hepatics), amyloplasts located immediately below the apical meris-

Figure 28. Fully differentiated chloroplasts in a mature thallus cell of *Aneura pinguis* (L.) Dum. Note the regular oval outline (cf. *Riccardia* Fig. 23–27) and the small compact grana, x15,000, scale bar = 1 μm.

Figure 29. Mature chloroplast of *Pallavicinia lyellii* (Hook.) Carruth. containing numerous starch grains (s) between the thylakoids, x25,500, scale bar = 1 μm.

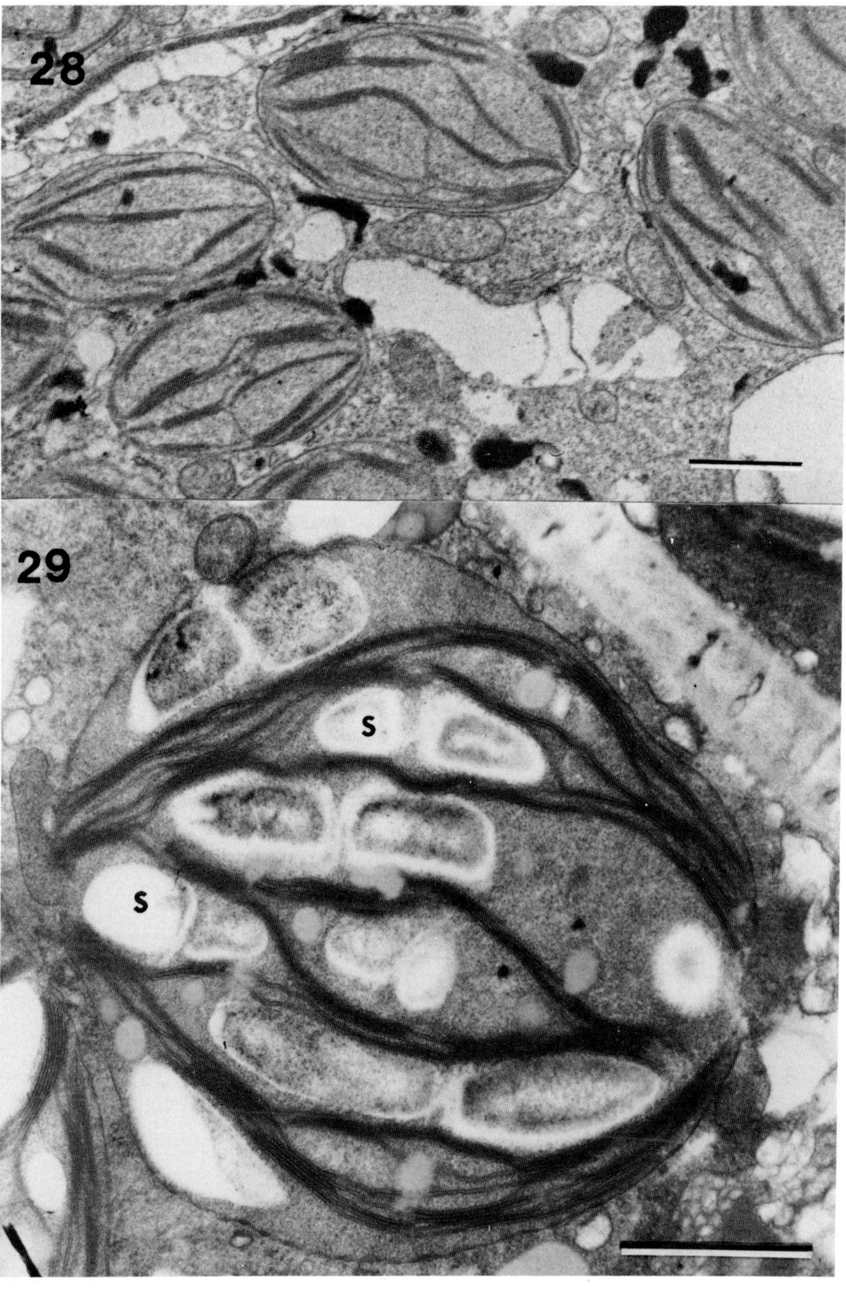
28
29
S
S

tem appear to play a vital role in the gravitropic responses of the stems of *Sphagnum* (Bismarck 1959) and similarly in the setae of moss and hepatic sporophytes.

Calobryales

Plastid ultrastructure in the stem apices and leaves of *Haplomitrium* is similar to that in *Blasia* (Renzaglia & Duckett, unpublished data). Each plastid contains several large grana and an extensive stroma-thylakoid system. In meristematic cells the thylakoid system comprises short grana and abundant intergranal thylakoids. Plastoglobuli and small starch grains are evident in the chloroplasts of young merophytes (Fig. 37).

Whereas all the meristematic cells in the gametophyte of *Haplomitrium* contain numerous plastids, those of *Takakia* (both stem and young leaves) possess only two large, starch-laden amyloplasts located at opposite poles (Crandall-Stotler 1986). In terms of size and shape these plastids are reminiscent of those in *Megaceros* (Burr 1970, Valentine et al. 1986). However, there is no hint of the vertical granal interconnections characteristic of hornwort plastids in the only published micrograph (Crandall-Stotler 1986, Fig. 30) that illustrates details of the thylakoid system (well-defined grana and numerous stroma lamellae). Crandall-Stotler (1986) cites the biplastidic condition as one of the criteria underlining the segregation of *Takakia* into its own division remote from all other archegoniates.

Experimental Studies

Although very few studies have explored the effects of experimental manipulation on the plastids of hepatics, some demonstrate ultrastructural changes (or lack of them) that parallel those in vascular plants, while others show differences. In the former category are the ultrastructural changes induced in chloroplasts of *Marchantia* by application of exogenous gibberellic acid (Albertine et al. 1976). Starch grains disappear, stromal density intensifies, and the grana increase in length and thickness until the

Figures 30, 31. *Trichocolea tomentella* (Ehrh.) Dum. **30,** Chloroplast in a young leaf cell with small grana and long intergranal thylakoids, x43,000; **31,** Chloroplast in a mature leaf cell with tall grana and an extensive intergranal thylakoid system, s, starch, x40,000. Scale bars = 1 μm.

Figures 32, 33. *Nardia scalaris* S. Gray. **32,** Probable division profile, grana the same size as in fully differentiated non-dividing chloroplasts, 1, lipid droplets x35,000; **33,** Fully differentiated chloroplasts typical of mature leaf and stem cells, note the small compact grana, lipid droplets (l) and approximately spherical outline, s, starch, x24,000. Scale bars = 1 μm.

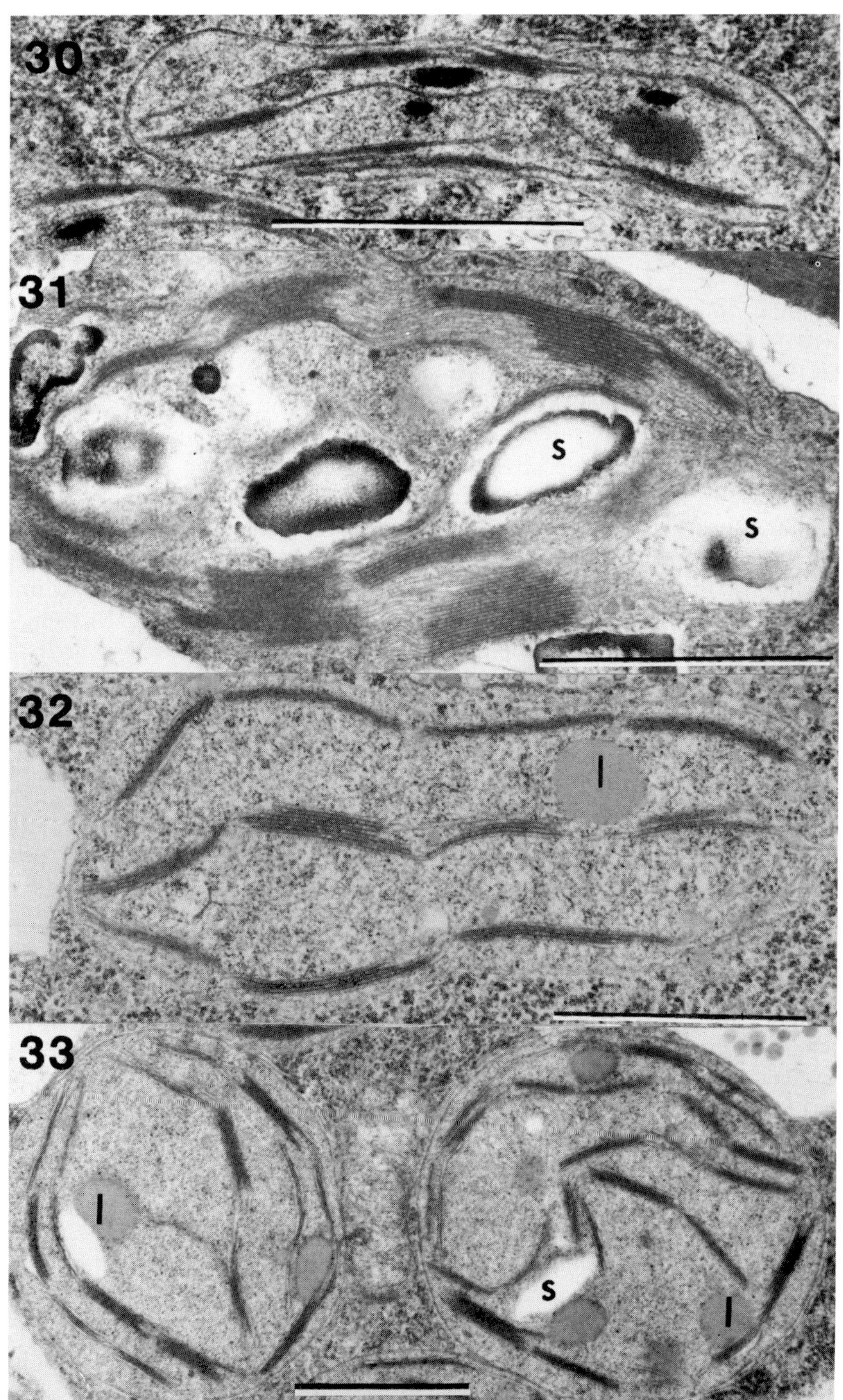

30
31
S
S
32
I
33
I
S
I

chloroplasts become packed with parallel arrays of thylakoids. Various growth retardants repress the ultrastructural changes associated with gibberellin.

Mache and Loiseaux (1973) were surprised to discover that chloroplasts of *Marchantia* were modified by an excess of light. After 33 days at 150 µmol photons m^{-2} s^{-1} (compared with the photosaturating intensity of 40 µmol photons m^{-2} s^{-1}), the small compact grana and extensive stromal network were replaced by elongate grana often extending from one edge of the plastid to the other. The intergranal thylakoids disappeared and the peripheral stroma contained numerous tubular and irregular vesicles derived from invaginations of the inner membrane of the envelope.

Chilling of *Pellia epiphylla* (L.) at -22^0 C produces some disorganization of the thylakoid system but results in much more profound changes in the oil-bodies and vacuoles (Pihakaski & Pihakaski 1979). The same authors (1980) trace the changes in ultrastructure and net photosynthesis following spraying of *Pellia* with N-phosphonomethyl glycine (glyphosate). In addition to ultrastructural alterations similar to those associated with senescence and herbicide treatment in monocotyledons and dicotyledons (e.g., swelling of the intergranal thylakoids followed by disintegration of the entire lamellar system), Pihakaski and Pihakaski (1980) recorded symptoms peculiar to *Pellia*, namely the appearance one day after treatment of vesicular and tubular structures on the chloroplast surface. Large granular bodies, apparently arising from the chloroplast stroma, were subsequently released into the cytoplasm. Swelling of the thylakoids has also been recorded in young (but not mature) leaves of *Hylocomium* following incubation in arsenate (Wells et al. 1987).

Figures 34–36. Plastids in subterranean axes of *Odontoschisma sphagni* (Dicks.) Dum. **34**, Amyloplasts (a) in the central cells of the stem immediately adjacent to the downwardly growing apex, arrow indicates the direction of the apex, x3500; **35**, Cluster of pleomorphic proplastids (p) in an epidermal cell, m, mitochondrion, n, nucleus, x22,500; **36**, Plastid from the apical region with single thylakoids (t) and a fibrillar region (f) in the stroma, x32,000. Scale bars = 1 µm, except **Fig. 34** = 10 µm.

Figure 37. *Haplomitrium hookeri* (Sm.) Nees. Chloroplasts in an immediate apical derivative with small grana, starch grains (s) and lipid droplets, x35,000. Scale bar = 1 µm.

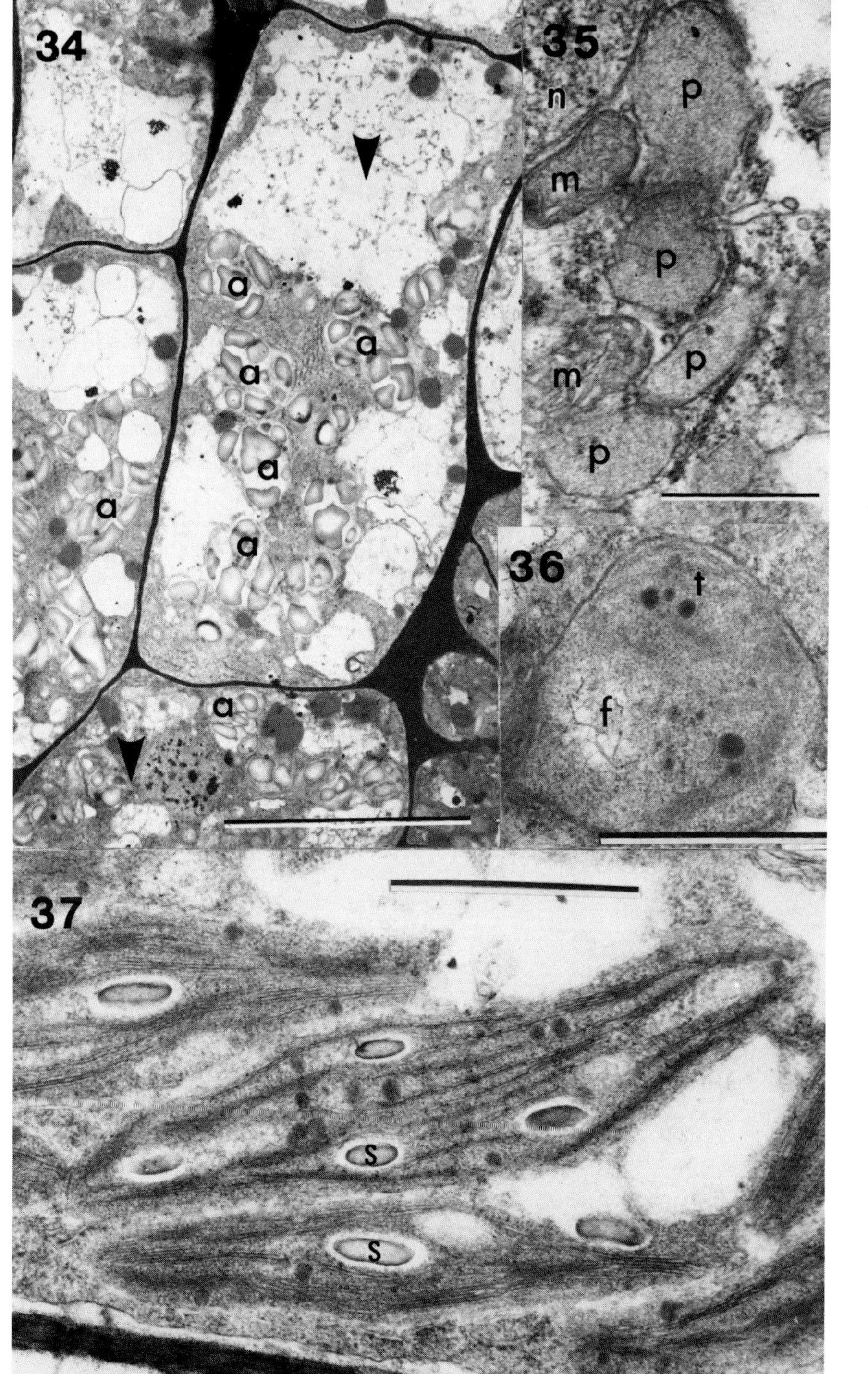

Bryophyta

Spores and Protonemata

The ripe spores of mosses exhibit considerable variation in plastid ultrastructure (Brown & Lemmon 1980b). According to the listings by Neidhart (1979) and Brown and Lemmon (1980b) the spores of most species are green, chlorophyllose, and presumably photosynthetically competent at the time of dispersal. The plastids of such spores contain well-organized grana and stroma-thylakoid systems (e.g., *Bryum, Polytrichum, Ceratodon, Leptodictyum*). Other mosses, however, produce spores that are non-green at maturity. Those of *Dawsonia* contain starch-filled amyloplasts with few or no internal lamellae (Stetler & DeMaggio 1976). The plastids in *Funaria* (Neidhart 1979) possess scattered vesicles and globules, while those of *Sphagnum* have small scattered grana (Brown et al. 1982).

As might be anticipated, ultrastructural studies have revealed more profound changes accompanying germination in the plastids of mosses with non-green rather than in those with chlorophyllose spores. The small elongate grana seen in the mature green spores of *Bryum capillare* Hedw. hardly change during germination (Chevallier & Nurit 1974). In *Ceratodon* plastids of mature spores possess a few elongate trilamellate grana. On germination each plastid contains numerous compact grana comprising up to 10 thylakoids (Valanne 1966). The well defined compact grana found in the mature spores of *Polytrichum* are generally smaller and more elongate after germination (Karunen 1972). In *Dawsonia* (Stetler & DeMaggio 1976) and *Funaria* (Monroe 1968, Chevallier 1975, Chevallier & Nurit 1974) germination sees the rapid elaboration of numerous elongate grana.

Irrespective of their condition in the spores, the chloroplasts of the young protonemata of mosses contain well defined grana and an extensive intergranal lamellar system. In view of the use of plastids as one of the diagnostic features of chloronema and caulonema [viz., numerous and rounded versus lens-shaped and scattered (Bopp 1984)], it is surprising that no critical comparison has yet been made of the plastids at the ultrastructural level. Virtually all electron microscope studies of moss protonemata consider either caulonema or chloronema (the latter almost fortuitously because these works are principally concerned with spore germination) but not both. It is only possible to gain limited insight into possible differences in plastid ultrastructure between the two types of protonemata. However, this should be viewed with some caution since it involves comparisons between authors whose culture regimes never correspond exactly.

The lens-shaped plastids in mature caulonemal cells of *Ceratodon* contain trilamellar grana closely similar to those in the leaves (Valanne 1984) but more elongate and smaller in terms of thylakoid number than those in protonemata immediately after germination (Valanne 1966). A similar picture emerges in *Funaria* when the micrographs of the chloronema in Monroe (1968) are compared with those of the caulonema in

Conrad et al.(1986). The latter authors also illustrate that thylakoid architecture remains unaltered during bud formation. Jensen and Jensen's study (1984) centers on the caulonemal apical cells in *Physcomitrium*. The chloroplasts in the basal region of these cells are like those previously described in *Funaria* and *Ceratodon*. Those near the tip, however, are approximately spherical with grana comprising two to four thylakoids radiating from a central point adjacent to the envelope. Unlike those in other mosses hitherto investigated, the plastids in the gametophyte cells of *Buxbaumia* (Fig. 46) are small and spheroidal. They contain a few single thylakoids, very small grana, numerous plastoglobuli, scattered phytoferritin particles, and occasionally prolamellar bodies (Ligrone et al. 1982b). These plastids in *Buxbaumia* are the only protonema plastids described thus far that show a level of organization that even approaches that seen in the proplastids of vascular plants. Judging from the frequent occurrence of umboshaped profiles in published micrographs (e.g., Jensen & Jensen 1984), we conclude that division by constriction of plastids with fully differentiated thylakoid systems is the rule in protonemata.

Gametophores

Mature leaves of the majority of mosses studied under the electron microscope contain chloroplasts similar to those found in the chlorenchyma of vascular plants (Schofield & Hébant 1984). Chloroplasts with numerous compact grana each containing up to 10 or more thylakoids have been noted in a range of mosses including *Sphagnum* (Fabré & Orcival 1982, Karunen & Kälviäinen 1985), the acrocarpous genera *Leucobryum* (Castaldo et al 1989), *Mnium* (Luttge & Krapf 1968), *Atrichum* and *Polytrichum* (Paolillo & Reighard 1967), and the pleurocarps *Hookeria* (Schofield & Hébant 1984), *Cratoneuron* (Krochko et al. 1978), and *Pleurozium* (Aro & Valanne 1979). Paolillo and Reighard (1967) note that in *Polytrichum* the lateral limits of the grana are not as clearly defined as in angiosperms because pairs of thylakoids often extend beyond the boundaries of the taller stacks. Sheets of osmiophilic globules between adjacent grana have been noted in *Atrichum* (Sitte 1963) and *Cratoneuron* (Aro & Valanne 1979), and central aggregations of such globuli are a characteristic feature of the chloroplasts in the mature chlorophyllose cells in *Sphagnum* leaves (Karunen & Kälviäinen 1985; Clymo & Duckett, unpublished data). We are aware of only one study that specifically considers plastid ultrastructure during leaf development in mosses, viz., Aro and Valanne's description (1979) of smaller grana in the young leaves of *Pleurozium* (but see *Tortula* considered below). Large chloroplasts closely similar to those in the leaves appear to be the norm in the cortex of the stems (Schofield & Hébant 1984), e.g., *Pleurozium* (Noailles 1978).

In contrast to the aforementioned genera chloroplasts in the mature leaf cells of *Ceratodon* (Valanne 1976) contain elongate grana each comprising only two or three thylakoids and very few stroma lamellae. Similar thylakoid architecture also occurs in the mature leaf chlorenchyma of *Funaria* (Wiencke & Schulz 1983) and in *Tortula*. Figure

42 shows grana of *Tortula* with up to five thylakoids and a very sparse stroma lamellar system.

The most comprehensive analysis of plastid ontogeny in moss gametophytes is probably that of Hébant (1974a), who has traced development in the hydroids from the apical cells in a variety of polytrichalean taxa. The apical cells contain a large number of pleomorphic plastids scattered throughout the cytoplasm. The plastids contain one to three arrays of membranes, each comprising single, double, or triple thylakoids. The peripheral stroma is packed with irregular vesicles derived from invaginations of the inner membrane of the envelope.

At an early stage in differentiation the plastids become associated with the nucleus. This situation has also been noted in *Polytrichum* by Bonnot (1967), in *Timmiella* by Ligrone (1985), and in *Pleurozium* by Noailles (1974, 1978). Some of the plastids remain in the vicinity of the nuclear envelope during elongation of the conducting elements, while others lie freely in the cytoplasm. Whatever their position, these plastids contain small elongate grana of up to three thylakoids and numerous plastoglobuli. Division appears to be by constriction in the Polytrichales, but plastid profiles in the differentiating stem parenchyma of *Pleurozium* (Noailles 1974, 1978) suggest partitioning is also likely.

Profound changes in the shape of the plastids accompany hydroid differentiation. In developing hydroids plastids are either highly elongate and cylindrical or attenuated plates with swollen margins. There is no further elaboration of the thylakoid system, and eventually the plastids as well as the surrounding cytoplasm degenerate. By contrast the hydrom sheath contains ovoid plastids with small grana very similar to those in the immediate derivatives of the apical cell.

The leaf leptoids of *Funaria* contain highly elongate plastids (up to 14 μm in length compared with 4–6 μm in leaf parenchyma) with small grana (Wiencke & Schulz 1983). Small grana with three to five thylakoids also characterize the plastids in the leptoids of *Polytrichum* leaves (Scheirer 1983). However, far more unusual forms occur in the stem leptoids of Polytrichales (Schofield & Hébant 1984, Hébant 1974b, 1977). Differentiation of these organelles involves regression of the thylakoid system to aggregations of vesicles and occasional prolamellar bodies. The stroma matrix contains arrays of crystalline and granular materials recalling chromoplast ultrastructure in angiosperms (Sitte et al. 1980, Knoth et al. 1986).

In general terms the patterns of plastid development in Polytrichales are broadly in line with those of vascular plants. Meristematic tissues contain less complex thylakoid arrays than mature photosynthetic tissues, whereas differentiation of specializied con-

Figures 38, 39. Transection of the stem apex of *Tortula fragilis* Tayl. **38,** The plastids (p) in the apical cell (ac) and its derivatives (d) contain massive grana, n, nucleus, x3600, scale bar = 10 μm; **39,** Massive grana in an apical cell plastid (serial section with Fig. 38), l, lipid globules, s, starch, x41,000. Scale bar = 0.5 μm.

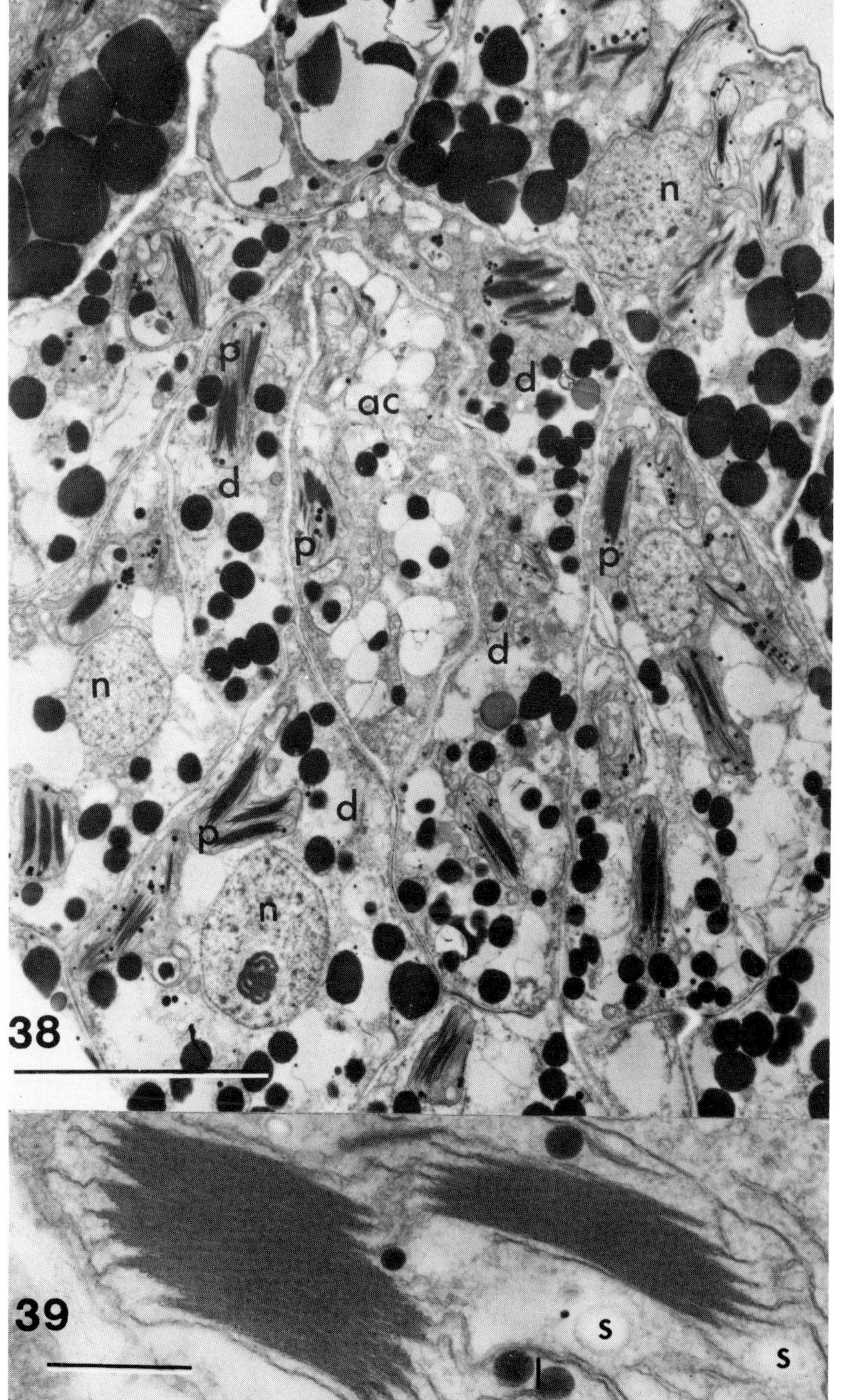

ducting tissues involves regression of the internal membrane systems. The most remarkable feature of the plastids is their intimate association with the nucleus during cell differentiation. Similar associations have not been reported elsewhere in bryophytes, but are apparently quite common in epidermal cells of angiosperm leaves (Duckett, unpublished observations).

As we have already illustrated for the hepatics, generalizations about plastid behavior based on just one moss can be highly misleading. Fabré and Orcival (1982) figure plastids with large grana and very few stroma thylakoids in apical regions of *Sphagnum* stems. Our observations (unpublished data) confirm the absence of rudimentary intergranal lamellar systems in the plastids of *Sphagnum* apices.

Investigation of the stem apices of *Tortula* (Fig. 38–42) reveals a sequence of plastid ontogeny certainly very different from any hitherto recorded in bryophytes and absolutely contrary to any pattern recorded by Whatley (1977, 1978, 1980) in vascular plants. The ovoid plastids in the apical cell contain one or two massive grana each with up to 50 thylakoids (Fig. 38, 39). It is not only in terms of size that these grana are highly unusual; their lateral margins have a regular serrated appearance. These serrations are the consequence of regular stepwise extension of the lateral boundaries of the thylakoids beyond the central mass of the granum. The thylakoids of the greatest diameter, i.e., those forming the tips of each serration, continue as single lamellae into the stroma.

In immediate derivatives of the apical cell the plastids are highly pleomorphic (Fig. 38, 40, 41), with cytoplasmic invaginations often containing large lipid droplets and mitochondria. These plastids contain grana of the same diameter and have serrated margins similar to those in the apical cells. However, grana number is greater, and each granum contains fewer thylakoids. From cells near the apex to mature leaf cells there is a progressive reduction in granal size (in terms of thylakoid number but not diameter) and a corresponding increase in granal number. The simplest explanation for these different plastid morphologies is that differentiation involves the progressive splitting and separation of the huge grana found in the apical cells. The most likely points of separation are the indentations between the marginal teeth (i.e., between the thylakoids of minimum diameter in the stacks). Each group of thylakoids forming a serration in the giant granum will eventually become a single granum of four to six thylakoids typical of those in mature leaf cells (Fig. 42).

It is now clearly of considerable importance to investigate plastid ontogeny in a

Figures 40–42. *Tortula fragilis* Tayl. **40, 41,** Highly pleomorphic plastids in the immediate derivatives of the apical cell. **40,** Several grana (g) probably derived from splitting of the massive grana found in the apical cell plastid, x15,500; **41,** The cytoplasmic invaginations in the plastid surface contain lipid globules (l) and a mitochondrion (m), g, grana, x25,500. **42,** Chloroplast in mature leaf cell with small grana and a poorly-developed intergranal thylakoid system, x22,000. All scale bars = 1 μm.

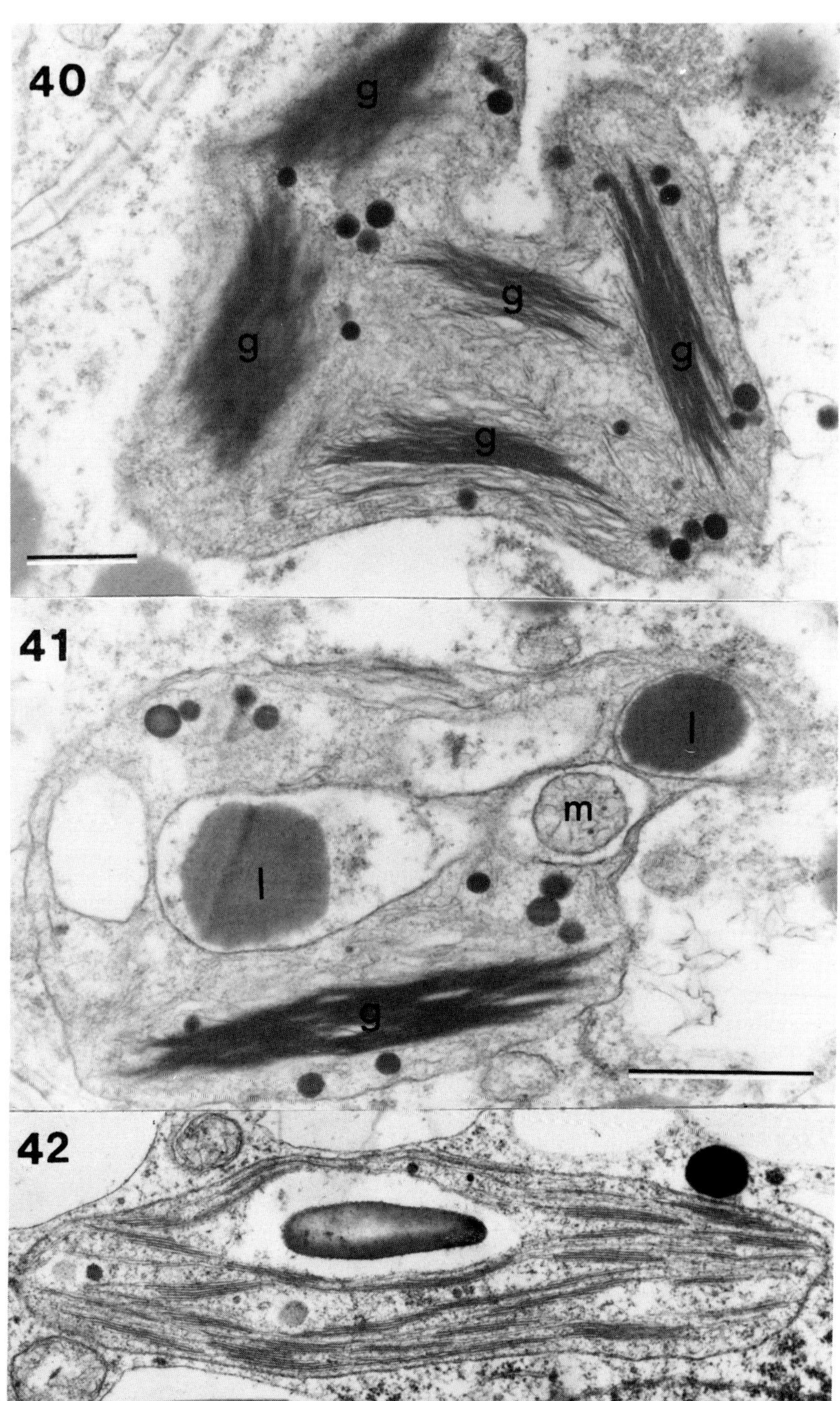

range of moss apices. It will be especially interesting to discover whether the occurrence of massive grana in apices correlates in any way with the presence of small elongate grana in mature leaf cells (e.g., *Funaria* and *Ceratodon*).

Extensive search of the literature on thylakoid ontogeny in vascular plants has failed to reveal any situation at all comparable with that in *Tortula*. However, the micrographs of the chloroplasts during stomatal development in *Selaginella* appear to show some similarities with those of *Tortula* (Brown & Lemmon 1985b). The single plastid in young guard cells of *Selaginella* contains two large granal masses with lateral splits between groups of thylakoids (Brown & Lemmon 1985a: Fig. 8). During subsequent division of the plastid one granal mass passes to each daughter organelle. The micrographs of maturing guard cells show plastids with more numerous smaller grana. Giant grana are frequent in gametophytic cells of *Equisetum* (Duckett, unpublished data), but plastid ontogeny has not been observed. Similarly, the origin of giant grana, occasionally observed in higher plants (Whatley & Price 1983), has yet to be explored.

Sporophytic Tissues

Micrographs that illustrate the range of forms of sporophytic plastids in mosses are much scarcer than those of gametophytes. Developmental interrelationships have not been explored, and it is impossible to draw any conclusions about the possible existence of significant differences between the two generations. Most details are available for the haustorium. Although hardly representative of the sporophyte generation as a whole, this region contains plastid types not described elsewhere in mosses.

Figures 43–47. Plastid diversity in *Buxbaumia piperi* Best. **43**, An inner parenchyma cell of the apophysis, small grana radiate from the vicinity of two prolamellar bodies (pb), microbodies (mi) are numerous in these cells, m, mitochondria. x27,000; **44, 45**, Plastids in sporophyte transfer cells, note abundance of plastoglobuli, membrane-bound vesicles (in 45) and single thylakoids orientated perpendicularly to the long axis (in 45), pb, prolamellar body, m, mitochondrion; 44, x29,000; 45, x15,000; **46**, Typical gametophyte plastid with very small grana, plastoglobuli, and small particles, probably pytoferritin (arrow) scattered through the stroma, x34,000; **47**, Detail of inner spore sac cell-plastid showing grana adjacent to the envelope and abundant fibrillar (f) material in the stroma, x21,000. All scale bars = 1 μm. (Micrographs courtesy of R. Ligrone et al. 1982b.)

Figure 48. Plastids with very regular grana in a sporophytic foot transfer cell of *Timmiella barbuloides* (Brid.) Moenk., pb, prolamellar body, mi, microbody, m, mitochondrion, x23,000. Scale bar = 1 μm. (Micrograph courtesy of R. Ligrone et al. 1982a.)

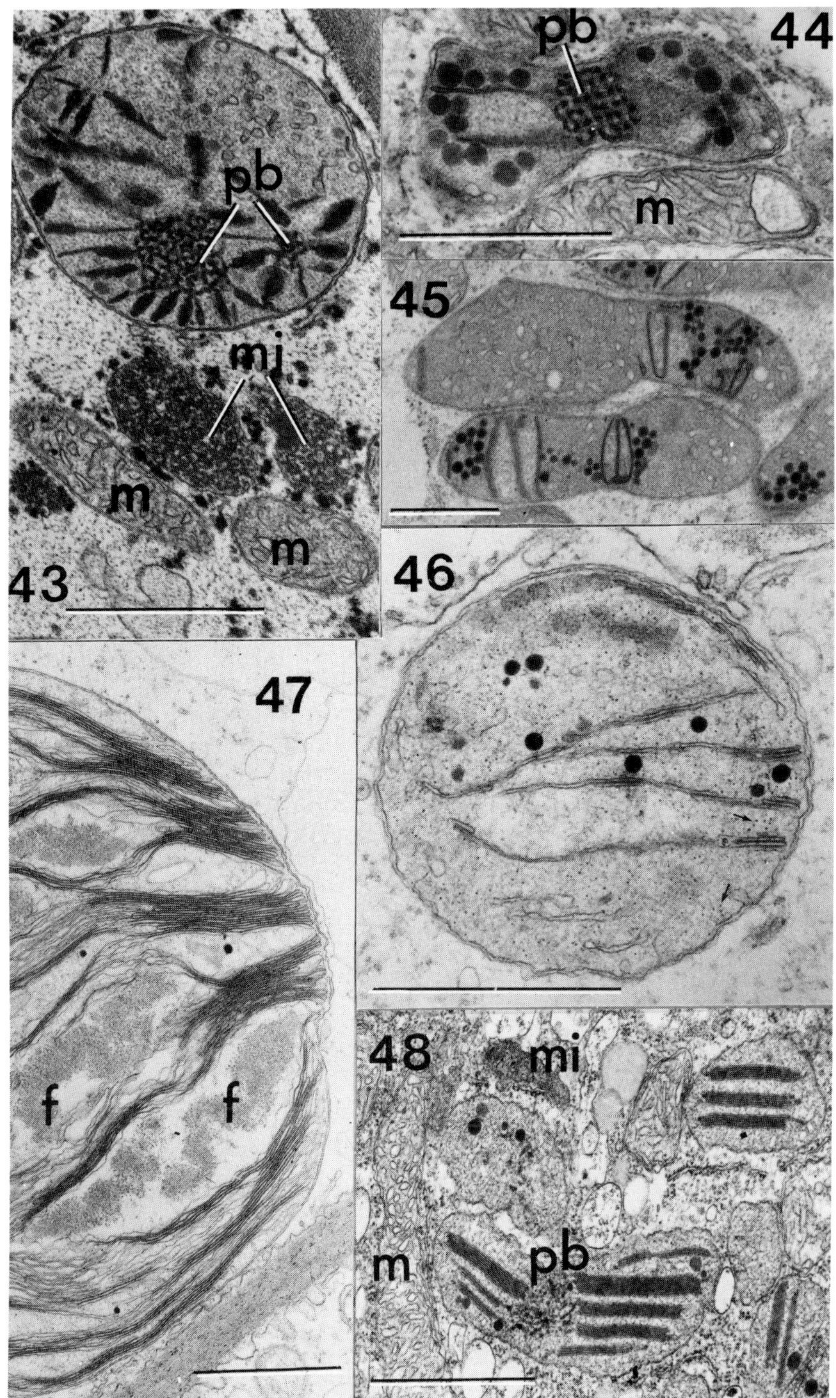

In the sporophytic transfer cells of *Buxbaumia* (Ligrone et al. 1982b) the plastids are highly pleomorphic. They contain numerous vesicles and plastoglobuli and occasionally small prolamellar bodies (Fig. 44). Thylakoids are either absent or single and perpendicular to the long axis of the organelle (Fig. 45). In *Mnium* (Eymé & Suire 1967) the transfer cell plastids are similary pleomorphic but in contrast to *Buxbaumia* contain swollen grana. Those in the middle region of the foot-vaginula complex of *Timmiella* (Ligrone et al. 1982a) contain large regular grana but no stroma lamellae and prolamellar bodies (Fig. 48).

During initial development of the transfer cells in *Funaria* (Browning & Gunning 1979, Wiencke & Schulz 1977) the plastids are pleomorphic to dumb-bell shaped and have swollen thylakoids and numerous vesicles. At a later stage they are cylindrical to highly elongate with well developed grana containing up to 10 thylakoids. By contrast the grana plastids of the apophysis contain one to three thylakoids similar to those in the leaf chlorenchyma (Sack & Paolillo 1983a, b, 1985). As in vascular plants, the stomatal guard cell plastids are packed with starch. Cells of the developing peristome in *Fissidens* (Mueller 1973) contain spherical or ovoid plastids with one to three grana each comprising one to three thylakoids.

The only comprehensive analysis of plastid diversity in moss sporophytes is that by Ligrone et al. (1982b) on *Buxbaumia*. These workers recognize a range of forms varying in size, shape, thylakoid architecture, stroma inclusions, and amounts of starch in 15 different sporophytic tissues, but as yet have not extended their work to consider ontogenetic interrelationships. Prolamellar bodies (Fig. 43), most abundant in the seta, inner cells of the apophysis, and transfer cells, are associated with rudimentary thylakoid systems. At the other extreme large grana and extensive intergranal networks are found in the apophysis, urn, spore-sac tissues, columella, and young operculum. Fibrillar material is abundant in the stroma of the outer spore-sac plastids (Fig. 47). A peculiar feature of many of the chloroplasts not noted by Ligrone et al. (1982b) is the abutment of the grana against and perpendicular to the plastid envelope (Fig. 47). In common with the spore-sac tissues of *Buxbaumia* the tapetal plastids of *Polytrichum* have a well developed granal and intergranal thylakoid network (Paolillo 1964), like those in the leaves of the gametophore (Paolillo & Reighard 1967).

Experimental Studies

Unlike in vascular plants there is little or no reduction in the membrane system of moss plastids when plants are subjected to extended periods of darkness (Valanne et al. 1979, Valanne 1984), nor is there the formation of prolamellar bodies (cf. Bradbeer 1981, Rascio et al. 1986). The size of the grana tends to increase in darkness at the expense of the intergranal system. These differences between tracheophytes and mosses have been attributed partly to the characteristics of the plastid envelopes but mainly to the ability of mosses to synthesize chlorophyll in the dark (Valanne 1984). Much of the basis for these generalizations derives from observations on the effects of the light regime on spore germination and protonemata. After two weeks in the dark germinating spores of *Funaria* contain well organized thylakoid systems, though not as highly developed as those kept in the light (Monroe 1968). Addition of 1 % glucose to the culture medium inhibits lamellar differentiation in dark-grown material. Valanne (1966, 1971) records degeneration of the lamellae and formation of amyloplasts in dark-grown sporelings of *Ceratodon* on a medium containing 0.1 % sucrose. When these sporelings were exposed to blue or red light (Valanne 1971), normal chloroplasts developed. There was no difference in thylakoid configuration in plants exposed to either of the two wavelengths, but the stroma was denser in red light. After one to two months in the dark, protonemata of *Ceratodon* (Valanne 1976) contain lenticular grana with up to six thylakoids (compared with the mainly trilamellar structures produced in light). When returned to light, these protonemata initially exhibit a transient swelling of the thylakoids. This is followed by a rapid increase in the number of grana and a decrease in the number of thylakoids. Within a day the plastids are indistinguishable from those in light-grown plants.

Under continuous illumination the plastids of *Ceratodon* and *Pleurozium* contain more lamellae per granum (Aro & Valanne 1979, Valanne 1977a, b), whereas an increase in quantum flux density from 100 to 500 µm photons m^{-2} s^{-1} leads to a reduction in the number of thylakoids (Aro 1982). Aro and Valanne (1979) also record the destruction of the stroma thylakoids (particularly in *Pleurozium*) in continuous light and note that many of the effects of continuous light on these mosses resembles those in the early phases of senescence in higher plants.

Chloroplasts with elongate grana containing small numbers of thylakoids and poorly developed fret systems (e.g., *Ceratodon*) are characteristic of sun-plants, whereas those with large grana and numerous intergranal lamellae (e.g., *Pleurozium, Marchantia*) resemble shade-type counterparts in vascular plants. However, it must be underlined that when other attributes of the photosynthetic apparatus are considered, bryophytes emerge as very different from vascular plants (Valanne 1984). The taxa investigated to date comprise one hepatic, *Marchantia* (Aro 1982, Mache & Loiseaux 1973), and fewer than 10 mosses (Valanne 1984). Irrespective of their ecology, they all contain large amounts of the light-harvesting chlorophyll *a/b* complex, low chlorophyll *a/b* ratios, and rapid *in vivo* fluorescence induction kinetics. These are characteristics of sciophytes, i.e., shade plants with an inherently low capacity for photosynthesis that are unable to respond to more favorable light regimes (Bjorkman 1981). There is now a pressing need for studies on a far greater range of bryophytes. Such investigations should include not only species from sunny habitats with sun-type chloroplasts (e.g., *Ceratodon*) and species from shady habitats with shade-type chloroplasts (e.g., *Pleurozium*) but also those from shady habitats with sun-type chloroplasts (e.g., *Metzgeria, Geocalyx*).

In the desiccation-sensitive moss, *Cratoneuron filicinum* (Hedw.) Spruce, rehydration following rapid drying causes rupture of the plastid envelopes and the accumulation of plastoglobuli (Krochko et al. 1978), whereas dehydration and rehydration have little or no irreversible effects on plastid ultrastructure in the drought tolerant species *Tortula ruralis* (Hedw.) Gaertn., Meyer & Scherb. (Tucker et al. 1975) and *Pleurozium schreberi* (Brid.) Mitt. (Noailles 1978). The plastids in *Pleurozium* after 34 days of desiccation still contain small grana but very few intergranal lamellae. Thylakoid architecture indistinguishable from that in control specimens is re-established 10 min after rehydration (Oliver & Bewley, 1982). As recently pointed out (Duckett 1986), there are no reports of water stress or other experimental treatments that lead to the precipitation of RUBISCO in the stroma of bryophyte plastids.

Aro et al. (1987) have shown that the lipid, protein, and chlorophyll contents of the thylakoids of *Ceratodon* protonema grown at 20° C are significantly different from those in plants grown at 4° C, however, chloroplast ultrastructure remains unchanged. The protein and lipid characteristics of the thylakoids are thought to be related to the ability of the plant to photosynthesize even at sub-zero temperatures.

Spermatogenesis

Since plastid ultrastructure during spermatogenesis in bryophytes, particularly systematic aspects, receives consideration in other recent reviews (Duckett 1975, Duckett et al. 1982, 1984), it is only pertinent in the present context to highlight aspects not previously explored in detail.

The monoplastidic condition of the spermatids of hornworts, mosses, and liverworts (Paolillo 1984, Duckett 1986) stands out as a clear-cut difference from the multiplastidic male gametes of *Chara* and *Nitella* (Pickett-Heaps 1975). The question of when antheridial plastids are reduced in number was originally addressed at the light microscope level (Weier 1931). This study, which indicates a rapid reduction in plas-

Figures 49–54. Plastids during spermatogenesis in bryophytes; **49**, Mid-generation spermatogenous cell of *Phaeoceros laevis* containing an irregular plate-like plastid (p) with swollen margins, n, nucleus, m, mitochondria, x18,500; **50**, Transection of the minute starch-filled plastid (p) in the mature spermatozoid of *Phaeoceros*, n, nucleus, f, flagellum, x29,500; 51, The biplastidic condition in an early generation spermatogenous cell of *Hypnum mamillatum* (Brid.) Loeske, p, plastid, n, nucleus, x8500; **52**, Peripherally located sheet-like plastid associated with mitochondria (m) along its inner surface in a spermatid mother cell of *Hypnum*, n, nucleus, x22,500; **53**, Plastid in a mid-spermatid of *Hypnum* associated with the mitochondrial *Nebenkern* (m) situated on the inner surface of the nucleus (n), plastid with centrally located starch grains (s) and the peripheral vesicles in the stroma closely invested by a profile of endoplasmic reticulum (er), x28,000; **54**, Transection through the *Fibrillenscheide* (f) in a mature spermatozoid of *Petalophyllum ralfsii* (Wils.) Nees & Gott., plastid tubules clearly smaller than the spline microtubules (mt), n, nucleus, r, cytoplasmic remnant, x62,000. All scale bars = 1 μm, except Fig. 50 and 54 = 0.5 μm.

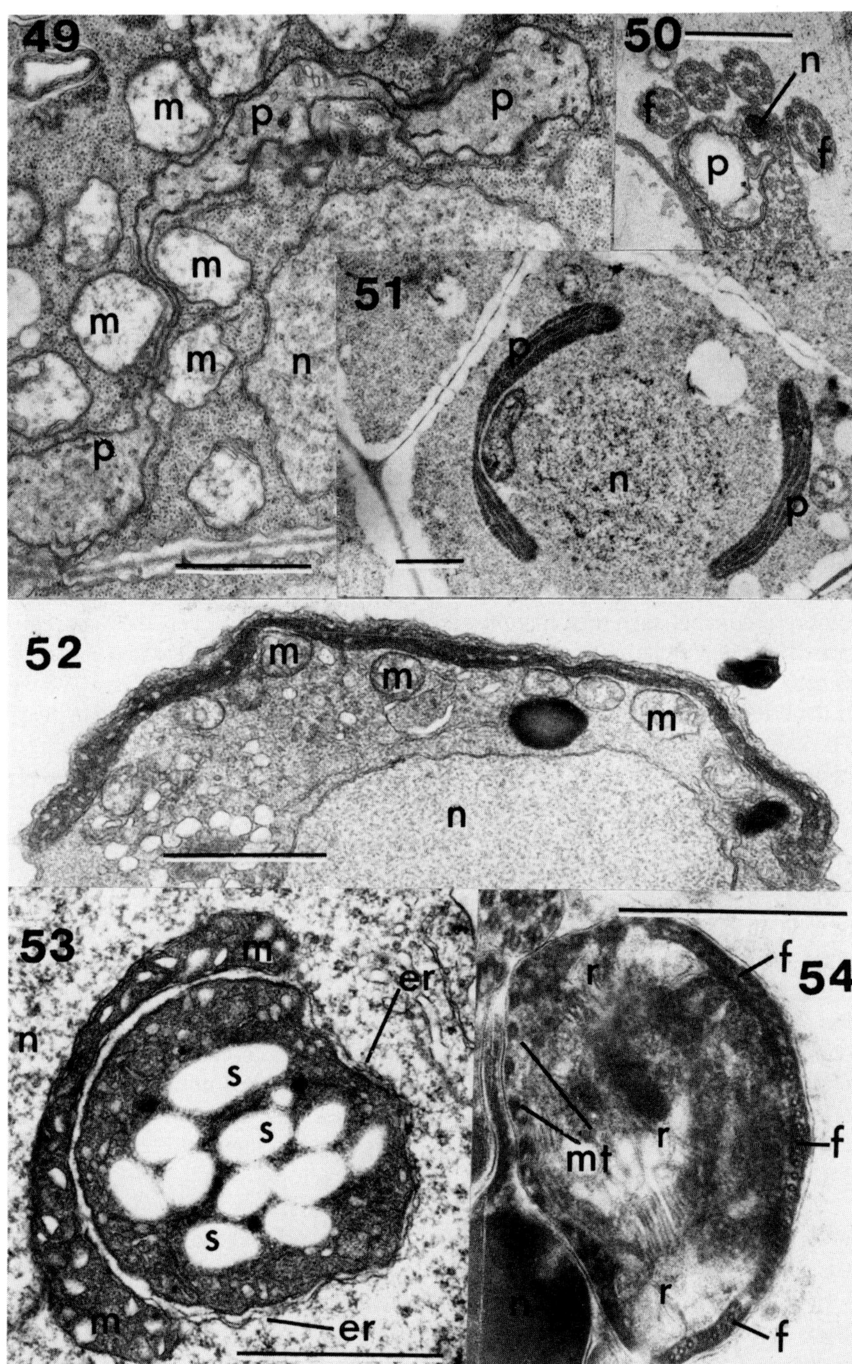

tid number early in the development of the antheridium of *Polytrichum*, receives confirmation from our EM observations of plastid behavior throughout antheridial ontogeny in *Polytrichum*, *Hypnum*, and the liverwort *Blasia*. In all three the earliest generations of spermatogenous (androgonial) cells contain two plastids (four in recently delimited spermatogenous cells of *Blasia*) (Fig. 52). The biplastidic condition is maintained through succeeding generations with plastid replication immediately following cytokinesis. Of the pair of plastids present in the spermatid mother cell one passes to each of the spermatids during the final antheridial mitosis. We conclude from these observations that the reduction in plastid number occurs very early and suddenly during antheridial development – most likely immediately following the delimitation of the spermatogenous tissue from the jacket. Whether this dramatic reduction in number involves plastid fusion or exclusion of the majority of the plastids from nascent spermatogenous cells by unequal partitioning has yet to be determined. It is clear, however, that plastid reduction in antheridia is not a gradual, step-wise process involving slowing down of plastid division relative to cytokinesis, as suggested by Duckett (1986) and Paolillo (1984) for sporogenous cells.

Consideration of the behavior of antheridial plastids also brings into focus the plastid reduction process leading to monoplastidity in the SMCs of mosses (Paolillo 1969, 1984). Weier's light microscope study (1930) of *Polytrichum* seems to suggest a more gradual change than that observed in antheridia. However, the data remain open to the interpretation that the transition takes place over relatively few generations of cells, but later in sporogonial differentiation than in antheridial ontogeny. Whenever the critical stage, several generations of sporogenous cells contain a single plastid. The exhaustive studies of sporogenesis in bryophytes by Brown and Lemmon (this volume) center almost exclusively on events starting immediately prior to mitosis, so it is impossible to say as yet whether or not the events in *Polytrichum* are ubiquitous in mosses. The control mechanisms maintaining the monoplastidic condition are unknown (Paolillo 1984, Butterfass 1979).

The reduction in plastid number in the spermatogenous (and sporogenous) tissues of bryophytes also involves progressive regression of the thylakoid system and changes in the shape of these organelles. In very young antheridia the spermatogenous cell plastids still contain rudimentary grana (Duckett 1975). Subsequently, these are reduced to lamellae (Fig. 51) and finally to scattered vesicles (Fig. 49, 52). Concomitantly, the shape of the plastids changes from pleomorphic to highly attenuated plates (Fig. 52), often with swollen margins (Fig. 49). The mitochondria become closely aligned at the surface of the plastid in latter generations of spermatogenous cells in hornworts, liverworts and more particularly in mosses (Fig. 52) (also see Fig. 12 in Paolillo 1969).

A plastid/mitochondrial association is not apparent in very young spermatids but is re-established by the time the multilayered structure (MLS) becomes positioned over the nucleus. Mitochondrial/plastid associations (the so-called *Nebenkern*) are a characteristic feature of the spermatids of mosses, hepatics, and hornworts. These have not been noted in other archegoniates nor in charalean motile cells. The *Nebenkern* appears to be most highly differentiated in *Polytrichum* (Paolillo et al. 1968). In this moss numerous mitochondrial profiles, closely investing virtually the entire surface of the plastid, are part of a labyrinthine mitochondrion extending to beneath the lamellar MLS strata. More typically a saucer-shaped mitochondrion lies between the plastid and the inner surface of the nucleus (Fig. 53). Parts of the plastid envelope free from the confines of the mitochondrion are closely invested by endoplasmic reticulum.

Starch begins to accumulate in the plastids of young spermatids until, at maturity, the stroma is completely packed with starch. In most mosses and liverworts the plastids contain numerous starch grains (Fig. 53), but only one is found in hornworts (Fig. 50) and *Sphagnum*. *Sphagnum* is also unusual in that it is the only archegoniate observed to date with spermatid plastids that contain prolamellar bodies (Duckett et al. 1982, Miller et al. 1983).

Amyloplasts are ubiquitous in the spermatozoids of archegoniates and charalean algae (Pickett-Heaps 1975). In all cases their functional significance is highly problematic since the starch is not utilized as an energy source during motility (see Duckett 1975 and Moestrup 1975 for discussions). It is more likely that starch accumulation is an inevitable consequence of the liberation of quantities of soluble carbohydrate during breakdown of excess cytoplasm in the spermatids (Miller & Duckett 1986).

As in the majority of plants (see Sears 1980 and Whatley 1982 for reviews), the male parent does not appear to contribute the plastid to the zygote in bryophytes. In *Sphagnum* (Manton 1957) and other mosses (Duckett, unpublished observations) the spermatozoid plastids are frequently jettisoned during motility. Diers (1967) reports that in *Sphaerocarpos* the cytoplasmic remnant containing a posterior mitochondrion and the plastid is generally lost by the time the spermatozoid enters the egg. In *Blasia* a weak, narrow region of cytoplasmic remnants at the postior end of the nucleus and between it and the plastid assures detachment of the elongate starch-filled plastid (and attached mitochondrion) at the time of spermatozoid liberation (Renzaglia & Duckett, unpublished data).

Two major differences have been detected between the spermatid plastids of mosses and hepatics. Firstly, the plastid is intimately associated with the posterior tip of the spline in liverworts (and hornworts, Carothers & Duckett 1980). In mosses the plastid lies freely in the cytoplasmic remnant. Secondly, spermatid plastids of the majority of liverworts, but not of *Haplomitrium* or mosses and hornworts, contain a band of peripheral tubules forming a partial sheath, the *Fibrillenscheide*. Variation in the morphology of the sheath in terms of the length of the tubules correlates closely with the shape of the plastid in the mature gametes (cf. Fig. 54 with Fig. 55, 56) and the disposition of the starch grains filling the stroma. The tubules are shorter in taxa with a clumped array of starch grains (e.g., *Petalophyllum*, less than 1.5 µm), than those where the starch is linearly disposed (e.g., *Marsupella*, 8.5 µm, and *Diplophyllum*, over 3 µm; Fig. 55, 56). In *Marchantia*, *Blasia*, *Sphaerocarpos*, and *Petalophyllum* (Fig. 54) the tubules extend forward from the main body of the plastids in a flap-like structure (see Carothers & Duckett 1980 for reconstruction in *Marchantia*). Similar structures have not been noted in jungermannialean spermatozoids. Since the tubules are associated with plastids with distinct shapes, it has been suggested that they have a supporting role in maintaining shape (Duckett et al. 1984). They should, however, not be confused with the spline microtubules. Although both elements are resistant to detergent extraction (Miller 1982), their size and staining properties are different (Fig. 54, 56). The narrower plastid tubules assemble in the presence of colchicine at concentrations that inhibit spline elongation (Miller 1982).

Oogenesis

The all too fragmentary details presently known about oogenesis in bryophytes are summarized in Duckett et al. (1984). Although only a handful of taxa have so far been investigated, it is already clear that during oogenesis plastids undergo changes different from those recorded elsewhere in the life cycle. Whereas in the neck and venter wall cells plastids contain well developed grana hardly differing from those elsewhere in vegetative tissues, plastids in the axial cells show regressed lamellar systems of irregular scattered vesicles. Starch accumulation in the plastids accompanies cytoplasmic degeneration in the neck canal cells. The young oocytes of *Phascum* (Richards, unpublished data) and *Marchantia* (Zinsmeister & Carothers 1974) contain attenuated plastids with single thylakoids and numerous plastoglobuli. Subsequently, the plastids are amoeboid and have vesiculate internal lamellae. In maturing eggs of *Phascum* and *Cryptothallus* (Fig. 61) some of the plastids become closely invested by mi-

Figures 55, 56. The amyloplast in the mature spermatozoid of *Diplophyllum albicans* (L.) Dum. **55,** Longitudinal section showing the linearly-arranged starch grains (s), x31,000; **56,** Transverse section showing the sheath of plastid tubules, specimen permeabilized by addition of the detergent Triton X-100 to the primary fixative and, after osmication, stained with lanthanum nitrate, lumina of the plastid tubules more electron-dense than the walls (cf. **Fig. 54**), note the different staining properties of the spline microtubules (mt), s, starch, x72,500. Scale bars = 1 µm.

Figures 57–63. Plastid diversity in *Cryptothallus mirabilis* Malmb. **57,** Pleomorphic profile in a meristematic cell of the thallus, note absence of thylakoids, the scattered starch grains (s), and large plastoglobuli (l), x21,000; **58,** Single thylakoids are occasionally found in plastids in the thallus meristem, s, starch, l, plastoglobuli, x28,000; **59,** Amyloplast in an inner thallus cell adjacent to the fungal zone, x13,000; **60,** Proplastid with single thylakoids in the seta of a young sporophyte, x36,000; **61,** Amyloplast (the starch grains are electron-dense following Thiery staining for carbohydrates) surrounded by mitochondrial (m) profiles in a young egg, x23,000; **62,** Degenerating plastid in a maturing egg, s, starch, l, lipid globules, pb, prolamellar bodies, x34,000; **63,** Aggregation of prolamellar body material in the cytoplasm of a maturing egg, x44,000. All scale bars = 1 µm, except **Fig. 56** = 0.25 µm.

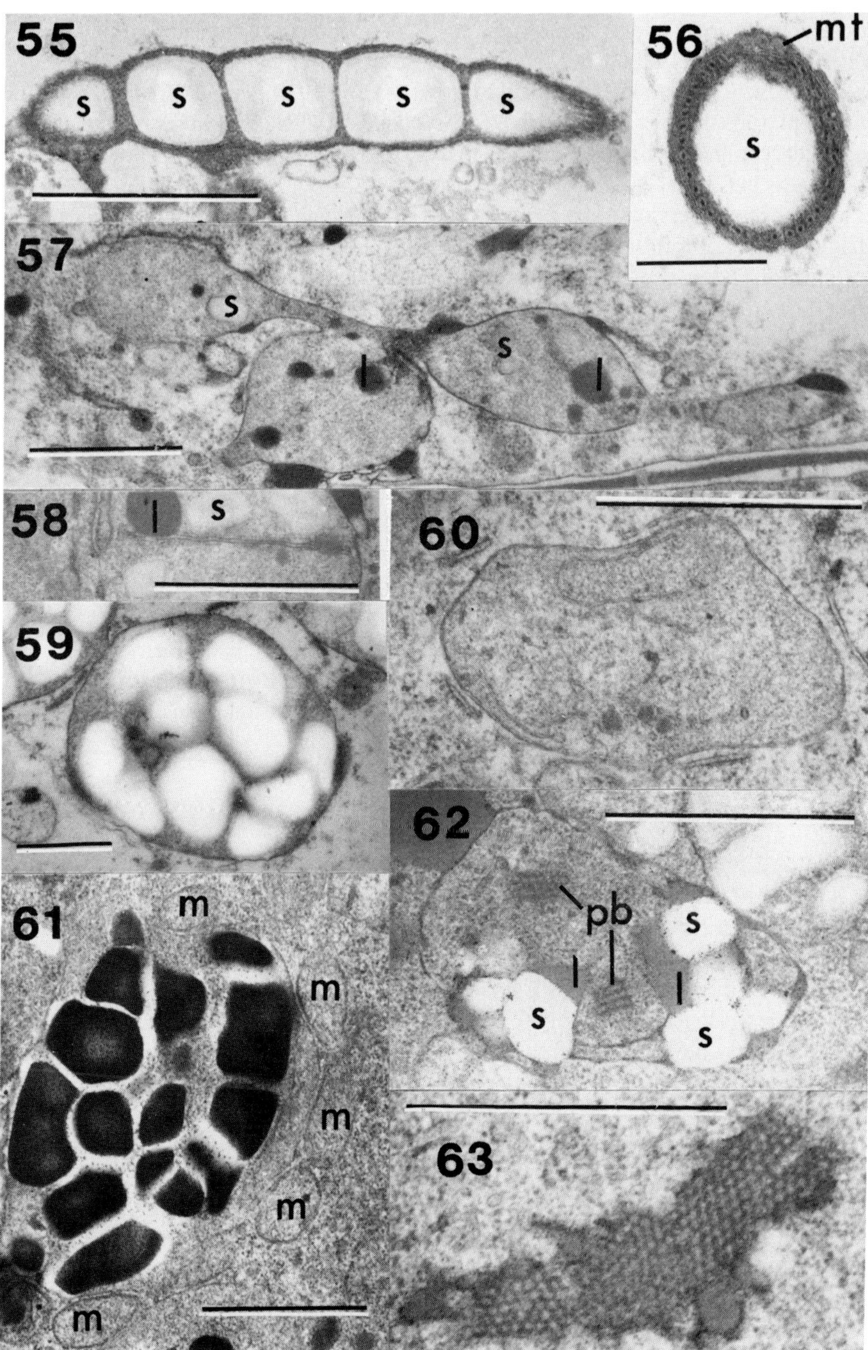

tochondrial profiles (cf. the *Nebenkern* of spermatids). Zinsmeister and Carothers (1974) and Diers (1970) could find no evidence in *Marchantia* and *Sphaerocarpos* of plastid degeneration similar to that in fern eggs (Bell 1979). However, differentiation of the plastids during oogenesis in both bryophytes and pteridophytes recalls parallel events accompanying female meiosis in angiosperms (Dickinson & Potter 1978) and oogenesis in *Chara* (Pickett-Heaps 1975).

Our recent observations on *Cryptothallus* provide additional insight into plastid behavior during oogenesis. Maturing eggs of *Cryptothallus* contain two distinct kinds of plastid. Those associated with mitochondria contain numerous starch grains (Fig. 61), but others possess smaller amounts of starch, irregular accumulations of lipid, and prolamellar bodies (Fig. 62). The envelopes of many of the latter appear to be discontinuous. Overall, these give the impression of organelles in the process of breaking down. The cytoplasm of the same eggs (Fig. 63) contains large irregular aggregations of material identical to the prolamellar bodies. Starch-filled plastids persist in the mature eggs, but those with prolamellar bodies are absent. The micrographic data strongly suggest that oogenesis in *Cryptothallus* involves the degeneration and elimination of at least some of the plastids. The two kinds of plastid have also been noted in the ventral canal cell (but never the neck canal cells) of *Cryptothallus*, but both kinds degenerate there. The stroma of amoeboid plastids in the young oocytes of *Chara* contains both starch grains and aggregations of tubular material (appreciably larger than microtubules and colchicine insensitive). Pickett-Heaps (1975) also records the proliferation of lipid bodies in these cells and their association with endoplasmic reticulum and the plastid envelopes, which may disintegrate during egg maturation. The parallels with *Cryptothallus* are most striking.

Sporogenesis

Of all the notable cytological events associated with sporogenesis in bryophytes perhaps the most striking is the reduction of the plastids to one during spore mother cell (SMC) formation in mosses. This dish-like plastid, appropriately called the "plastefoliace" by the French authors Eymé and Suire (1969, 1971), may reach 8–12 μm in diameter, while remaining only 0.2–0.5 μm thick in cross section. The thylakoid system is poorly developed, starch is rare, but phytoferritin inclusions are sometimes present (Neidhart 1979). This huge plastid divides into four parts immediately before meiosis. The four daughter organelles are distributed into the four meiospores where they rapidly undergo further divisions. Mature moss spores are multiplastidic (Neidhart 1979).

Although the behavior of the plastid in the SMCs and subsequently during spore maturation has been beautifully documented in a variety of mosses (see Brown & Lemmon, this volume, and references cited therein), the ontogeny of the plastids throughout archesporial development has been traced only in *Polytrichum* (Paolillo 1969). Gradual simplification of the internal lamellae and flattening of the plastids characterizes early stages in the development of the spore-sac. As the sporogenous cells multiply, the number of plastids is maintained as one flattened structure cupped around the interphase nucleus. How and when monoplastidy is achieved is not identified. Following division into four in the spore mother cells, the daughter plastids round up immediately prior to meiosis. The single plastid in each nascent spore is nearly spherical and contains rudimentary lamellae. After elaboration of the exine the plastid becomes extensively lobed. The main body is traversed by elongate and predominantly single thylakoids, while the swollen ends contain aggregations of vesicles and one or two small prolamellar bodies. Fully differentiated chloroplasts are developed before the spores are shed, elaboration of the grana-fretwork system apparently taking place over the course of several divisions. Paolillo's comment (1969) that his observations on *Polytrichum* indicate a shut down in vegetative plastid activities during sporogenesis and a return to the vegetative state before the spores are shed immediately brings to mind the dedifferentiation and redifferentiation of the plastids associated with meiosis, microsporogenesis, and megasporogenesis in angiosperms (Dickinson 1981, Dickinson & Potter 1978, Dickinson & Willson 1983). Undifferentiated plastids are associated with meiosis in *Selaginella* (Brown & Lemmon 1985c) in contrast to the prominent grana in those of mitotic cells (Brown & Lemmon 1985b).

Thylakoid regression has also been noted in archesporial tissues of *Mnium* followed by the appearance of prolamellar bodies in the meiocytes (Eymé & Suire 1971). Prolamellar bodies are similarly associated with the post-meiotic elaboration of the thylakoid network in *Leptodictyum* (Brown & Lemmon 1980b). Unusual parallel thylakoid arrays invest expanding starch grains in these plastids. Profiles of endoplasmic reticulum frequently lie adjacent to the plastid envelopes in sporocytes [e.g., *Mnium* (Eymé & Suire 1971), *Amblystegium* (Brown & Lemmon 1984b), *Rhynchostegium* (Brown & Lemmon 1982a)], but intimate mitochondrial-plastid associations are not evident (cf. section on spermatogenesis). By contrast to bryopsid mosses, minimally differentiated plastids have not been described during sporogenesis in *Sphagnum*. During the post-meiotic cytokinesis well developed grana are present (Brown et al. 1982).

Rudimentary plastid states are unknown during sporogenesis in hepatics (the SMCs are not monoplastidic), but it is possible that the critical phases in the process may have been missed in all the studies performed to date. Among the hepatics only *Blasia* has been found to possess four large, starch-filled plastids in its SMCs. The resulting four spores are each supplied with a single amylochloroplast (Renzaglia & Duckett, unpublished data). In all the stages of spore development in *Aneura* figured by Horner et al. (1966) the multiple plastids contain prominent grana; indeed, these have more thylakoids than in mature thallus tissues. The sporocytes and mature spores of *Haplomitrium* contain amyloplasts (Brown & Lemmon 1986). Archesporial plastids of *Pellia* have one to three small grana. Those in the sporocytes are amylochloroplasts, while the multicellular spores contain well developed grana and extensive intergranal networks (Suire 1970). The spores of *Radula* possess large stacks of swollen thylakoids but no fret system (Suire 1970). Well defined grana are illustrated by Neidhart (1979) in *Frullania* and *Lophocolea*; the latter also possesses prolamellar bodies. The spores of *Sphaerocarpos* contain amyloplasts and small grana, a situation that remains unaltered from the early stages in the differentiation of the intracapsular cells (Kelley & Doyle 1975). The only unusual feature is the aggregation of plastotubules, the nature, fate, and function of which are unknown. Oltmann (1974) describes etioplasts and proplastids in the spores of the Jungermanniales.

The elaters in *Pellia* contain ovoid plastids with loose aggregations of elongate thylakoids, whereas the thylakoid system in elaters of *Radula* comprises irregular vesicles (Suire 1970).

Associations with other Organelles

Microbodies have frequently been identified in a variety of tissues in all three groups of bryophytes (Ligrone et al. 1982b). As in vascular plants (Richardson 1974, Tolbert 1971), these are spheroidal or irregular in outline and delimited by a single membrane. They contain a finely granular matrix frequently with a paracrystalline core (Fig. 27). The nature and function of these heterogeneous organelles have not been investigated in detail in bryophytes. By analogy with vascular plants and on the basis of appropriate cytochemical tests it is suggested that microbodies in cells with well developed chloroplasts are peroxisomes with a role in photorespiration. Microbodies of this type are found in the urn wall and spore-sac tissue of *Buxbaumia* (Ligrone et al. 1982b), leaf lamellae in *Polytrichum* (Hébant & Marty 1972), protonemata of *Ceratodon* (Valanne 1971, Idzikowska & Szweykowska 1978), leaves of *Leucobryum* and *Timmiella* (Gambardella et al. 1979a), and photosynthetic lamellae in *Conocephalum* (Gambardella et al. 1979b).

In other situations, such as in antheridial jacket cells (Duckett 1973) and associated with dividing chloroplasts in hornworts (Brown & Lemmon 1985a), their functions are less certain. Their high frequency of occurrence in the hydrom sheath (Hébant 1974a) and especially in sporophytic foot transfer cells (Fig. 43, 48) (Ligrone et al. 1982a, b), intimately associated with lipid globules in the absence of photosynthetic plastids and starch, suggests glyoxysomal functions involving the mobilization of lipids. This is almost certainly the case in differentiating oil-body cells in *Marchantia* (Galatis & Apostolakos 1976, Galatis et al. 1978b). However, we have found microbodies to be surprisingly rare in the gametophytic tissues of liverworts. The only place where they occur in any abundance is in cells adjacent to the fungal zones in *Cryptothallus*, in *Riccardia* (Fig. 27), and in various members of the Jungermanniales.

Pais and Carrapiço (1979) describe two phases of microbody activity in germinating spores of *Bryum capillare* Hedw. An initial glyoxysomal phase (indicated by malate synthetase activity), associated with lipid mobilization, is subsequently replaced by peroxidase activity, which corresponds with the development of the photosynthetic apparatus.

Plastid-mitochondrial associations similar to those noted here in various situations in bryophytes have often been described in all groups of algae. In fact they are so common they rarely receive detailed comment (Wellburn 1982, Ledoigt & Lefort-Tran 1985). In contrast they are rare in vascular plants. A further similarity between bryophytes and algae that is also shared by fern gametophytes (Cran 1979) and the sporophytic tissues of a variety of lower vascular plants (Whatley 1977) is the frequency with which endoplasmic reticulum completely or partially surrounds the plastids. By contrast plastid-ER associations in angiosperms are apparently transitory and confined to immature or specialized cells (Whatley 1977). The functional significance of both the ER and mitochondrial associations is not understood.

Although bryophytes are good material for observations on plastid division and replication cycles (reviewed by Paolillo 1984), the mechanisms that actually control their numbers remain obscure. Lying at the heart of the problem are complex interactions between plastid and nuclear genomes. Further understanding of the regulation of plastid numbers in bryophytes now awaits detailed information on nuclear and plastid DNA replication. Equally fundamental, but for the immediate future far less intractable experimentally, are questions of how plastids are partitioned into daughter cells at cytokinesis and how their position is maintained and regulated during cellular differentiation at interphase.

The discovery of quadripolar microtubular arrays associated with the plastids and independent of the meiotic spindles in the SMCs of mosses immediately focuses attention on the involvement of the cytoskeleton in plastid apportionment at cytokinesis, not to mention the possible roles of cytoskeletal elements associated with plastids in the establishment of polarity (Brown & Lemmon 1980a, 1982a, b, c, d, 1983, 1984b, Eymé & Suire 1971, Lambert 1970, 1978). Preprophasic microtubule arrays are also intimately associated with dividing plastids in hornworts (Brown & Lemmon 1985a) and monoplastidic lycopods (Brown & Lemmon 1984a, 1985b, c). But in multiplastidic vascular plants microtubules are seldom seen in the immediate vicinity of plastids, and the same is true in hepatics and mosses. However, Apostolakos and Galatis (1985a, b) illustrate protuberances of the plastids extending into the polar microtubule organizing centers (MTOCs) at preprophase in *Marchantia*. At prophase the plastids are concentrated around the MTOCs, with microtubules running parallel to their envelopes. In caulonemal tip cells of *Funaria* microtubules are associated with the chloroplasts (Schnepf 1982: Fig. 2). Thus there is ultrastructural evidence for links between microtubules and plastids in all three groups of bryophytes.

Immunofluorescence studies have clearly established the involvement of microtubules in nuclear positioning in moss protonemata (Doonan et al. 1985). Since microtubule networks, visualized by monoclonal antibodies, also extend around the plastids, it is probably that these cytoskeletal elements are important in the siting of plastids. However, it is unlikely that all of the complex processes involving anchorage, separation, and migration of the plastids that are characteristic of bryophyte development can be explained solely in terms of microtubule dynamics (assembly, disassembly, sliding, and the forces thereby generated). There are many instances of morphogenetic plastid migrations not associated with microtubules, for example the actin-mediated plastid movements in charalean algae (review by Williamson 1986). It is not without significance in this context that Brown and Lemmon (1982b) point out the abundance of fibrillar elements associated with microtubules and the plastid envelope in the SMCs of mosses. With precise positioning of the plastids (not to mention their associations with mitochondria, nucleus, and ER), a key feature of morphogenesis, bryophytes may be singled out among multiplastidic plants as offering perhaps the greatest potential for the elucidation of how cytoskeletal elements determine the location of cytoplasmic organelles. The fact that morphogenesis, plastid duplication, and cytoskeletal dynamics may all be modified experimentally (Dyer & Duckett 1984) is a further bonus accruing from the use of bryophyte systems. We predict that evidence for the involvement of actin in plastid migrations in bryophyte morphogenesis will not be long forthcoming and that intermediate filament proteins, recently discovered in plants (Parke et al. 1987), will be found to be intimately involved in the positioning of these organelles.

Conspectus

This review reveals that plastid ultrastructure in bryophytes is not only extremely diverse but also that the developmental pathways follow courses very different from those in vascular plants. It is impossible to extrapolate about plastid ultrastructure on the basis of observations on one species: even closely allied genera (e.g., *Riccardia, Aneura, Cryptothallus*) may be very different. In the 1960s generalizations about the cell structure (and biochemistry) of a whole group from data on just one or two taxa were commonplace. Plastid ultrastructure in bryophytes amply illustrates why this should not be the case today. Some three decades after the advent of biological electron microscopy, the deduction of chloroplast gene organization in *Marchantia* (Ohyama et al. 1986) from the complete sequence of chloroplast DNA (determined by cloning techniques) heralds the dawn of bryophyte molecular biology. In the light of the lessons from cell biology we must not now assume that *Marchantia* is any more representative of liverworts (indeed bryophytes as a whole) at the level of organelle genome organization than it is at the level of chloroplast ultrastructure.

Whereas in vascular plants it is generally assumed that the meristems contain proplastids, plastids with minimally differentiated internal lamellar systems are uncommon in the same locations in bryophytes. In the gametophyte generation of the majority of mosses, hepatics, and hornworts, few ultrastructural differences are detectable between the chloroplasts in dividing cells and those in fully expanded tissues. However, in *Tortula* small elongate grana in mature leaf cells derive from the progressive fragmentation of massive grana found in the apical cell and its immediate derivatives. Not even the ripe spores of most bryophytes contain proplastids. A similar absence of proplastids is characteristic of the gametophytes (but not spores) of ferns (Cran 1979, Kotenko 1986). Here, as in bryophytes, division of fully differentiated chloroplasts by constriction (and less frequently by partition) is the norm. In vascular plants the stage of plastid development seems to bear little or no relationship to their capacity to divide (Whatley 1986, 1987); proplastids undergo replication during early organogenesis, while chloroplasts in different stages of differentation divide freely in more mature cells.

Although this review highlights the relative uniformity in plastid ultrastructure in meristems of bryophytes, it should be underlined that in no single case have plastids been studied in detail throughout the cell cycle. Thus the possibility that the plastids in the apical cell or its derivatives undergo rapid changes before cell division cannot be totally ruled out.

Thylakoid architecture in mature chloroplasts varies greatly, sometimes even between closely related taxa. Vertical intergranal links are absolutely diagnostic of the hornworts. Similar structures occur in the charalean alga *Coleochaete*. Some liverworts and mosses have tall compact grana and extensive fretwork systems, while in others the stroma is transversed by elongate grana and containing small numbers of thylakoids, and the intergranal network is poorly differentiated. An unusual feature of some metzgerialean chloroplasts is the predominantly peripheral location of elongate thylakoids. These impinge on the contours of the envelope, resulting in plastids that are polyhedral in shape. As such, they bear a striking resemblance to some of the chloroplasts of *Chara*. Even in chlorenchyma the shape of the plastids in bryophytes rarely approaches the regular plano-convex lens found in seed plants. Angular contours and elongated grana in mature chloroplasts, and occasional grana in proplastids can all be found in lower vascular plants, e.g., *Equisetum* (Whatley 1971) and *Pilularia* (Whatley 1975). It could well be that as in bryophytes they are not that uncommon but have just not been properly described (Whatley, personal communication).

Regression rather than elaboration of the thylakoid network is a striking feature of highly differentiated cells in bryophytes (e.g., transfer cells, conducting elements, and spermatids). Specific thylakoid configurations and stromal inclusions such as phytoferritin are often peculiar to cell types. Prolamellar bodies are rare and, except for a recent report on the sporophytes of hornworts, their formation is not induced by extended periods of darkness nor by changes in light quality [cf. their transient appearance when fern gametophytes are transferred from red to blue light (Cran 1979)]. Chloroplast ultrastructure in bryophytes is far more stable under different light regimes than it is in vascular plants. Although sun and shade-type chloroplasts are found in bryophytes, other features of their photosynthetic machinery are very different. Variable amounts of starch in the chloroplasts of photosynthetic tissues are short-term stores of reserve carbohydrate, but amyloplasts in the setae of mosses and hepatics, immediately below the stem apex in *Sphagnum*, and in underground axes of some leafy hepatics, probably function in gravity perception.

The monoplastidic cell lineages characteristic of sporogenesis in mosses and spermatogenesis in all bryophytes are established rapidly during the initial delimitation of the sporogenous and spermatogenous tissues, respectively, and not as a gradual process as previously assumed. Dedifferentiation and subsequent redifferentiation of the plastids during sporogenesis recalls similar behavior associated with meiosis in angiosperms. There are no clear-cut differences between the plastids of sporophyte and gametophyte generations.

There is evidence of degeneration and elimination of part of the plastid complement during oogenesis in *Cryptothallus*. Spermatid plastids are excluded from the eggs at fertilization in all bryophytes.

Associations of the bryophyte plastids with other organelles, viz. mitochondria, nucleus and endoplasmic reticulum, have more in common with algae and lower vascular plants than with seed plants. Microtubule-plastid links seen in spore mother cells of mosses, during mitosis in liverworts and hornworts, and in caulonemata suggest bryophytes as promising material for experimental studies on the role of cytoskeletal elements in plastid positioning, a phenomenon fundamental to plant morphogenesis.

Acknowledgments

K.S.R. acknowledges a Grant-in-Aid from the State of Tennessee enabling her to work at Queen Mary College, University of London, in 1985 and 1986. The authors gratefully thank Dr. R. Ligrone for permission to reproduce the following micrographs: Fig. 3, 6–10, from Ligrone and Fioretto 1987; Fig. 43–47, from Ligrone et al. 1982b; Fig. 48, from Ligrone et al. 1982a. We would also like to thank the Pathology Department, Quillen-Dishner College of Medicine, East Tennessee State University, for the use of their TEM, and Dr. Brent Mishler of Duke University for contributing live specimens of *Tortula*.

Literature Cited

Albertine, K. H., Maravolo, N. C. & Kaustinen, H. 1976. Effects of gibberellin and several growth retardants on plastid ultrastructure in the hepatic *Marchantia polymorpha*. The Bryologist 79: 22–34.

Apostolakos, P. & Galatis, B. 1985a. Studies on the development of air pores and air chambers of *Marchantia paleacea*. II. Ultrastructure of the initial microtubule organizing centres. Canadian Journal of Botany 63: 744–756.

—— 1985b. Studies on the development of air pores and air chambers of *Marchantia paleacea*. III. Microtubule organization in preprophase-prophase initial aperture cells – formation of incomplete preprophase microtubule bands. Protoplasma 128: 120–135.

—— 1985c. Studies on the development of air pores and air chambers of *Marchantia paleacea*. IV. Cell plate arrangement in initial aperture cells. Protoplasma 128: 136–146.

Aro, E.-M. 1982. A comparison of the chlorophyll-protein composition and chloroplast ultrastructure in two bryophytes and two higher plants. Zeitschrift für Pflanzenphysiologie 108: 97–105.

Aro, E.-M., Somersalo, S. & Karunen, P. 1987. Membrane lipids in *Ceratodon purpureus*. Physiologia Plantarum 69 : 65–72.

Aro, E.-M. & Valanne, N. 1979. Effect of continuous light on CO_2 fixation and chloroplast structure of the mosses *Pleurozium schreberi* and *Ceratodon purpureus*. Physiologia Plantarum 45: 460–466.

Bell, P. R. 1979. Gametogenesis and fertilization in ferns, pp. 471–503. In: Dyer, A. F. (Ed.), The Experimental Biology of Ferns. London. Academic Press.

Berrie, G. K. & Webster, P. M. 1982. Ultrastructure of plastids and mitochondria in gemmae of *Marchantia polymorpha* L. Annals of Botany 50: 199–206.

Bismarck, R. von 1959. Über den Geotropismus der Sphagnen. Flora, Jena 148: 23–83.

Bjorkman, O. 1981. Comparative studies on photosynthesis in higher plants. Encyclopedia of Plant Physiology, New Series 12A: 57–109.

Bonnot, E.-J. 1967. Relations structurales nucléo-plastidales dans les cellules méristématiques du gamétophyte feuille de *Polytrichum formosum*. Comptes Rendus Hebdomadaire des Séances de L'Académie des Sciences, Paris, Série D, 264: 2555–2558.

Bopp, M. 1984. Developmental physiology of bryophytes, pp. 276–324. In: Schuster, R. M. (ed.), New Manual of Bryology, Vol. 1. Nichinan, Japan. Hattori Botanical Laboratory.

Bradbeer, J. W. 1981. Development of photosynthetic function during chloroplast biogenesis, pp. 423–472. In: Stumpf, T. K. & Hatch, M. D. (eds.), The Biochemistry of Plants, Vol. 8. London. Academic Press.

Brett, D. W. & Somerard, A. P. 1986. Ultrastructural development of plastids in the epidermis and starch layer of glossy *Ranunculus* petals. Annals of Botany 58: 903–910.

Brown, R. C. & Lemmon, B. E. 1980a. Ultrastructure of sporogenesis in a moss, *Ditrichum pallidum*. I. Meiotic prophase. The Bryologist 83: 137–152.

— — 1980b. Ultrastructural aspects of chloroplast development in spores of the moss *Leptodictyum riparium* (Hedw.) Warnst. The Bryologist 83: 545–554.

— — 1982a. Ultrastructural aspects of moss meiosis: review of nuclear and cytoplasmic events during prophase. Journal of the Hattori Botanical Laboratory 53: 29–30.

— — 1982b. Ultrastructural aspects of moss meiosis: cytokinesis and organelle apportionment in *Rhynchostegium serrulatum*. Journal of the Hattori Botanical Laboratory 53: 41–50.

— — 1982c. Ultrastructure of meiosis in the moss *Rhynchostegium serrulatum*. I. Prophasic microtubules and spindle dynamics. Protoplasma 110: 23–33.

— — 1982d. Ultrastructure of sporogenesis in the moss *Amblystegium riparium*. I. Meiosis and cytokinesis. American Journal of Botany 69: 1096–1107.

— — 1983. Microtubule organization and morphogenesis in young spores of the moss *Tetraphis pellucida* Hedw. Protoplasma 116: 115–124.

— — 1984a. Plastid apportionment and preprophase microtubule bands in monoplastidic root meristem cells of *Isoetes* and *Selaginella*. Protoplasma 123: 95–103.

— — 1984b. Ultrastructure of sporogenesis in the moss *Amblystegium riparium*. II. Spore wall development. Journal of the Hattori Botanical Laboratory 57: 139–152.

— — 1985a. Preprophasic establishment of division polarity in monoplastidic mitosis of hornworts. Protoplasma 124: 175–183.

— — 1985b. Development of stomata in *Selaginella*: division polarity and plastid movements. American Journal of Botany 72: 1914–1925.

— — 1985c. A cytoskeletal system predicts division plane in meiosis of *Selaginella*. Protoplasma 127: 101–109.

— — 1986. Spore wall development in the liverwort *Haplomitrium hookeri*. Canadian Journal of Botany 64: 1174–1182.

Brown, R. C., Lemmon, B. E. & Carothers, Z. B. 1982. Spore wall development in *Sphagnum lescurii*. Canadian Journal of Botany 60: 2394 – 2409.

Browning, A. J. & Gunning, B. E. S. 1979. Structure and function of transfer cells in the sporophyte haustorium of *Funaria hygrometrica* Hedw. I. The development and ultrastructure of the haustorium. Journal of Experimental Botany 30: 1233–1246.

— — 1970. Phylogenetic transitions in the chloroplasts of the Anthocerotales. I. The number and ultrastructure of the mature plastids. American Journal of Botany 57: 97–110.

Butterfass, T. 1979. Patterns of chloroplast development. A developmental approach to protoplasmic plant anatomy. Cell Biology Monographs 6. Vienna. Springer-Verlag.

Carothers, Z. B. & Duckett, J. G. 1980. The bryophyte spermatozoid: a source of new phylogenetic information. Bulletin of the Torrey Botanical Club 107: 281–297.

Castaldo, R., Ligrone, R. & Gambardella, R. 1979. A light and electron microscope study on the phylloids of *Leucobryum candidum* (P. Beauv.) Wils. Revue Bryologique et Lichénologique 45: 345–360.

Chevallier, D. 1975. Effets d'un sejour à l'obscurité sur le pouvoir germatif et sur l'évolution de l'appareil photosynthétique des spores de *Funaria hygrometrica*. Physiologia Plantarum 34: 216–220.

Chevallier, D. & Nurit, F. 1974. Evolution structurale et fonctionnelle de l'appareil photosynthétique des spores de *Funaria hygrometrica* et de *Bryum capillare* au cours de leur germination. Canadian Journal of Botany 52: 927–934.

Conrad, P. A., Steucek, G. L. & Hepler, P. K. 1986. Bud formation in *Funaria*: organelle redistribution following cytokinin treatment. Protoplasma 131: 211–223.

Cox, G. 1986. Comparison of *Prochloron* from different hosts. I. Structural and ultrastructural characteristics. New Phytologist 104: 429–445.

Cran, D. G. 1979. The ultrastructure of fern gametophyte cells, pp. 171–212. In: Dyer, A. F. (Ed.), The Experimental Biology of Ferns. London. Academic Press.

Crandall-Stotler, B. J. 1981. Morphology/anatomy of hepatics and anthocerotes. Advances in Bryology 1: 315–398.

— 1986. Morphogenesis, developmental anatomy and bryophyte phylogenetics: contraindications of monophyly. Journal of Bryology 14: 1–23.

Dickinson, H. G. 1981. The structure and chemistry of plastid differentation during male meiosis in *Lilium henryi*. Journal of Cell Science 52: 223–241.

Dickinson, H. G. & Potter, U. 1978. Cytoplasmic changes accompanying the female meiosis in *Lilium longiflorum* Thunb. Journal of Cell Science 29: 147–169.

Dickinson, H. G. & Willson, C. 1983. Two stages in the redifferentiation of amyloplasts in the microspores of *Lilium*. Annals of Botany 52: 803–810.

Diers, L. 1967. On the behaviour of the plastids during the development of the archegonium and the egg cell in an archegoniate plant, pp. 306–320. In: Sironval, C. (Ed.), Le Chloroplaste; Croissance et Vieillissement. Paris. Masson.

— 1970. Origin of plastids: cytological results and interpretations including some genetical aspects, pp. 129–145. In: Miller, P. (Ed.), Control of Organelle Development. Society for Experimental Biology Symposium 24. Cambridge. University Press.

Doonan, J. H., Cove, D. J. & Lloyd, C. W. 1985. Immunofluorescence microscopy of microtubules in intact cell lineages of the moss *Physcomitrella patens*. I. Normal and CIPC-treated tip cells. Journal of Cell Science 75: 1–17.

Duckett, J. G. 1973. Wall ingrowths in the jacket cells of the antheridia of *Anthoceros laevis* L. Journal of Bryology 7: 405–412.

— 1975. An ultrastructural study of antheridial plastids in *Anthoceros laevis* L. Cytobiologie 10: 432–448.

— 1986. Ultrastructure in bryophyte systematics and evolution: an evaluation. Journal of Bryology 14: 25–42.

Duckett, J. G. & Renzaglia, K. S. 1987. Transmission electron microscopy of oil bodies in hepatics: games with systematics and cytochemistry. Bulletin of the British Bryological Society 49: 17–19.

Duckett, J. G., Carothers, Z. B. & Miller, C. C. J. 1982. Comparative spermatology and bryophyte phylogeny. Journal of the Hattori Botanical Laboratory 53: 107–125.

— — 1984. Gametogenesis, pp. 232–275. In: Schuster, R. M. (Ed.), New Manual of Bryology, Vol. I. Nichinan, Japan. Hattori Botanical Laboratory.

Duckett, J. G. & Prasad, A. K. S. K., Davies, D. A. & Walker, S. 1977. A cytological analysis of the *Nostoc*-bryophyte relationship. New Phytologist 79: 349–362.

Dyer, A. F. & Duckett, J. G. (Eds.) 1984. The Experimental Biology of Bryophytes. 292 pp. London. Academic Press.

Eymé, J. & Suire, C. 1967. Au sujet de l'infrastructure des cellules de la région placentaire de *Mnium cuspidatum* Hedw. (Mousse bryale acrocarpe). Comptes Rendus Hebdomadaire des Séances de l'Académie des Sciences, Paris, Série D, 265: 1788–1791.

— — 1969. Ultrastructure des cellules sporogènes des Mousses: observations sur le plastidome et le chondriome. Comptes Rendus Hebdomadaire des Séance' de l'Académie des Sciences, Paris, Série D, 268: 290–293.

— — 1971. Recherches sur l'évolution ultrastructurale des constituants cellulaires durant la sporogénèse des mousses. Le Botaniste 54: 109–151.

Fabré, M.-C. & Orcival, J. 1982. Données ultrastructurales sur les apex végétatif et sexualise de *Sphagnum flexuosum* Dozy et Molk. Comptes Rendus Hebdomadaire des Séances de l'Académie des Sciences, Paris, Série D, 294: 1073–1080.

Fowke, L. C. & Pickett-Heaps, J. D. 1978. Electron microscope study of vegetative cell division in two species of *Marchantia*. Canadian Journal of Botany 56: 467–475.

Frederick, S. E., Gruber, P. J. & Tolbert, N. E. 1973. The occurrence of glycolate dehydrogenase and glycolate oxidase in green plants. An evolutionary survey. Plant Physiology, Lancaster 52: 318–323.

Galatis, B. & Apostolakos, P. 1976. Associations between microbodies and a system of cytoplasmic tubules on oil-body cells of *Marchantia*. Planta 131: 217–221.

— — 1977. On the fine structure of differentiating mucilage papillae of *Marchantia*. Canadian Journal of Botany 55: 772–795.

Galatis, B., Apostolakos, P. & Katsaros, C. 1978a. Ultrastructural studies on the oil bodies of *Marchantia paleacea* Bert. I. Early stages of oil-body cell differentiation: origination of the oil body. Canadian Journal of Botany 56: 2252–2267.

Galatis, B., Katsaros, C. & Apostolakos, P. 1978b. Ultrastructural studies on the oil bodies of *Marchantia paleacea* Bert. II. Advanced stages of oil-body cell differentiation: synthesis of lipophilic material. Canadian Journal of Botany 56: 2268–2285.

Gambardella, R., Castaldo, R. & Ligrone, R. 1979a. Presence of microbody-like organelles in Bryophyta. Caryologia 32: 117.

Gambardella, R., Ligrone, R. & Castaldo, R. 1979b. First ultrastructural demonstration of catalase activity with diaminobenzidine in a liverwort *Conocephalum conicum* (L.) Dum. (Marchantiales). Delpinoa 20: 67–73.

— — 1981. Ultrastructure of the sporophyte foot of *Phaeoceros*. Cryptogamie Bryologie et Lichénologie 2: 23–45.

Graham, L. E. 1982. The occurrence, evolution and phylogenetic significance of parenchyma in *Coleochaete* Breb. (Chlorophyta). American Journal of Botany 69: 447–454.

Graham, L. E. & McBride, G. E. 1975. The ultrastructure of multilayered structures associated with flagellar bases in motile cells of *Trentepohlia aurea*. Journal of Phycology 11: 86–96.

—— 1979. The occurrence and phylogenetic significance of a multilayered structure in *Coleochaete* spermatozoids. American Journal of Botany 66: 887–894.

Griffiths, D. J. 1970. The pyrenoid. Botanical Review 36: 19–58.

— 1980. The pyrenoid and its role in algal metabolism. Science Progress, Oxford 66: 537–553.

Gruber, P. J. & Frederick, S. E. 1977. Cytochemical localization of glycolate oxidase in microbodies of *Klebsormidium*. Planta 135: 45–49.

Hébant, C. 1974a. Studies on the development of the conducting tissue-system in the gametophytes of some Polytrichales. II. Development and structure at maturity of the hydroids of the central strand. Journal of the Hattori Botanical Laboratory 38: 565–607.

— 1974b. Polarized accumulations of endoplasmic reticulum and other ultrastructural features of leptoids in *Polytrichadelphus magellanicus* gametophytes. Protoplasma 81: 373–382.

— 1977. The conducting tissues of bryophytes. Bryophytorum Bibliotheca 10. Vaduz. Cramer.

Hébant, C. & Marty, F. 1972. Fine structural identification of peroxisomes in the cells of the photosynthetic lamellae of the leaf of *Polytrichum commune* gametophytes. Journal of Bryology 7: 195–199.

Hensel, W. 1986. Cytodifferentiation of polar plant cells. Use of anti-microtubular agents during the differentiation of statocytes from cress roots (*Lepidium sativum* L.). Planta 169: 293–303.

Hölzl, J. 1969. Elektronenmikroskopische Beobachtungen an den Blattzellen des Lebermooses *Mylia taylori*. Österreichische Botanische Zeitschrift 117: 203–304.

Honegger, R. 1980. Zytologie der Blaualgen-Hornmoos-Symbiose bei *Anthoceros laevis* aus Island. Flora, Jena 170: 290–302.

Horner, H. T., Jr., Lersten, N. R. & Bowen, C. C. 1966. Spore development in the liverwort *Riccardia pinguis*. American Journal of Botany 53: 1048–1064.

Idzikowska, K. & Szweykowska, A. 1978. The ultrastructural aspects of the cytokinin-induced bud formation in *Ceratodon pupureus*. Protoplasma 94: 41–52.

Jensen, L. C. W. & Jensen, C. G. 1984. Fine structure of protonemal apical cells of the moss *Physcomitrium turbinatum*. Protoplasma 122: 1–10.

Karunen, P. 1972. Studies on moss spores. I. The triglycerides of *Polytrichum commune* spores and their mobilization and degradation in relation to the germination phases. Annales Universitatis Turkuensis, Series A, 51: 1–70.

Karunen, P. & Kälviäinen, E. 1985. Senescence and post-mortem changes in the ultrastructure of *Sphagnum fuscum* (Klinggr.) Schlieph. leaf cells. New Phytologist 100: 419–427.

Kelley, C. B. & Doyle, W. T. 1975. Differentiation of intracapsular cells in the sporophyte of *Sphaerocarpos donnellii*. American Journal of Botany 62: 547–559.

Knoth, R., Hansmann, P. & Sitte, P. 1986. Chromoplasts of *Palisota barteri*; and the molecular structure of chromoplast tubules. Planta 168: 167–174.

Kotenko, J. L. 1986. Antheridium formation on *Onoclea sensibilis* L: cytoplasmic polarity and determination of wall positions. Botanical Gazette 147: 28–39.

Krochko, J. E., Bewley, J. D. & Pacey, J. 1978. The effects of rapid and very slow speeds of drying on the ultrastructure and metabolism of the desiccation-sensitive moss *Cratoneuron filicinum* (Hedw.) Spruce. Journal of Experimental Botany 29: 905–917.

Lambert, A.-M. 1970. Étude de structures clinétique en rapport avec la rupture de la membrane nucléaire, en début de méiose chez *Mnium hornum* L. Organisation des centromères. Comptes Rendus Hebdomadaire des Séances de l'Académie des Sciences, Paris, Série D, 270: 481–484.

—— 1978. La méiose chez lez bryophytes: ultrastructure et dynamique du fuseau chez une mousse, *Mnium hornum* Hedw. Bryophytorum Bibliotheca 13: 113–145.

Ledoigt, G. & Lefort-Tran, M. 1985. Regulation of plastid development in *Euglena gracilis*. Relationship between cell compartments. Biologie Cellulaire 55: 87–96.

Lehmann, H. & Jaster, B. 1981. Feinstrukturelle Untersuchungen zur Entwicklung der Olzellen bei dem Lebermoos *Riella*. Protoplasma 106: 109–119.

Lehmann, H. & Schulz, D. 1969. Elektronenmikroskopische Untersuchungen von Differenzierungsvorgängen bei Moosen. I. Probleme zur Darstellung der Feinstruktur. Planta 84: 57–67.

—— 1982. Die Verwendung der Semisimultan-Fixierung zur Darstellung der Feinstruktur spezialisierter Mooszellen. Mikroskopie 39: 285–291.

Ligrone, R. 1985. Massive ribonuclear protein bodies in gametophyte vegetative cells of the moss *Timmiella barbuloides* (Brid.) Moenk. Protoplasma 127: 204–211.

Ligrone, R. & Fioretto, A. 1987. Chloroplast development in light- and dark-grown sporophytes of *Phaeoceros laevis* (L.) Prosk. (Anthocerotophyta). New Phytologist 105: 301–308.

Ligrone, R., Gambardella, R. & Sposito, L. M. L. 1982a. Ultrastructure of the sporophyte foot-gametophyte vaginula complex in *Timmiella barbuloides* (Brid.) Moenk. Planta 154: 414–425.

Ligrone, R., Gambardella, R., Castaldo, R., Glordano, S. & de Lucia Sposito, M. L. 1982b. Gametophyte and sporophyte ultrastructure in *Buxbaumia piperi* Best (Buxbaumiales, Musci). Journal of the Hattori Botanical Laboratory 52: 465–499.

Lüttge, U. & Krapf, G. 1968. Die Ultrastruktur der Blattzellen junger and alter *Mnium*-Sprosse und ihr Zusammenhang mit der Ionenaufnahme. Planta 81: 132–139.

Mache, R. & Loiseaux, S. 1973. Light saturation of growth and photosynthesis of the shade plant *Marchantia polymorpha*. Journal of Cell Science 12: 391–401.

Manton, I. 1957. Observations with the electron microscope on the cell structure of the antheridium and spermatozoid of *Sphagnum*. Journal of Experimental Botany 8: 382–400.

—— 1962. Observations on plastid development in the meristem of *Anthoceros*. Journal of Experimental Botany 13: 325–333.

Marchant, H. J. & Pickett-Heaps, J. D. 1973. Mitosis and cytokinesis in *Coleochaete scutata*. Journal in Phycology 9: 461-471.

Mattox, K. R. & Stewart, K. D. 1984. Classification of the green algae: a concept based on comparative cytology, pp. 29–72. In: Irvine, D. E. G. & John, D. M. (Eds.), Systematics of the Green Algae. London & Orlando. Academic Press.

Menke, W. 1961. Über die chloroplasten von *Anthoceros punctatus*. 5. Mitteilung zur Entwicklungsgeschichte der Plastiden. Zeitschrift für Naturforschung 16: 334–336.

Miller, C. C. J. 1982. Experimental studies on spermatogenesis in Bryophytes. Ph. D. dissertation. London University. England.

Miller, C. C. J. & Duckett, J. G. 1986. Cytoplasmic deletion processes during spermatogenesis in mosses. Gamete Research 13: 253–270.

Miller, C. C. J., Duckett, J. G., Sheterline, P. & Carothers, Z. B. 1983. Immunofluorescence microscopy of the flagella and multilayered structure in two mosses: *Sphagnum palustre* L. and *Polytrichum juniperinum* Hedw. Journal of Cell Science 61: 71–86.

Moestrup, Ø. 1974. Ultrastructure of the scale-covered zoospores of the green alga *Chaetosphaeridium*, a possible ancestor of the higher plants and bryophytes. Biological Journal of the Linnean Society 6: 111–125.

— 1975. Some aspects of sexual reproduction in eukaryotic algae, pp. 23–35. In: Duckett, J. G. & Racey, P. A. (Eds.), The Biology of the Male Gamete. Biological Journal of the Linnean Society 7 (Supplement 1). London. Academic Press.

Monroe, J. H. 1968. Light- and electron-microscopic observations on spore germination in *Funaria hygrometrica*. Botanical Gazette 129: 247–258.

Mueller, D. M. J. 1973. The peristome of *Fissidens limbatus*. University of California Publications in Botany 63: 1–34.

Neidhart, H. V. 1979. Comparative studies of sporogenesis in bryophytes, pp. 251–280. In: Clarke, G. C. S. & Duckett, J. G. (Eds.), Bryophyte Systematics. London. Academic Press.

Noailles, M.-C. 1974. Étude ultrastructurale comparée de la différenciation du parenchyma et des cellules périphériques de la tige d'une Mousse pleurocarpe: *Pleurozium schreberi* (Willd.) Mitt. (Hypnobryale). Comptes Rendus Hebdomadaire des Séances de l'Académie des Sciences, Paris, Série D, 441–444.

— 1978. Étude ultrastructurale de la récupération hydrique après un periode de sécheresse chez une Hypnobryale: *Pleurozium schreberi* (Willd.) Mitt. Annales des Sciences Naturelles Botanique et Biologie Végétale, Série 12, 19: 249–265.

Ohyama, K., Fukuzawa, H., Kohchi, T., Shirai, H., Sano, T., Sano, S., Umesono, K., Shiki, Y., Takeuchi, M., Chang, Z., Aota, S., Inokuchi, H. & Ozeki, H. 1986. Chloroplast genome organization deduced from complete sequence of liverwort (*Marchantia polymorpha*) chloroplast DNA. Nature 322: 572–574.

Oliver, M. J. & Bewley, J. D. 1982. Desiccation and ultrastructure in bryophytes. Advances in Bryology 2: 91-132.

Oltmann, O. 1974. Licht- und electronenmikroskopische Untersuchungen zur Sporogenese von *Lophocolea heterophylla* (Schrad.) Dum. Die Entwicklung von der Sporenmutterzelle bis zur Tetradenstadium. Pollen et Spores 16: 5–25.

Osborn, B. A. & Raven, J. A. 1986. Light absorption by plants and its implications for photosynthesis. Biological Reviews 61: 1–61.

Pais, M. S. & Carrapiço, F. 1979. Localisation cytochimique de la malate synthétase et de la glycolate oxydase au niveau des microbodies des spores chlorophylliennes de la mousse *Bryum capillare*. Comptes Rendus Hebdomadaire des Séances de l'Académie des Sciences, Paris, Série D, 288: 395–398.

Paolillo, D. J. Jr. 1964. The plastids of *Polytrichum commune*. I. The capsule at meiosis. Protoplasma 58: 667–680.

— 1969. The plastids of *Polytrichum*. II. The sporogenous cells. Cytologia 34: 133–144.

— 1984. Cell and plastid cycles, pp. 117–142. In: Dyer, A. F. & Duckett, J. G. (Eds.), The Experimental Biology of Bryophytes. London. Academic Press.

Paolillo, D. J. Jr. & Reighard, J. A. 1967. Ultrastructural features of some polytrichaceous moss leaves. The Bryologist 70: 62–69.
Paolillo, D. J. Jr., Kreitner, G. L. & Reighard, J. A. 1968. Spermatogenesis in
Polytrichum juniperinum. I. The orgin of the apical body and the elongation of the nucleus. Planta 78: 226–247.
Parke, J. M., Miller, C. C. J., Cowell, I., Dodson, A., Dowding, A.,
Downes, M., Duckett, J. G. & Anderton, B. H. 1987. Monoclonal antibodies
against plant proteins recognise animal intermediate filaments. Cell Motility and the Cytoskeleton. 8: 312–323.
Pickett-Heaps, J. D. 1972. Cell division in *Klebsormidium subtilissimum* (formerly
Ulotrix subtilissima), and its possible phylogenetic significance. Cytobios 6: 167–183.
— 1975. Green Algae: Structure, Reproduction and Evolution in Selected Genera. Sunderland, Massachusetts. Sinauer Associates.
Pickett-Heaps, J. D. & Marchant, H. J. 1972. The phylogeny of the green algae: a
new proposal. Cytobios 6: 255–264.
Pihakaski, K. & Pihakaski, S. 1979. Effects of chilling on the ultrastructure and net
photosynthesis of *Pellia epiphylla*. Annals of Botany 43: 773–781.
Pihakaski, S. & Pihakaski, K. 1980. Effects of glyphosate on ultrastructure and photosynthesis of *Pellia epiphylla*. Annals of Botany 46: 133–141.
Pocock, K. & Duckett, J. G. 1985. On the occurrence of branched and swollen rhizoids
in British hepatics: their relationships with the substratum and associations with fungi.
New Phytologist 99: 281–304.
Rascio, N., Mariani, P., Chitano, P. & Vecchia, F. D. 1986. An ultrastructural
study of maize leaf etioplasts throughout their entire life-cycle. Protoplasma 130: 98–
107.
Richardson, M. 1974. Microbodies (glyoxysomes and peroxisomes) in plants. Science
Progress, Oxford. 61: 41–61.
Rogers, C. E., Mattox, K. R. & Stewart, K. D. 1980. The zoospore of *Chlorokybus atmosphyticus*, a charophyte with sarcinoid growth habit. American Journal of Botany 67: 774–783.
Sack, F. D. & Paolillo, D. J. Jr. 1983a. Protoplasmic changes during stomatal development in *Funaria*. Canadian Journal of Botany 61: 2515–2526.
— — 1983b. Stomatal pore and cuticle formation in *Funaria*. Protoplasma 116: 1–13.
— — 1985. Incomplete cytokinesis in *Funaria* stomata. American Journal of Botany 72:
1325–1333.
Sack, F. D., Suyemoto, M. M. & Leopold, A. C. 1986. Amyloplast sedimentation
and organelle saltation in living corn columella cells. American Journal of Botany 73:
1692–1698.
Scheirer, D. C. 1983. Leaf parenchyma with transfer cell-like characteristics in the moss,
Polytrichum commune Hedw. American Journal of Botany 70: 987–992.
Schimper, A. F. W. 1885. Untersuchungen über die chlorophyll Körner und die ihnen
homologen Gebilde. Jahrbuch für Wissenschaftliche Botanik 16: 1–220.
Schnepf, E. 1982. Morphogenesis in moss protonemata, pp. 321–344. In: Lloyd, C. W.
(Ed.), The Cytoskeleton in Plant Growth and Development. London. Academic Press.
Schofield, W. B. & Hébant, C. 1984. The morphology and anatomy of the moss gametophore, pp. 627–695. In: R. M. Schuster (Ed.), New Manual of Bryology, Vol. II.
Nichinan, Japan. Hattori Botanical Laboratory.

Schulz, D. & Lehmann, H. 1969. Wachstum und Aufbau der Zellwand bei Moosen. Cytobiologie I. 343–356.

Sears, B. B. 1980. Elimination of plastids during spermatogenesis and fertilization in the plant kingdom. Plasmid 4: 233–255.

Sigee, D. C. 1969. The fine structure of plastids in the apical region of the gametophyte of *Cryptothallus mirabilis* Malmb. Transactions of the British Bryological Society 5: 820–822.

Sitte, P. 1963. Hexagonale Anordnung der Globuli in Moos-Chloroplasten. Protoplasma 56: 198–201.

Sitte, P., Falk, H. & Liedvogel, B. 1980. Chromoplasts, pp. 117–148. In: Czygan, F.-C. (Ed.), Pigments in Plants. Stuttgart & New York. G. Fischer.

Sluiman, H. J. 1983. The flagellar apparatus of the zoospore of the filamentous green alga *Coleochaete pulvinata*: absolute configuration and phylogenetic significance. Protoplasma 115: 160–175.

Stead, A. D. & Duckett, J. G. 1980. Plastid ontogeny in the corolla cells of *Digitalis purpurea* L. cv Foxy. Annals of Botany 46: 549–555.

Steer, M. W. 1985. Mitosis in bryophytes. Advances in Bryology 2: 1–63.

Stetler, D. A. & DeMaggio, A. E. 1976. Ultrastructural characteristics of spore germination in the moss *Dawsonia superba*. American Journal of Botany 63: 438–442.

Stewart, K. D. & Mattox, K. R. 1975. Comparative cytology, evolution and classification of the green algae with some considerations of the origin of other organisms with chlorophylls a and b. Botanical Review 41: 104–135.

—— 1978. Structural evolution in the flagellated cells of green algae and land plants. BioSystems 10: 145–152.

Suire, C. 1970. Recherche cytologiques sur deux Hépatiques: *Pellia epiphylla* (L.) Corda (Metzgériale) et *Radula complanata* (L.) Dum. (Jungermanniale). Ergastrome, Sporogénèse et spermatogénèse. Le Botaniste 53: 125–392.

Tolbert, N. E. 1971. Microbodies, peroxisomes and glyoxysomes. Annual Review of Plant Physiology 22: 45–74.

Tucker, E. B., Costerton, J. W. & Bewley, J. D. 1975. The ultrastructure of the moss *Tortula ruralis* on recovery from desiccation. Canadian Journal of Botany 53: 94–101.

Valanne, N. 1966. The germination phases of moss spores and their control by light. Annales Botanici Fennici 3: 1–60.

— 1971. The effects of prologed darkness and light on the fine structure of *Ceratodon purpureus*. Canadian Journal of Botany 49: 547–554.

— 1976. Development of chloroplast structure and photosynthetic competence in dark-adapted moss protonemata after exposure to light. Protoplasma 89: 359–369.

— 1977a. The combined effects of light intensity and continuous light on the CO_2 fixation, chlorophyll content and chloroplast structure of the protonemata of *Ceratodon purpureus*. Zeitschrift für Pflanzenphysiologie 83: 275–283.

— 1977b. Effect of continuous light on CO_2 fixation, chlorophyll content, growth and chloroplast structure in *Ceratodon purpureus*. Zeitschrift für Pflanzenphysiologie 81: 347–357.

— 1984. Photosynthesis and photosynthetic products in mosses, pp. 257–273. In: Dyer, A. F. & Duckett, J. G. (Eds.), The Experimental Biology of Bryophytes. London. Academic Press.

Valanne, N., Pennanen, A. & Vapaavuori, E. 1979. Comparison between the presentation of chloroplast structure in the dark and the turnover rate of chlorophyll-protein complexes in a moss and two varieties of pea. Plant & Cell Physiology 20: 1511–1522.

Valentine, L. J., Campbell, E. O. & Hopcroft, D. H. 1986. A study of chloroplast structure in three *Megaceros* species and three *Dendroceros* species (Anthocerotae) indigenous to New Zealand. New Zealand Journal of Botany 24: 1–8.

Weier, T. E. 1930. A study of moss plastid after fixation by mitochondrial, osmium, and silver techniques. I. The plastid during sporogenesis in *Polytrichum commune*. La Cellule 40: 261–289.

— 1931. A study of the moss plastid after fixation by mitochondrial, osmium and silver techniques. II. The plastid during spermatogenesis in *Polytrichum commune* and *Catherinaea undulata*. La Cellule 41: 51–84.

Wellburn, A. R. 1982. Bioenergetic and ultrastructural changes associated with chloroplast development. International Review of Cytology 80: 133–191.

Wells, J. M., McNally, S. F. & Richardson, H. S. 1987. The effects of arsenate, selenite and distilled water on the physiology and fine structure of *Hylocomium splendens*. Journal of Bryology. 14: 551–563.

Whatley, J. M. 1971. The chloroplasts of *Equisetum telmateia* Erhr: a possible developmental sequence. New Phytologist 70: 1095–1102.

— 1975. Chloroplast structure in coiled and uncoiled croziers of *Pilularia globulifera*. New Phytologist 74: 413–420.

— 1977. Variations in the basic pathway of chloroplast development. New Phytologist 78: 407–420.

— 1978. A suggested cycle of plastid developmental interrelationships. New Phytologist 80: 489–502.

— 1980. Plastid growth and division in *Phaseolus vulgaris*. New Phytologist 86: 1–16.

— 1982. Ultrastructure of plastid inheritance: green algae to angiosperms. Biological Reviews 57: 527–569.

— 1983. Ultrastructure of plastids in roots. International Review of Cytology 85: 175–220.

— 1986. Patterns of plastid replication during plant development, pp. 3–36. In: Mantell, S. H., Chapman, G. P. & Street, P. F. S. (Eds.), The Chondriome-Chloroplast and Mitochondrial Genomes. London. Longman.

— 1987. Plastids in a changing environment. Festschrift. Herbert Baker. Botanical Society of America (in press).

Whatley, J. M. & Price, D. N. 1983. Do lemon cotyledons green in the dark? New Phytologist 94: 19–27.

Wiencke, C. & Schulz, D. 1977. The development of transfer cells in the haustorium of the *Funaria hygrometrica* sporophyte. Bryophytorum Bibliotheca 13: 147–167.

— — 1983. The fine structure of symplasmic and apoplasmic transport in the "nerve" of the *Funaria* leaflet. Zeitschrift für Pflanzenphysiologie 112: 337–350.

Williamson, R. E. 1986. Review. Organelle movement along actin filaments and microtubules. Plant Physiology, Bethesda 82: 631–634.

Wilsenach, R. 1963. Differentiation of the chloroplast in *Anthoceros*. Journal of Cell Biology 18: 419–428.

Zinsmeister, D. D. & Carothers, Z. B. 1974. The fine structure of oogenesis in *Marchantia polymorpha*. American Journal of Botany 61: 499–512.

Advances in Bryology 3: 95 – 134 (1988).

Comparative Morphology of the Bryophyte Blepharoplast

by

Zane B. Carothers

Department of Plant Biology, University of Illinois
Urbana, Illinois 61801, U.S.A.

and

Ann E. Rushing

Department of Botany and Microbiology, Auburn University
Auburn, Alabama 36849, U.S.A.

With 16 Figures and 2 Tables

Abstract: This paper emphasizes the more recent advances in bryophyte blepharoplast morphology and comprises both original research and a review of pertinent literature. The bryophytes on which the review focuses include representatives of 14 liverwort, two hornwort, and six moss genera. In order to facilitate comparison, seven different blepharoplasts – those of the hepatics *Haplomitrium gibbsiae*, *Bazzania trilobata*, and *Marchantia polymorpha*, the hornwort *Notothylas orbicularis*, and the mosses *Funaria hygrometrica*, *Archidium tenerrimum*, and *Thuidium delicatulum* – are diagrammatically represented to show only morphologically equivalent structures and at the same scale of magnification. A detailed analysis of the blepharoplast of *Haplomitrium gibbsiae* is reported for the first time.

Of the dozen or so characters derived from structural analysis, the features that appear to be most useful for comparative purposes are 1) shape of the spline anterior, 2) number of spline microtubules, 3) spline aperture, 4) lamellar strip, and 5) basal bodies with their associated transition zones. The hornworts, represented by *Phaeoceros* and *Notothylas*, have a commonly held group of blepharoplast characters not found elsewhere in the bryophytes, thus emphasizing the unity of the group. These characters include an inaperturate, 12-microtubule spline with a rectangular anterior, a transversely elongated lamellar strip, mono-

morphic basal bodies evenly aligned with the spline's leading edge, and flagellar transition
zones devoid of stellate patterns. Within the liverworts and mosses, however, a wide range of
variation exists. In general the moss blepharoplast has a somewhat spatulate spline with either a closed or open aperture, a longitudinally elongated lamellar strip with a truncated posterior margin, and dimorphic basal bodies oriented with the anterior one parallel to the spline
axis and the posterior one divergent. In addition to these few shared features the moss
blepharoplast exhibits variations between all taxa except *Thuidium* and *Hypnum* in the Hypnobryales. Within the liverworts, the blepharoplasts of most taxa have a bilaterally asymmetrical spline of the closed-aperturate type, a bluntly rounded MLS-anterior, a relatively
long-tapered lamellar strip, and dimorphic basal bodies that are quite strongly subapical in
position. The most conspicuous departure from this generalized form is that of *Haplomitrium gibbsiae*. Its distinguishing features include 1) an obliquely asymmetrical spline anteri-
or, the right margin of which widens in two steps, 2) a spline consisting of up to 90 parallel
microtubules, 3) a spline aperture of the open type situated to the left of the spline's midline,
4) a lamellar strip with an elongated, curved drop-shape, 5) both anterior and posterior basal
bodies situated to the left of the spline's midline, 6) three ventrally positioned microtubular
triplets extending forward from each basal body, and 7) stellate patterns measuring only 100
nm in length.

Except for *Haplomitrium* the subclass Marchantiidae exhibits relatively uniform blepharoplast morphology. Species representing *Marchantia* and the sphaerocarpalean genera *Riella*,
Sphaerocarpos, and *Geothallus* share such features as a closed aperture of three microtubule-
diameters in width and a relatively short lamellar strip with a notch-like offset on its right
side. In contrast, the Jungermanniidae is a diverse subclass displaying many variations in
blepharoplast morphology. In general the representatives of this group have narrower spline
apertures, longer lamellar strips, and longer posterior basal bodies than those of the Marchantiidae. Within the Jungermanniales, as determined from four representatives (*Marsupella*,
Chiloscyphus, *Bazzania*, and *Cephalozia*), the lamellar strip is elongated with a gradually
sloped, lateral offset that extends forward beyond the spline's anterior margin. Structural diversity of the blepharoplasts in the Metzgeriales exceeds that of the Jungermanniales. Studies
of three representatives (*Pellia*, *Pallavicinia*, and *Blasia*) show that each possesses major
characters that, in combination, distinguish them from each other. Thus, beyond citing such
common features as a bilaterally spatulate spline anterior and a closed spline aperture, it is
not possible to characterize the metzgerialean blepharoplast within narrow limits. *Blasia*, for
example, shares features with members of the Marchantiidae, while *Pallavicinia* has an
elongated strip similar in shape to that found in the Jungermanniales. With the exception of
the hornworts and sphaerocarpalean liverworts, which possess unique blepharoplast features,
it is not yet possible to characterize most bryophyte groups. Much of value will be learned
by continuing the survey and determining the ranges of morphological variation within major bryophyte taxa.

Zusammenfassung: Dieser Beitrag betont die neueren Fortschritte der Kenntnisse über die
Morphologie des Bryophyten-Blepharoplasten und enthält sowohl Resultate eigener For-
schung wie auch aus der dazugehörigen Literatur. 14 Lebermoos-, 2 Hornmoos- und 6 Laub-
moosgattungen wurden untersucht. Zur besseren Vergleichbarkeit wurden 7 verschiedene Blepharoplasten – jene der Lebermoose *Haplomitrium gibbsiae*, *Bazzania trilobata* und *Marchantia polymorpha*, das Hornmoos *Nothothylas orbicularis* und die Laubmoose *Funaria hygrometrica*, *Archidium tenerrimum* und *Thuidium delicatulum* – graphisch dargestellt, um die

morphologisch entsprechenden Strukturen im selben Maßstab zu zeigen. Der Blepharoplast von *Haplomitrium gibbsiae* wurde zum ersten Mal ausführlich beschrieben. Ein gutes Dutzend Merkmale wurden für die Strukturanalysen verwendet. Die folgenden 5 scheinen die nützlichsten für Vergleichzwecke zu sein: 1) Form des Vorderteils des mikrotubularen Bandes (spline anterior), 2) Anzahl der Mikrotubuli in diesem Band, 3) Apertur des mikrotubularen Bandes, 4) Lamellarstreifen und 5) Basalkörper mit den dazugehörigen Übergangszonen. Die Hornmoose, vertreten durch *Phaeoceros* und *Notothylas*, zeigen gemeinsame Blepharoplastenmerkmale, die nirgends sonst bei den Bryophyten gefunden werden und somit die Einheitlichkeit dieser Gruppe hervorheben. Zu diesen Merkmalen gehören ein Band aus 12 Microtubuli ohne Apertur mit einem rechteckigen Vorderteil, ein quer verlaufender verlängerter Lamellarstreifen, monomorphe Basalkörper, die gleichmäßig entlang der Führungskante des Bandes angeordnet sind und die Übergangszonen zur Geißel ohne Sternmuster. Innerhalb der Leber- und Laubmoose hingegen besteht eine große Mannigfaltigkeit. Im allgemeinen besitzt der Laubmoosblepharoplast ein etwas spatelförmiges Band mit einer entweder geschlossenen oder einer offenen Apertur, einem in Längsrichtung verlängerten Lamellarstreifen mit gestutztem Hinterrand und dimorphen Basalkörpern, deren Vorderteile parallel zur Bandachse gerichtet sind und deren Hinterteile davon abstehen. Außer diesen wenigen gemeinsamen Merkmalen unterscheiden sich aber die Blepharoplasten aller Laubmoostaxa untereinander mit Ausnahme von *Thuidium* und *Hypnum* in den Hypnobryales. Die Blepharoplasten der meisten Lebermoostaxa zeigen ein bilateral-asymmetrisches Band mit geschlossener Apertur, ein stumpf abgerundetes Vorderteil der „vielschichtigen Struktur" (MLS), einen recht lang zugespitzten Lamellarstreifen und dimorphe Basalkörper in subapikaler Position. Die auffälligste Abweichung von dieser allgemeinen Form findet sich bei *Haplomitrium gibbsiae* mit folgenden Unterscheidungsmerkmalen: 1) ein schräg asymmetrisches Vorderteil des Bandes, dessen rechter Rand sich zweimal ausweitet, 2) ein Band mit bis zu 90 parallelen Mikrotubuli, 3) eine offene Apertur des Bandes auf der linken Seite von dessen Mittellinie, 4) ein verlängerter, gebogen-tropfenförmiger Lamellarstreifen, 5) sowohl vordere wie hintere Basalkörper links von der Mittellinie des Bandes, 6) drei ventral angelegte Mikrotubuli–Tripletts, die von jedem Basalkörper vorwärts weisen und 7) Sternmuster, die nur 100 nm lang sind. Mit Ausnahme von *Haplomitrium* zeigt die Unterklasse Marchantiidae eine relativ einheitliche Blepharoplastmorphologie. Arten welche *Marchantia*, sowie die Gattungen *Riella*, *Sphaerocarpos* und *Geothallus* (Sphaerocarpales) repräsentieren, haben als gemeinsame Merkmale eine geschlossene Apertur von den drei Microtubuli-Breitendurchmessern und einen ziemlich kurzen Lamellarstreifen mit einer etwas eingekerbten Ausbuchtung auf der Seite. Im Gegensatz dazu ist die Morphologie des Blepharoplasten in der Unterklasse Jungermanniidae vielfältig. Im allgemeinen besitzen die Vertreter dieser Gruppe eine engere Apertur des Bandes, längere Lamellarstreifen und längere hintere Basalkörper als jene der Marchantiidae. Wie an vier Vertretern der Jungermanniales (*Marsupella, Chiloscyphus, Bazzania, Cephalozia*) festgestellt wurde, sind hier die Lamellarstreifen verlängert mit einem allmählich geneigten, seitlichen Ausbuchtung, der über den Vorderrand des Bandes hinausreicht. Die strukturelle Diversität des Blepharoplasten der Metzgeriales übersteigt noch jene der Jungermanniales. Untersuchungen an drei Vertretern (*Pellia, Pallavicinia, Blasia*) zeigen, daß bei jedem wichtige Kriterien vorhanden sind, welche, kombiniert, die Taxa voneinander abtrennen. Es ist deshalb nicht möglich, eine eng umgrenzte Charakterisierung des Blepharoplasten der Metzgeriales zu geben, die über eine Aufzählung von verbreiteten Merkmalen wie bilaterale spatelförmige Bandvorderseite und eine geschlossene Apertur hinausgeht. *Blasia* z. B. zeigt gemeinsame Züge mit Vertretern der Marchantiidae, während *Pallavicinia* einen verlängerten Streifen ähnlich jenem der Jungermanniales aufweist. Die Charakterisierung der meisten Bryophytcngruppen anhand von Blepharoplastenmerkmalen ist bis jetzt noch nicht möglich mit Ausnahme der Hornmoose und Lebermoose der Ordnung Sphaerocarpales. Die Fortsetzung der Untersuchungen einschließlich der Bestimmung der morphologischen Variationsbreite innerhalb der Hauptgruppen der Bryophyten wird weitere wertvolle Ergebnisse liefern.

Keywords: *Archidium*, basal body, *Bazzania*, blepharoplast, bryophyte, *Funaria*, *Haplomitrium*, *Marchantia*, morphology, multilayered structure, *Notothylas*, spermatid, spline, *Thuidium*, ultrastructure.

Contents

Introduction

Comparative blepharoplast morphology is a newcomer among the various lines of inquiry into the systematics and evolution of bryophytes. Reckoning from the first detailed investigations of *Marchantia polymorpha* L. (Carothers & Kreitner 1968) and *Polytrichum juniperinum* Hedw. (Paolillo et al. 1968a, b), studies of blepharoplast structure have now been reported for representatives of some 22 genera, 11 of which were published between 1984 and 1987. The species chosen for inclusion here constitute a broad selection including representatives of 14 liverwort, two hornwort, and six moss genera. Not unexpectedly, these studies range widely also in the kind and amount of structural details they report and in the interpretations on which the diagrammatic reconstructions (when given) are based. A consequence of the variation among different sets of data is that their comparison is often made difficult and sometimes impossible.

With the intent of facilitating the presentation and use of blepharoplast data a simplified style of graphic reconstruction was introduced in a recent study of *Funaria hygrometrica* Hedw. (Carothers & Brown 1985). Since then, that same graphic style has been used in several other reports, including the present account. Shown here are seven comparable reconstructions depicting only morphologically equivalent structures and at the same scale of magnification. The seven species (and their data sources) include the liverworts *Haplomitrium gibbsiae* (Steph.) Schust. (present paper), *Bazzania trilobata* (L.) S. Gray (Rushing & Carothers 1986a), and *Marchantia polymorpha* (Carothers & Kreitner 1968), the hornwort *Notothylas orbicularis* (Schwein.) Sull. (Renzaglia & Carothers 1986), and the mosses *Funaria hygrometrica* (Carothers & Brown 1985), *Archidium tenerrimum* Mitt. (Brown & Carothers 1986), and *Thuidium delicatulum* (Hedw.) B.S.G. (Rushing & Carothers 1986b).

Our emphasis in the present review is on recent advances and includes a documented analysis of the *Haplomitrium gibbsiae* blepharoplast. Selected earlier studies are cited when pertinent; however, no attempt is made to review all published data regarding the topic. Much of the older literature is incorporated in studies of spermatozoid development and can be accessed readily through appropriate reviews (e.g., Carothers & Duckett 1980, Duckett et al. 1982, 1983). In addition to having the present work provide a research update it is intended to serve as an introduction to the topic for those having little or no familiarity with blepharoplast ultrastructure. To that end we have added a concise background statement identifying some of the structures and events associated with male gamete development and an explanation of the terms used in describing them. Lastly, it is our intent to show how this information on blepharoplast morphology is serving as a useful supplement to the more traditional evidence used in studies of bryophyte systematics and evolution.

Background

The long history of inquiry concerning bryophyte blepharoplasts is intimately associated with that of male gamete formation. Early studies of cell lineages within the antheridium revealed the sequential appearance of spermatogenous (androgonial) cells,
spermatid mother cells, and spermatids, which differentiate to become motile spermatozoids (e.g., Durand 1908, see also Fig. 1 in Carothers & Duckett 1980). Careful observation of these cells revealed the presence in the developing spermatid of a minute
cytoplasmic body that elongated and gave rise to a pair of flagella (see, for example,
Taf. III in Ikeno 1903). The term "blepharoplast" was soon adopted and became
widely used for these flagellum-producing structures (Wilson 1911). Because the
small size of the blepharoplast made it extremely difficult to study (and even to demonstrate its presence) with light microscopy, such topics as the origin, structure, and
homologies of this entity remained speculative and controversial for more than half a
century (Lepper 1956).

With the development of suitable ancillary procedures for use with transmission
electron microscopy (TEM), investigators were finally able to characterize the subcellular components and follow the various series of ultrastructural changes attending the
formation of spermatozoids. The greatest amount of attention has been given to spermatids. Produced by the final cell duplication, these are the cells in which occurs most
of the structural transformation into a functional male gamete. They also exhibit the
greatest structural complexity of any protoplasts produced by the plant. Detailed TEM
analyses (e.g., Carothers & Kreitner 1967, 1968) have shown that the bryophyte
blepharoplast is an assemblage of cytoplasmic components, the functions of which, in
part, are associated with cell support and motility.

The principal blepharoplast component affording support is the spline, a band of
parallel microtubules that in mature spermatozoids extends the length of the gamete.
This microtubular cytoskeleton typically has a somewhat widened or spatulate outline
toward its anterior end and a longer, narrower shank toward the posterior. Appressed
beneath the spline at its anterior end is the lamellar strip (LS) consisting of many,
closely parallel, diagonally oriented lamellae and having a variously shaped outline according to the species. Unlike the spline, the bryophyte LS is a transitory structure
with only a small remnant (Carothers 1975: Plate 6B) or, perhaps, none at all in the
mature gamete. A large mitochondrion is situated immediately beneath the LS and persists in that position under the spline after reduction of the LS. Collectively, the spline
and lamellar strip constitute the so-called multilayered structure (MLS). The spline
constitutes the uppermost layer of the dorsiventrally flattened MLS, while the LS typically comprises three lower strata (Carothers & Kreitner 1967).

Affixed above the spline are the two locomotory structures (flagella) characteristic
of bryophyte spermatozoids. Each consists of three structurally dissimilar regions: a
proximally located basal body, a short transition zone and, distally, the long flagellar
shaft. Of these three, only the basal body, which is derived from a centriole, is interpreted as a component of the blepharoplast. Typically, the basal bodies of each

blepharoplast are dimorphic, longer than their precursor centrioles, and of different lengths. Owing to unequal growth at their proximal ends, the posterior basal body is longer than the anterior one. Dimorphism is also evident in the differing amounts of proximal extension undergone by the central hubs and individual microtubular triplets. (For diagrammatic illustrations showing the major components and their spatial relationships in the blepharoplast and the mature gamete, see Carothers & Duckett 1980).

The Blepharoplast of *Haplomitrium gibbsiae*[1]

The multilayered structure (MLS) of *Haplomitrium gibbsiae* has the typical, dorsiventrally 4-layered organization. Its lamellar strip (LS) is comprised of closely spaced, vertical lamellae (Fig. 3) oriented with their long axes at about 45° to that of the spline juxtaposed above them (Fig. 2). MLS microanatomy of *H. gibbsiae* is essentially identical to that of *H. hookeri* (Carothers & Duckett 1979) and need not be repeated here.

A composite reconstruction (Fig. 14) based on many sections such as those illustrated in this paper reveals that the anterior portion of the spline is asymmetrical. From a rounded apex, its left margin gradually curves back (Fig. 2) to a level near the distal end of the posterior basal body (PBB). In contrast, the right margin widens in two stages. First, from the rounded apex, it reaches a width equivalent to about 29 parallel microtubules to the right of the spline aperture (SA); it extends posteriorly at this width to a position about 1.3 µm behind the apex, then gradually curves outward (Fig. 14). At their widest regions the splines shown here (Fig. 11–14) comprise 84 microtubules; the highest number we have observed in *H. gibbsiae* is 90. The aperture is located to the left of the spline's midline (Fig. 14). It is open at its anterior end (Fig. 4, 5) and two microtubule-diameters wide (Fig. 6, 7) over most of its 1.5 µm length. At its posterior end the SA is closed by convergence of the two groups of microtubules it laterally separates (Fig. 14). This closure occurs at approximately the level (ca. 1.3 µm behind the apex) where the spline's right margin curves outward.

The overall outline of the LS has a form resembling that of a curved, elongated drop (Fig. 14). Its left margin is coincident with that of the spline except at the spline's apex where the LS extends anteriorly for a short distance. In overall length and width the LS measures about 3.1 and 1.9 µm, respectively. The distal end of the LS and that of the posterior transition zone (PTZ) are located at about the same level (Fig. 13, 14).

[1]The materials and methods pertaining to the study of *Haplomitrium gibbsiae*, together with explanations concerning taxonomic usage, tabular data, and diagrammatic representations contained in this paper, are presented in Appendix A.

The locomotory structures, like the SA, are situated to the left of the spline's midline (Fig. 8–14). The PBB is 3 µm long and follows closely the curvature of the spline's left margin (Fig. 14). The ABB is 1 µm long and straight, diverging about 10° to the left of the spline axis with its distal end above the SA (Fig. 7). Nearly all of the ABB is overlapped by the PBB. Both basal bodies are subapical, the apex of the ABB slightly nearer (ca. 0.1 µm) the spline apex than is that of the PBB. The posterior portion of each basal body consists of a circle of nine triplet microtubules arranged around a central structure that, in transverse section, resembles the hub and spokes of a wheel – the so-called cartwheel configuration (Fig. 7, 11). The longer, anterior portion of each basal body commonly consists of three extended, often splayed ventral triplets and the associated cartwheel structure (Fig. 6, 8).

Immediately posterior to each basal body is a transition zone (TZ). The anterior (ATZ) and posterior (PTZ) zones are unequal in length, the former about 0.74 µm long and the latter about 0.56 µm. Within each is a 100 nm-long stellate pattern (SP; Fig. 8, 12) situated about 0.15 µm distal to the TZ's anterior end. The portions of the TZ located proximal and distal to the SP are seen in transverse section to have a circle of nine doublet microtubules surrounding an apparently structureless interior (e.g., Fig. 9, 13). The SPs are longitudinally separated by about 2 µm (Fig. 1), a distance reflecting the unequal lengths of the basal bodies. Extending posteriorly from each TZ is the flagellum with its familiar axoneme, i.e., the circular configuration comprising nine doublet microtubules and two central singlets (e.g., Fig. 13). The level where the flagellum projects above the body of the cell typically coincides with that of the SP (Fig. 14).

Figures 1–7. Blepharoplast anatomy of *Haplomitrium gibbsiae*. See Appendix A for key to labeling. **1**, Oblique-longitudinal section intersecting both anterior and posterior stellate patterns in their respective transition regions, x42,100; **2**, Tangential section showing the anterior and left-marginal curvature of the multilayered structure and diagonally oriented S_2 lamellae of the subtending lamellar strip, x49,800; **3**, Diagonal section of the multilayered structure showing the microtubular S_1 layer, or spline, and the subtending tristratose lamellar strip (bracket), x52,900; **Figures 4–7**. Transverse sections of the blepharoplast, proceeding from anterior to posterior. (See Fig. 14 for the longitudinal locations of these four levels.) **4**, At this level (the anterior end of the spline aperture) the spline comprises 26 parallel microtubules, the tubule at left (arrow) marks the right-hand boundary of the aperture. (See Fig. 5 for consecutive serial sections.), x50,300; **5**, Section adjacent to that in Fig. 5 shows 29 microtubules to the right of the spline aperture and three to the left, the S_1 tubule indicated by the arrow is similarly identified in Fig. 4, x50,300; **6**, The aperture is here flanked by 9 S_1 tubules on the left and 29 on the right, the extended triplets are associated with the posterior basal body, x50,300; **7**, At this level (near the aperture's posterior end) the widening spline has 20 tubules on the left side of the aperture and 30 on the right, and the ABB appears as a circle of nine imbricated triplets, extended triplets of the posterior basal body are evident at the left, x50,300. Scale bars = 0.2 µm.

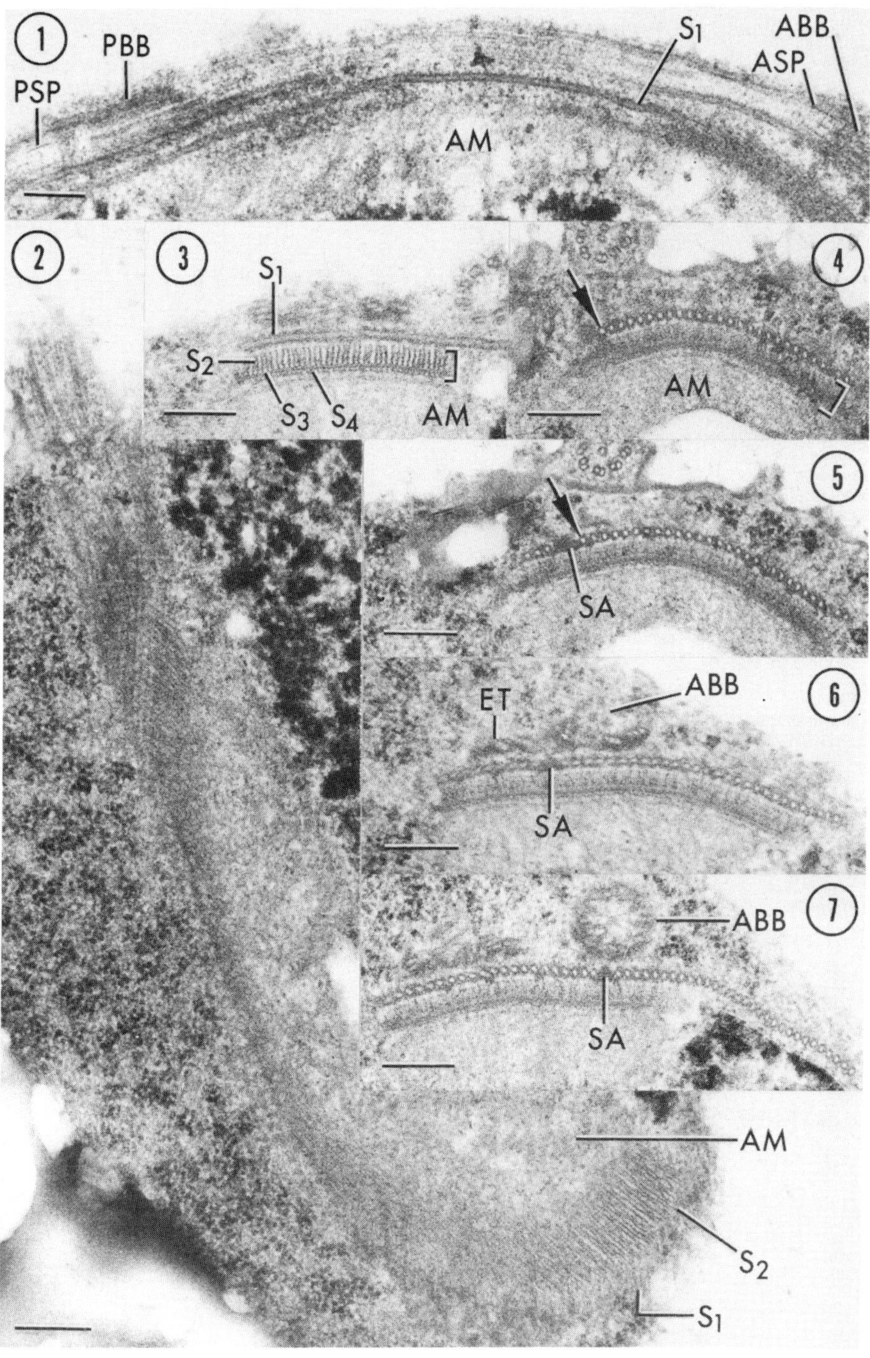
1
PSP
PBB
S1
ABB
ASP
AM
2
3
S1
S2
S3
S4
AM
4
AM
5
SA
6
ET
ABB
SA
7
ABB
SA
AM
S2
S1

Table 1. Summary of blepharoplast characteristics of the seven bryophytes depicted in Figs. 15 and 16.

	Spline type	Most commonly observed width at spline's widest region (parallel S_1 tubules)	Spline aperture position (spline tubules to right of aperture)	Predominant width of spline aperture (tubule-diameters)	Spline aperture length (μm)	Divergent spline microtubules	Lamellar strip outline in relation to spline anterior	Lamellar strip length (μm)	Lamellar strip overall width (μm)
Haplomitrium gibbsiae (Calobryales)	open aperturate	84 (2.6 μm)	29	2	1.5	0	anterior extension	3.9	1.9
Bazzania trilobata (Jungermanniales)	closed aperturate	15 (0.45 μm)	3	1	0.75	0	anterior extension	2.9	0.45
Marchantia polymorpha (Marchantiales)	closed aperturate	17 (0.5 μm)	3	3	1.5	0	coincident	1.75	0.5
Notothylas orbicularis (Anthocerotales)	inaperturate	12 (0.4 μm)	–	–	–	0	right- and left-lateral extensions	0.4	0.7
Funaria hygrometrica (Funariales)	open aperturate	13 (0.5 μm)	4	3	2.0	1	anterior extension	1.1	0.5
Archidium tenerrimum (Archidiales)	open aperturate	24 (0.75 μm)	6	2	1.15	0	anterior and right-lateral extension	1.5	0.9
Thuidium delicatulum (Hypnobryales)	open aperturate	14 (0.5 μm)	4	3	3.0	1	anterior extension	2.6	0.5

Basal body length, inc. extended triplets and hub (µm)		Distance between basal body apex and spline apex (µm)		Transition zone length (µm)		Stellate pattern length (nm)		Distance from anterior stellate pattern to spline apex (µm)	Distance between anterior and posterior stellate pattern apices (µm)	Source of data
ABB	PBB	ABB	PBB	ATZ	PTZ	ASP	PSP			
1.0	3.0	0.4	0.5	0.75	0.55	100	100	1.5	2.0	Present paper, Fig. 14
0.7	2.3	0.45	0.85	0.17	0.17	150	140	1.15	2.0	Rushing & Carothers (1986a), Fig. 15
0.65	1.0	0.5	0.65	0.3	0.3	230	230	1.15	0.6	Carothers & Kreitner (1968), Fig. 9
0.4 (right BB)	0.4 (left BB)	0.0	0.0	0.1	0.1	–	–	–	–	Renzaglia & Carothers (1986), Fig. 11
0.8	3.5	0.2	0.5	0.15	0.2	110	150	1.0	3.0	Carothers & Brown (1985), Fig. 16
1.3	5.0	0.15	0.15	0.2	0.2	140	130	1.5	3.5	Brown & Carothers (1986), Fig. 15
2.1	7.0	0.2	0.3	0.1	0.1	45	45	2.3	5.0	Rushing & Carothers (1986b), Fig. 22

A summary including the above and other measurements pertaining to the *Haplomitrium gibbsiae* blepharoplast is presented in Table 1 along with comparable sets of data from other bryophytes.

The *Haplomitrium*-type Blepharoplast

The relevance of *Haplomitrium* to the comparative study of blepharoplast morphology stems from its recognition as a representative genus in a morphologically remote (and putatively primitive) order of liverworts (Schuster 1971). The genus consists of 12 species (Engel 1981); until now, however, *H. hookeri* has been the sole source of our information. With the fortunate acquisition of *H. gibbsiae* antheridia, we have been able to broaden that data-base to include representatives of two subgenera (Schuster 1967): subgen. Haplomitrium (*H. hookeri*) and subgen. Calobryum (*H. gibbsiae*). Furthermore, the work on *H. gibbsiae* led to a reexamination of *H. hookeri* and the modification of an earlier interpretation of blepharoplast morphology for that species (Carothers & Duckett 1979: Fig. 20).

The new data for *Haplomitrium hookeri* indicate that its morphology is closely similar to that of *H. gibbsiae* (Fig. 14). In particular we have observed that the spline has an open-type aperture (though only one microtubule-diameter in width), and that both the left margin of the spline and the PBB gradually curve back as the spline widens. A revised reconstruction of the *H. hookeri* blepharoplast would show that this curved form also brings the LS outline into closer agreement with that of *H. gibbsiae*. Differences in such quantitative characters as the number of microtubules constituting the spline (84 in *H. gibbsiae* versus 55 in *H. hookeri*) likely reflect interspecific variation, with a negligible effect on overall blepharoplast morphology. Some of the other values cited for *H. hookeri* (Carothers & Duckett 1979: Table II), e.g., overall width of the lamellar strip, should now be disregarded because they would necessarily be superseded in a revised reconstruction for that species.

Figures 8–13. Blepharoplast anatomy of *Haplomitrium gibbsiae*, all figures are shown at x49,500. See Appendix A for key to labeling. **8,** An oblique-transverse section showing the stellate pattern within the anterior transition zone. **Figures 9–13.** Transverse sections of the blepharoplast, proceeding from anterior to posterior. (See Fig. 14 for the longitudinal locations of these five levels.) **9,** The spline is shown here to include about 59 microtubules while the LS has narrowed to subtend about 18 of them; **10,** At this level the spline consists of about 71 microtubules while the LS subtends only 11, note the 9 doublet-2 singlet configuration of the flagellum; **11,** This spline has reached its greatest width of about 84 microtubules, the LS width is reduced to subtend only 5 of the S_1 tubules; **12,** The LS at this level is narrowed to about 3 microtubule-diameters, the stellate pattern of the posterior transition zone is shown at left; **13,** At its extreme posterior extent the LS is reduced in width to subtend only 2 of the spline's 84 microtubules. Scale bars = 0.2 μm.

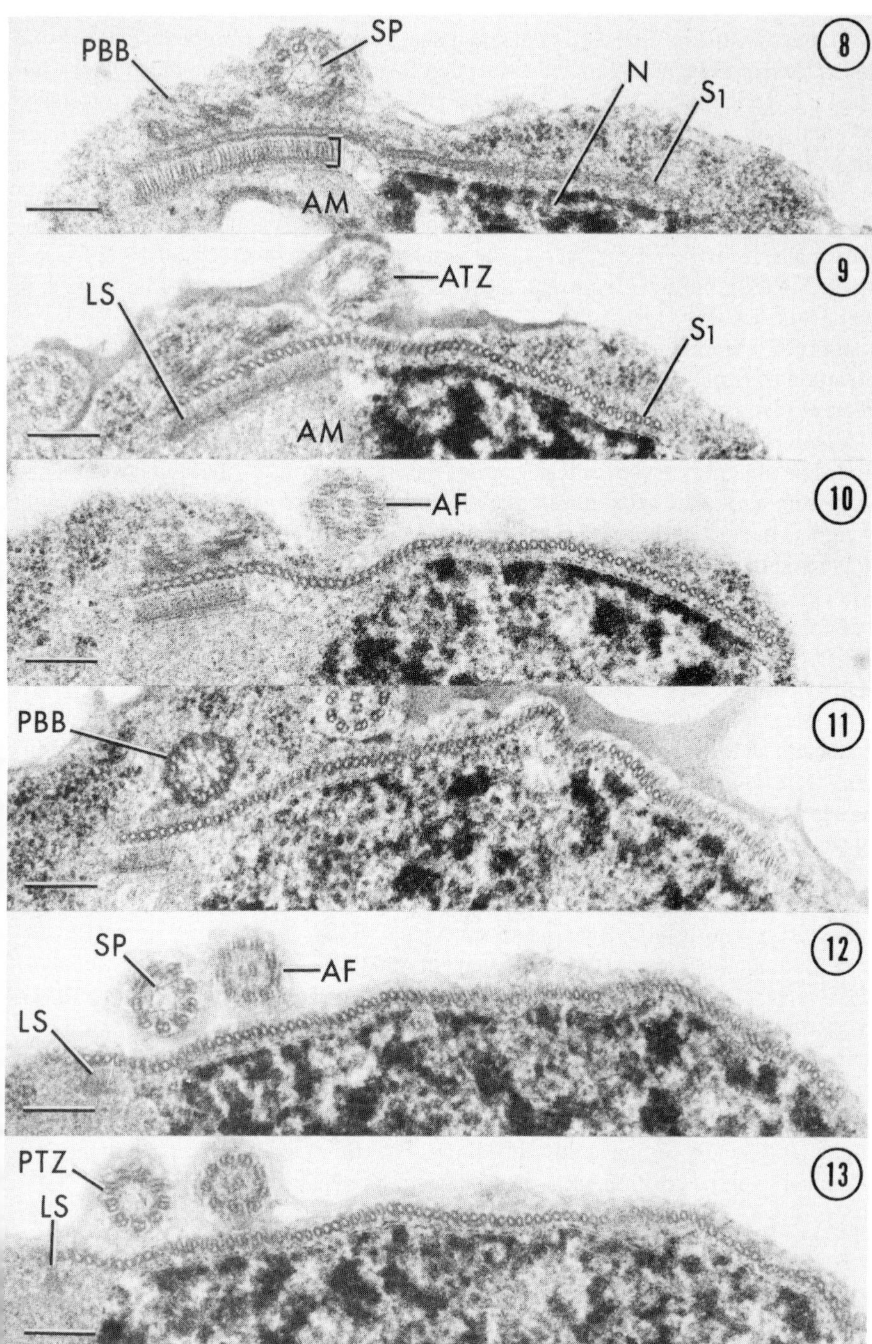
PBB
SP
8
N
S₁
9
LS
ATZ
S₁
AM
AM
10
AF
PBB
11
SP
12
AF
LS
PTZ
13
LS

Characters that may be used in distinguishing between the *Haplomitrium gibbsiae* and *H. hookeri* blepharoplast and associated flagellar structures are mostly quantitative (e.g., spline width and basal body length), but an unusual qualitative difference between their transition zones occurs also. In addition to the differences in TZ lengths (in *H. hookeri*, both TZs are ca. 0.2 μm long; in *H. gibbsiae*, the PTZ is ca. 0.55 μm and the ATZ ca. 0.75 μm long), the stellate patterns occupy different positions within the TZ. In *H. hookeri*, which exhibits the position typically found in bryophytes, the proximal ends of the SP and TZ occur at virtually the same level (Carothers & Duckett 1979: Fig. 20), but in *H. gibbsiae*, the SP's proximal end is about 0.15 μm distal to that of the TZ (Fig. 14).

It is our opinion that, except for the TZ differences noted above, the graphic interpretation of *Haplomitrium gibbsiae* (Fig. 14) is also appropriate for representing the general blepharoplast morphology of *H. hookeri*. Further, a characterization of the *Haplomitrium*-type blepharoplast must include the following distinguishing features:

1. The obliquely asymmetrical shape of the spline anterior – the spline widens distally from a rounded apex, its left margin curves in a near-parabolic outline, its right margin is parallel to the axis for a short distance and then is abruptly dilated with additional microtubules. This "delayed" accumulation of tubules on the right side has been reported only for *Haplomitrium*.

2. The exceptional number of spline microtubules – with maximum numbers of 57 and 90 observed for *H. hookeri* and *H. gibbsiae*, respectively. The *Haplomitrium*-type spline is the widest reported thus far for any bryophyte.

3. The spline aperture is of the open type. It is from one to two microtubule-diameters wide according to the species and is situated to the left of the spline center – thus far, *Haplomitrium* is the only genus of liverworts in which the open-type aperture has been reported, and the only bryophyte shown to have its aperture located left of center.

4. The lamellar strip outline has an elongated, curved drop-shape. Its left-lateral margin is coincident with that of the superposed spline, and its right-lateral margin narrows posteriorly from right to left beneath the spline.

5. Both the PBB and ABB are situated left of the spline's midline – the *Haplomitrium*-type blepharoplast is the only one reported to have an ABB so placed. The PBB is consistently left-of-center in all bryophytes.

6. In the anterior basal body, the trio of proximally extended microtubular triplets is ventrally situated – this feature has been reported also for *Pallavicina lyellii* (Hook.) Carruth. (see Renzaglia et al. 1985: Fig. 4); in other hepatics the extended triplets of the ABB typically occupy a dorsal position. The trio of extended triplets in the PBB occupies a ventral position, the typical situation for hepatics.

Figure 14. Diagrammatic representation of blepharoplast structure in *Haplomitrium gibbsiae*. The figure numbers at right refer to transverse sections taken at the levels indicated, x42,900; the scale bar at lower left represents 0.25 μm.

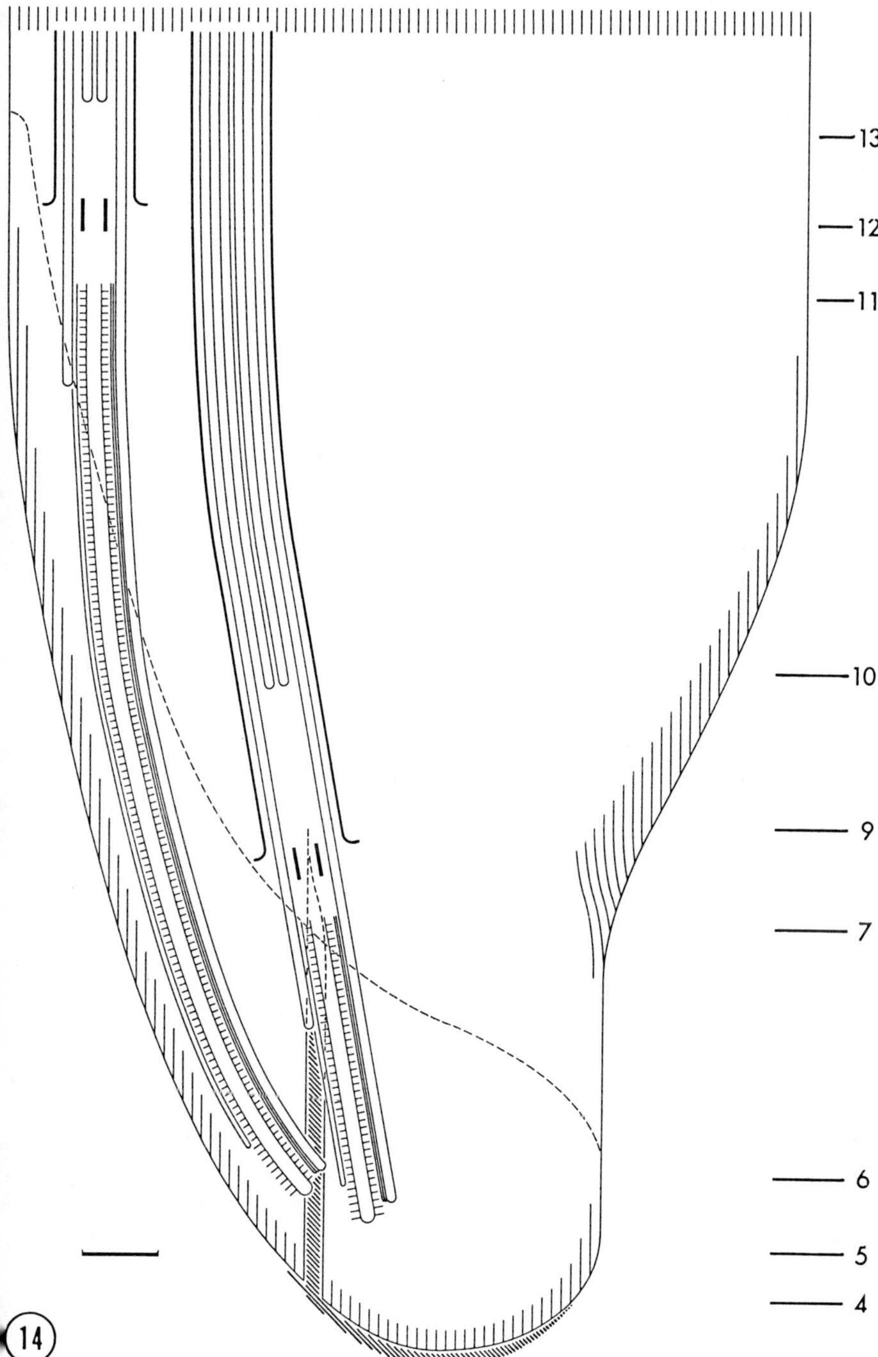

7. The lengths of the stellate patterns in *H. gibbsiae* and *H. hookeri* (100 nm and 120 nm, respectively) are at the low end of the overall range for liverworts – *H. gibbsiae* and *Pellia*, also measuring 100 nm, have the shortest SPs thus far reported for liverworts.

Other Blepharoplast Morphologies

In this section are compared selected major features of blepharoplast structure for representative species within each of the three bryophyte subdivisions: liverworts, hornworts, and mosses. The illustrative species are, for the most part, those with blepharoplast morphology depicted diagrammatically here or elsewhere. The data summarized below will help characterize blepharoplast morphology within each group and also provide evidence concerning the nature and range of structural variation.

Liverworts

The examples cited here are selected from orders other than the Calobryales – the *Haplomitrium*-type blepharoplast having been described in the preceding section. However, the basic characters enumerated below include those given for *Haplomitrium*.

1. Spline shape. In all reported cases the spline's anterior end is bluntly rounded, and its width is somewhat greater than that of its longer, posteriorly extended shank. The anterior portion is broadened overall by one or more shorter microtubules delimiting the front of the spline aperture and by other shorter microtubules situated laterally. In consequence the outline of the spline's dorsal surface is spatulate to a greater or lesser degree. Depending upon the numbers and dispositions of these shorter tubules, the outline of the wide anterior portion may be unilateral or bilateral, and, if the latter, the anterior portion may have equal or unequal amounts of lateral expansion (Table 2). *Marsupella emarginata* (Ehrh.) Dum. has a unilaterally expanded spline; the other liverworts in Table 2 are unequally bilateral (e.g., *Marchantia*, Fig. 15) except for *Bazzania*, which is classified as equally bilateral (Fig. 15). Further variation is evident in the asymmetry of all but equally bilateral splines, i.e., whether the greater amount of the lateral expansion occurs on the spline's left or right side.

2. Number of spline microtubules. In view of the spline's spatulate outline, the number of constituent tubules is usually best determined from a transverse section taken at a short distance in front of the (closed) aperture. The data of this kind given in Table 1 in Carothers & Duckett (1980) and here in Table 2 show that for some 21 non-calobryalean hepatics of broad taxonomic distribution the average number of microtubules is 19. These same data reveal a fairly narrow range in the number of microtubules and, therefore, in spline width. The overall range in tubule number reported for these liverworts is 13 for *Nowellia curvifolia* (Dicks.) Mitt., cited in Lee

(1974, p. 45) to 25 for *Chiloscyphus pallescens* (Ehrh. ex Hoffm.) Dum. (Rushing et al. 1984) – about twice as many as in the former. Both belong to the Jungermanniales. The ranges for other orders include 20 in, for example, *Pallavicinia lyellii* (Renzaglia et al. 1985) to 15 for *Pellia neesiana* (Gott.) Limpr. (Duckett et al. 1983) in the Metzgeriales, 20 for *Sphaerocarpos donnellii* Aust. (Carothers et al. 1983) to 24 for *Geothallus tuberosus* Campb. (Brown et al. 1983) in the Sphaerocarpales, and 17 for *Marchantia polymorpha* (Carothers & Kreitner 1968) to 22 for *Dumortiera hirsuta* (Sw.) Nees, cited in Carothers and Duckett (1980), for the Marchantiales.

3. Spline aperture. In all reported cases spline apertures of non-calobryalean liverworts are located to the right of the spline's midline and are of the closed type. The aperture is typically closed at its anterior end by shorter microtubules intercalated between longer ones that delimit the sides of the slot-like opening; for example, see *Marchantia* (Fig. 15). An exceptional type of closed aperture has been reported for *Marsupella* (Duckett et al. 1983) in which the spline aperture in mature spermatozoids is closed anteriorly by convergence of the same microtubules that delimit its sides, i.e., the aperture represents a discontinuous local separation of adjacent tubules. (For a graphic comparison of these and other types of bryophyte spline morphology, see Fig. 17 in Carothers & Brown 1985.)

The overall widths of the apertures (measured in microtubule-diameters) vary according to the taxon. As indicated in Table 2, representatives of the Jungermanniales and Metzgeriales exhibit a range of from one to three. Representatives of the three genera of Sphaerocarpales have aperture widths of three microtubule-diameters, as does *Marchantia polymorpha*. A second representative marchantialean species, *Reboulia hemisphaerica* (L.) Raddi, is shown in the dissertation of Turner (1966) to have a spline aperture width of three microtubule-diameters (Turner: Fig. 35c), an aperture of the closed type (cf. the two blepharoplasts in Turner: Fig. 33), and a spline tubule number of at least 19 (Turner: Fig. 35d).

4. Lamellar strip. Microanatomy of the LS is remarkably uniform among the liverworts, virtually all reported species having tristratose, dorsiventral organization (and, consequently, the 4-layered MLS) with the vertical lamellae oriented at approximately 45° to the spline axis, e.g., Fig. 2. Further, the LS is characteristically longer than wide. The range in length for liverworts is exhibited by *Sphaerocarpos donnelli*, measuring 1.1 µm (Carothers et al. 1983), and *Pallavicinia lyellii*, measuring 7.5 µm, the longest recorded for any bryophyte (Renzaglia et al. 1985). The length (0.9 µm) given for *Pellia neesiana* (Duckett et al. 1983, Fig. 36a) is taken from a later-stage blepharoplast in which maturational shortening has probably occurred (see also Planche XXXIV, Fig. 4, on p. 377 and Schema 9a on p. 261 in Suire 1970). Excluding the two extremes, the average LS length of the non-calobryalean hepatics cited in Table 2 is 2.1 µm. The posterior limit of the LS coincides more or less closely with the PBB's distal end.

Although the shape of the LS varies among species, there are some commonalities, viz., a bluntly rounded leading edge and a unilateral "offset", i.e., a rather abrupt reduction in width, typically on the right-hand side, as seen in *Bazzania* and *Marchantia*

Table 2. Summary of selected characters for the splines and lamellar strips (collectively, the so-called multilayered structures) of 22 representative liverworts, hornworts, and mosses. For "Spatulate spline anterior", LL (left lateral) or RL (right lateral) indicates the side having the greater amount or all of the lateral expansion. "Spline tubules" is expressed as the most commonly observed number of microtubules at the widest region of the spline's anterior end. "SA type" includes open aperturate (OA), closed aperturate (CA), and inaperturate (IA) splines. "SA width" is the predominant aperture width expressed as microtubule-diameters. "LS outline" implies longitudinal elongation of all lamellar strips exceptqv those of hornworts.

Taxon	Spatulate spline anterior	Spline tubules	SA type	SA width	LS outline	LS outline relative to spline anterior	Source of data
A. Liverworts Calobryales							
Haplomitrium gibbsiae (Steph.) Schust.	not applicable	84	OA	2	curved drop-shape	ant. extension	present paper
Haplomitrium hookeri (Smith) Nees	not applicable	55	OA	1	(undetermined)	ant. extension	Carothers & Duckett (1979), emended
Metzgeriales (Pelliineae)							
Pellia neesiana (Gott.) Limpr.	bilateral, unequal, RL	15	CA	1	inclined drop-shape	ant. & LL extensions	Duckett et al. (1983, Fig. 36 – a later stage)
(Pallaviciniineae)							
Pallavicinia lyellii (Hook.) Carruth.	bilateral, unequal, RL	20	CA	1	w/gradual RL offset	coincident	Renzaglia et al. (1985)
(Blasiineae)							
Blasia pusilla L.	bilateral, unequal, LL	20	CA	3	w/RL offset (notch)	coincident	Renzaglia & Duckett (1987)

Taxon	Symmetry						Reference
Jungermanniales (Jungermanniineae) *Marsupella emarginata* (Ehrh.) Dum.	unilateral, RL	14	CA	1	w/RL offset	ant. & RL extensions	Duckett et al. (1983) and Rushing et al. (1984, Table 1)
(Geocalycineae) *Chiloscyphus pallescens* (Ehrh. ex Hoffm.) Dum.	bilateral, unequal, LL	25	CA	2	w/gradual RL offset	ant. extension	Rushing et al. (1984)
Lophocolea heterophylla (Schrad.) Dum.	bilateral, unequal, LL	23	CA	3	w/gradual RL offset	(undetermined)	Lee (1974) Fig. 56; Fig. 45, 63
(Lepidoziineae) *Bazzania trilobata* (L.) Gray	bilateral, equal	15	CA	1	w/gradual RL offset	ant. extension	Rushing & Carothers (1986a)
(Cephaloziineae) *Cephalozia lunulifolia* (Dum.) Dum.	bilateral, RL	17	CA	2	w/gradual RL offset	ant. extension	Rushing et al. (1984), emended
Sphaerocarpales (Riellineae) *Riella americana* Howe et Underwood	bilateral, unequal, LL	24	CA	3	w/RL offset (notch)	coincident	Carothers et al. (1983)
(Sphaerocarpineae) *Sphaerocarpos donnellii* Aust.	bilateral, unequal, LL	20	CA	3	w/RL offset (notch)	coincident	Carothers et al. (1983)
Geothallus tuberosus Campb.	bilateral, unequal, LL	24	CA	3	w/RL offset (notch)	coincident	Brown et al. (1983)
Marchantiales (Marchantiineae) *Marchantia polymorpha* L.	bilateral, unequal, LL	17	CA	3	w/RL offset (notch)	coincident	Carothers & Kreitner (1968)

Taxon	Spatulate spline anterior	Spline tubules	SA type	SA width	LS outline	LS outline relative to spline anterior	Source of data
B. Hornworts Anthocerotales *Phaeoceros laevis* (L.) Prosk.	bilateral, equal	12	IA	–	trans. elongated, approx. rhomboidal	LL & RL extensions	Carothers et al. (1977)
Notothylas orbicularis (Schwein.) Sull.	bilateral, equal	12	IA	–	trans. elongated, approx. rhomboidal	LL & RL extensions	Renzaglia & Carothers (1986)
C. Mosses Sphagnales *Sphagnum palustre* L.	unilateral, RL	14	CA	2	post. narrowing from R to L	coincident	Duckett & Carothers (1979)
Archidiales *Archidium tenerrimum* Mitt.	bilateral, unequal, LL	24	OA	2	post. truncate from L to R	ant. & RL extensions	Brown & Carothers (1986)
Funariales *Funaria hygrometrica* Hedw.	unilateral, LL	13	OA	3	w/LL offset	ant. extension	Carothers & Brown (1985)
Hypnobryales *Thuidium delicatulum* (Hedw.) B.S.G.	bilateral, unequal, RL	14	OA	3	post. narrowing from R to L	ant. extension	Rushing & Carothers (1986b)
Hypnum mamillatum (Brid.) Loeske	"oblong, acute"	14	CA	2	post. narrowing from R to L	ant. & RL extensions	Duckett & Renzaglia (1986)
Polytrichales *Polytrichum juniperinum* Hedw.	unilateral, RL	10	CA	2	post. narrowing from R to L	coincident	Paolillo et al. (1968a)

(Fig. 15). In later-stage blepharoplasts, the outline of the LS changes concomitantly with the strip's acropetal shortening as depicted for *Marsupella* and *Pellia* (Duckett et al. 1983: Fig. 19 and 36a, respectively). As indicated in Table 2, a portion of the LS may or may not extend beyond the spline margin. In the four jungermannialean species that have been investigated, the LS projects forward of the spline's anterior margin for a short distance (e.g., *Bazzania*, Fig. 15), while the more mature LS of *Marsupella* extends beyond the right side as well (Duckett et al. 1983: Fig. 19). The later-stage blepharoplast of *Pellia* is unusual in that its LS extends far enough beyond the spline margin so that the area of its dorsal surface subtending the spline is less than that which is "uncovered" anteriorly (Duckett et al. 1983: Fig. 36a). In about half of the non-calobryalean liverworts listed in Table 2, the LS margin does not extend beyond that of the spline; that relationship of the LS margin to the spline anterior is termed coincident.

5. Basal bodies and transition zones. The basal bodies have a common origin; their immediate precursor is a centrosome comprised of two coaxial centrioles. After separation these now-discrete centrioles appear to be identical in structure but soon become dimorphic as a result of differential growth. Basal body dimorphism is a constant feature of hepatics and, in fact, of all bryophytes. Usually, the most conspicuous dissimilarity between a pair of these organelles is their different lengths. *Pellia* stands alone as a variant exhibiting the greatest extremes and range reported (Duckett et al. 1983) for any bryophyte: the length of the ABB is 0.4 μm and that of the PBB is 10.5 μm – about 26 times longer. A more representative expression of ABB and PBB lengths may be taken from the studies of non-calobryalean species listed in Table 2. Excluding *Lophocolea heterophylla* (Schrad.) Dum. (for which no comparable data are available) and *Pellia*, the ABB averages 0.7 μm with a range of 0.6 μm (*Sphaerocarpos*) to 1.2 μm (*Pallavicinia*), while the PBB averages 1.7 μm, with a range of 0.8 μm (*Sphaerocarpos*) to 2.6 μm (*Marsupella*). The difference between the PBB and ABB lengths in individual species ranges from the 1.3 times-longer PBB (0.8 vs. 0.6 μm) in *Sphaerocarpos* to the 4 times-longer PBB (2.6 vs. 0.65 μm) in *Marsupella*.

Another constant feature of basal body dimorphism is the different amounts of acropetal growth exhibited by the microtubular triplets. Arranged in a cylindrical array, the nine triplets enclose a distal, apparently structureless region, and more proximally, a central core that, in transverse section, resembles a hub and spokes – the cartwheel configuration (e.g., Fig. 11). By differential growth, some of the triplets (and the central core) extend farther forward, thus increasing the overall length of the basal body. However, the ABB and PBB usually differ as to the positions that their extended triplets occupy in the 9-membered arrays. In the ABB a number of dorsally positioned triplets usually project beyond the others, though in *Haplomitrium* and *Pallavicinia* (Renzaglia et al. 1985) it is the ventral ones that do. In the PBB there is typically a ventral trio of extended triplets. Further, in most reported cases, the basal body's central core projects forward for some distance beyond the extended triplets, as illustrated in *Bazzania* (Fig. 15) and the ABB of *Pallavicinia* (Renzaglia et al.

1985). However, the reverse also occurs, as seen in *Marchantia* (Fig. 15) and in the PBB of *Pallavicinia*.

The morphology of the non-calobryalean blepharoplast is affected also by the position of the basal bodies relative to the spline and to each other. The ABB lies above the aperture, which, as noted earlier, is to the right of the spline's centerline, while the PBB is left of center. Both BBs are "subapical" in that their apices are located rearward from the leading edge of the spline. Further, the apex of the ABB typically lies at some level within the widest region of the spline's expanded anterior. Except in *Pallavicinia* (Renzaglia et al. 1985) and *Pellia* (Duckett et al. 1983), the BBs overlap each other, i.e., the anterior end of the PBB lies forward of the ABB's posterior end. The blepharoplast is further characterized by basal body orientation, that is, how (if at all) the BB's longitudinal axes diverge relative to that of the subtending spline. Non-calobryalean hepatics exhibit at least four different combinations of BB orientation: 1) both ABB and PBB parallel, e.g., *Blasia* (Renzaglia & Duckett 1987) and *Pellia* (Duckett et al. 1983); 2) PBB parallel and ABB right-divergent, e.g., *Marsupella* (Duckett et al. 1983); 3) posterior portion of PBB divergent and ABB parallel, e.g., *Bazzania* (Fig. 15) and *Pallavicinia* (Renzaglia et al. 1985); and 4) both PBB and ABB divergent, e.g., *Marchantia* (Fig. 15).

The transition zone is continuous with the posterior end of the basal body, although, in a strict sense, the zone is not a part of it. Rather, the zone is a specialized region between the BB and the distal axoneme with its configuration of 9 doublet + 2 singlet microtubules. The TZ's inclusion here with the blepharoplast reflects its convenient utility in providing additional characters that help distinguish the locomotory structures in spermatids of various bryophytes. A characteristic feature of the TZ is the intricate stellate pattern, so-called because it may be seen in transverse section to contact each of the encircling doublets (Fig. 8, 11). The anterior end of the SP essentially coincides with that of the TZ; its posterior end terminates some distance forward of the TZ's distal limit, the latter being coincident with the anterior end of the pair of singlet microtubules in the axoneme. For the non-calobryalean liverworts listed in Table 2 (excluding *Lophocolea* for lack of data) the average length of the SP is about 155 nm, the range extending from 100 nm (*Pellia*) to 230 nm (*Sphaerocarpos* and *Marchantia*). In most instances the SPs of anterior and posterior TZs in a given spermatid have similar lengths or nearly so.

Hornworts

Knowledge of blepharoplast morphology in hornworts is derived from studies of two representative species: *Phaeoceros laevis* (L.) Prosk. (Moser 1970; Moser et al. 1977; Carothers et al. 1977) and *Notothylas orbicularis* (Renzaglia & Carothers 1986). These investigations have revealed a high degree of similarity in most of the structural characteristics.

At the maximally structured stage of MLS development, the anterior region of the

12-microtubule spline has a rectangular outline owing to its transverse leading edge (Fig. 15, *Notothylas*). More posteriorly the spline narrows bilaterally and equally to form the distal shank. In both species the spline is inaperturate and subtended by a transversely elongated LS of approximately rhomboidal outline, the greater amount of lateral extension occurring on the right side. The *Phaeoceros* LS is slightly longer (0.5 vs. 0.4 μm) and wider (0.9 vs. 0.7 μm) than that of *Notothylas*. The LS is tristratose, and its vertical lamellae are diagonally oriented at about 45° to the spline axis.

In both species the basal bodies (including their short triplet extensions) are about 0.4 μm long. Their position with respect to the spline's anterior margin is "apical", i.e., the BB apices lie directly above the spline's leading edge. However, the BBs of *Notothylas* are oriented parallel to the spline axis, while those of *Phaeoceros* diverge outward slightly at their distal ends. A further distinguishing characteristic found in both hornworts is the transition zone's lack of a stellate pattern (Fig. 15).

Mosses

Knowledge of comparative blepharoplast morphology in mosses is derived principally from studies of six species: *Polytrichum juniperinum* (Paolillo et al. 1968a, b), *Sphagnum palustre* L. (Duckett & Carothers 1979), *Funaria hygrometrica* (Carothers & Brown 1985), *Archidium tenerrimum* (Brown & Carothers 1986), *Thuidium delicatulum* (Rushing & Carothers 1986b), and *Hypnum mamillatum* (Brid.) Loeske (Duckett & Renzaglia 1986). In spite of such a small and nonrepresentative sample, some generalized observations can be advanced toward a fuller, future characterization.

1. Spline shape. As shown in Fig. 16, the outline of the moss spline's apical end is asymmetrical and varies from bluntly curved (*Funaria*) to rounded (*Archidium*) to pointed (*Thuidium*). Only *Sphagnum* has a nearly symmetrical leading edge (Duckett & Carothers 1979: Fig. 67). The elongated, wide anterior region where the greatest number of microtubules occurs extends rearward, usually to some distance behind the PBB, before substantially narrowing into the shank portion of the spline. In *Polytrichum*, however, the spline begins to narrow from the right side just behind the anterior transition zone; there it tapers gradually toward the left to a level about midway along the PBB where it continues distally as a narrow shank (Paolillo et al. 1968: Fig. 36c).

Published studies show that a divergent or "stray" left-marginal microtubule occurs in four of the six mosses specified above, viz., *Polytrichum*, *Funaria*, *Thuidium*, and *Hypnum*. Duckett and Renzaglia (1986) also cite unpublished data that add *Ditrichum*, *Bryum*, and *Mnium* to the list of those having divergent tubules.

The rearward course of the divergent microtubule has received two different interpretations. One of these, illustrated by *Funaria* and *Thuidium* (see diagrams in Fig. 16) and by *Polytrichum* (Paolillo et al. 1968a: Fig. 36C, D and 1968b: Fig. 1A, B), has the divergent tubule rejoining the spline at approximately the level terior transition zone. The other, described for *Hypnum*, holds that the tubule remains separate from the spline except at its anterior end (Duckett & Renzaglia 1986). Direct (but incomplete) evidence in the form of transverse sections has been drawn upon in all four of these studies, while indirect evidence has also been used in the *Hypnum* work. In *Funaria*, for example, most of the specimens had a spline comprising 13 microtubules. Transverse sections commonly showed the divergent tubule beside a 12-tubule portion of spline extending back to the level of the PSP; other transverse sections commonly showed a more posterior level having a spline comprised of 13 adjacent tubules (Carothers & Brown 1985: Fig. 13–15). These were major factors leading to the conclusion that the divergent tubule had rejoined the spline. Unfortunately, serial sections were not used in *Funaria*, *Thuidium*, *Polytrichum*, or *Hypnum* to trace the continuity of a divergent tubule either to its presumed reunion with the other spline members or to its distal limit if it remains separate. As matters stand, the published evidence is inconclusive and neither interpretation stands as proven. Accordingly, we have indicated the tentativeness of the interpretation given for *Funaria* and *Thuidium* by placing a question mark near the divergent microtubule's putative reunion with the rest of the spline (Fig. 16).

2. Number of spline microtubules. The data given here in Table 2, in Table 2 of Carothers and Duckett (1980), and in the report on *Timmiella barbuloides* (Brid.) Moenk. by Gambardella and de Lucia Sposito (1982) show that for 27 species representing 23 genera in 13 orders the average number of microtubules at the spline's widest region is 17. The lowest recorded number of microtubules for mosses (indeed, for all bryophytes) is 10 in *Polytrichum juniperinum* (Paolillo et al. 1968a); the upper limit of the range is 44 in *Timmiella barbuloides* – about 4.5 times greater. Actually this range is misleading, for it includes two species with unusually large numbers of spline microtubules: *Tetraphis pellucida* Hedw. with 38 (Carothers & Duckett 1980) and *Timmiella barbuloides* with 44. Of more significance are the facts that 11 of the 27 species have splines with 14 tubules and that 20 of the 27 have tubule numbers ranging from 10 to only 15.

3. Spline aperture. A consistent feature of the aperturate spline is the aperture's distal closure by convergence of the flanking microtubules. Proximally, the aperture extends all the way (open type) or only part of the way (closed type) to the spline's leading edge. Of the six mosses listed in Table 2, three (*Archidium*, *Funaria*, *Thuidium*) have open apertures; the others (*Sphagnum*, *Hypnum*, *Polytrichum*) have the closed type. Other mosses for which published evidence suggests closed apertures include *Andreaea rothii* Web. & Mohr (Duckett & Carothers 1979: Fig. 36–38) and, probably, *Tetraphis* (Carothers & Duckett 1978: Fig. 10). Duckett and Renzaglia (1986) also cite unpublished data that add *Ditrichum* and *Plagiothecium* to those having closed apertures.

Table 2 shows that the predominating widths of the moss spline apertures (expressed as microtubule-diameters) range from two to three. *Andreaea* has an aperture width measuring at least two diameters (Carothers & Duckett 1978: Fig. 9), and, according to unpublished data cited in Duckett and Renzaglia (1986), *Tetraphis* also has two, while aperture width in *Bryum* is only one. However, in the absence of data showing the longitudinal extent of the microtubules delimiting the apertures, these last two measurements cannot be construed as predominant widths. At its extreme anterior end, the open SA of *Archidium* widens for a very short distance to three tubule-diameters. Conversely, the open aperture in *Thuidium* progressively narrows anteriorly along the left margin in a stepwise manner from three to two to one tubule-diameters (Fig. 16).

Anterior closure of the aperture in *Polytrichum* is effected by intercalation of two short microtubules, the distal limits of which mark the proximal boundary of the aperture (Paolillo et al. 1968a: cf. Fig. 1, 2). Similar closure has been documented also for *Sphagnum* at an equivalent stage of development (Duckett & Carothers 1979: cf. Fig. 12, 16, 67). In contrast a different mode of anterior closure was later reported for the splines of mature *Sphagnum* spermatozoids (Duckett et al. 1983). The diagrammatic interpretation given in that paper (Fig. 51) shows the three microtubules on the right of the aperture converging proximally with the tubules on the aperture's left side. More recently, Duckett and Renzaglia (1986) reported that the closed aperture of *Hypnum* extends anteriorly to near the tip of the spline and that it is one tubule-diameter in width at its extreme anterior end and two over most of its length. They suggest that aperture closure may occur either by the anterior addition of microtubules (*Marchantia*) or by the anterior convergence of the microtubules (*Hypnum*) as the lamellar strips attain their final lengths.

The occurrence of the spline aperture's anterior closure via intercalated shorter microtubules has been well documented (e.g., Carothers & Kreitner 1968, Kreitner & Carothers 1976). A second mode of anterior closure, i.e., by proximal convergence of the microtubules delimiting the aperture, was earlier depicted for mature spermatozoids of *Marsupella* and *Sphagnum* (Duckett et al. 1983: Fig. 19 and 51) and more recently described for spermatids of *Hypnum* (Duckett & Renzaglia 1986). Our scrutiny of the micrographic evidence (for *Marsupella* and *Sphagnum*, see Fig. 10–14 and 40–44, respectively, in Duckett et al. 1983; for *Hypnum*, see Fig. 1C–E in Duckett & Renzaglia 1986) offered in support of the anterior-convergence interpretation finds it inadequate. In each of the three cases the sections are taken from different blepharoplasts and can be interpreted differently. In order to confirm the postulated convergence, micrographic evidence in the form of consecutive-serial sections parallel to the plane of the spline's anterior end and/or serial (preferably consecutive-serial) transverse sections through the same blepharoplast are needed.

4. Lamellar strip. In nearly all reported observations, the microanatomy of the LS in mosses appears to be the same as that exhibited by other bryophytes. In a few cases, however, horizontal stratification of the LS has been interpreted as consisting of two rather than three layers. The latter observations, notably on *Polytrichum juniperinum* (Paolillo et al. 1968a), *Physcomitrium coorgense* Broth. (Lal & Bell 1975), *Plagiothecium sylvaticum* (Brid.) B.S.G. (Carothers & Duckett 1978), and *Leptodictyum trichopodium* (Schultz) Warnst (Treeful 1978), show a dorsal stratum referable to the S_2 of a tristratose LS and a ventral stratum equivalent to the S_4 – the presuma-

bly missing layer is the thin, lightly staining S_3-equivalent. Some structural evidence has been cited (Carothers & Duckett 1978, 1980) as showing that the apparent absence of a differentiated S_3 may actually reflect a transitory developmental or physiological state of the 3-layered LS. However, more information will be needed before this or another interpretation can be confirmed.

At the stage of its maximum development the lamellar strip in mosses is characteristically longer than wide. The two pleurocarpous representatives among the six species listed in Table 2 have the greater lengths: the LS of *Thuidium* and *Hypnum* is 2.6 and 2.8 μm long, respectively, the latter being the longest reported for any moss. The LSs of the four acrocarpous representatives are appreciably shorter (*Sphagnum*, 1.3 μm; *Archidium*, 1.5 μm; *Funaria*, 1.1 μm; and *Polytrichum*, 1.0 μm), averaging only 1.2 μm in length. As evidenced by these six species, the shape of the moss LS may vary considerably, yet there is a degree of commonality for some of the features. For example, the posterior margin may narrow obliquely from right to left as in *Sphagnum*, *Thuidium*, *Hypnum*, and *Polytrichum*, or from left to right in a truncate or curved manner as in *Archidium* and *Funaria*, respectively (Fig. 16). The anterior portion of the LS margin also exhibits variation in outline according to the species. *Sphagnum* (Duckett & Carothers 1979: Fig. 67) and *Archidium* (Fig. 16) are bluntly rounded, the former symmetrically so, the latter somewhat asymmetric. *Thuidium*, *Hypnum*, and *Funaria* have more sharply rounded to acute anterior margins, while that of *Polytrichum* is depicted as rectangular or nearly so (Paolillo et al. 1968a: Fig. 36).

Considerable variation is also seen in the relationship between the LS and the anterior margin of the spline (Table 2). The relevant portions of these margins coincide in *Sphagnum* (Duckett & Carothers 1979: Fig. 67) and in *Polytrichum* (Paolillo et al. 1968a: Fig. 1–8). In *Funaria* and *Thuidium* the LS projects forward beyond the leading edge of the spline, while in *Archidium* and *Hypnum* the LS extends both forward and laterally on the right-hand side (Fig. 16). The distal limit of the moss LS approximately coincides with the level of the anterior transition zone's proximal end.

5. Basal bodies and transition zones. Fully developed basal bodies of the moss blepharoplast are characteristically dimorphic and of conspicuously different lengths (including the extended hubs and microtubular triplets). For the six species listed in Table 2, the range in ABB length extends from 0.8 μm in *Funaria* to 3 μm in *Hypnum*. *Thuidium* (2.1 μm) approaches the maximum, while the other three (*Polytrichum*, 1 μm; *Sphagnum*, 1.25 μm; and *Archidium*, 1.3 μm) are grouped near the minimum. The average ABB length is 1.6 μm. In comparison the range in PBB length extends from 3.5 μm in *Funaria* to 9 μm in *Hypnum*. *Sphagnum* (3.5–4.0 μm) is at or close to the minimum, while the others are rather evenly distributed (*Polytrichum*, 8 μm; *Thuidium*, 7 μm; and *Archidium*, 5 μm). The average PBB length of these six mosses is 6 μm. The greatest difference in length within a pair of basal bodies occurs in *Polytrichum* in which the PBB is eight times longer than the ABB. The least difference occurs in *Sphagnum* and *Hypnum*. In these the PBBs average three times longer than the ABBs.

Another feature contributing to basal body dimorphism is the different lengths of the microtubular triplets. At the distal end of the basal body all nine triplets are typically arrayed to form a cylinder. Serial transverse sections show that the longer triplets commonly occupy preferential positions in the array. The three longest triplets of the PBB typically occupy a ventral position, i.e., are closest to the upper surface of the spline. Transverse sections also show that the longer triplets of the ABB typically form a dorsal arc extending part way down the sides of the central hub, the number of constituent triplets increasing in the more distal levels. The longest of the triplets commonly extends to a level near the proximal end of the central hub. However, the hub may far exceed the length of the triplets as in the PBB of *Thuidium* and *Funaria* and in the ABB of *Archidium* (Fig. 16).

The positions of the basal bodies relative to the spline and to each other in mosses have much in common with those of liverworts. The ABB is superposed partially or wholly above the spline aperture, and the apices of both the ABB and PBB lie a short distance behind the spline's leading edge. Except in *Archidium*, the proximal tip of the ABB is closer to the spline apex than is the tip of the PBB (Fig. 16). Also, the anterior portion of the PBB overlaps the posterior end of the ABB. In *Archidium* and *Thuidium* the short, proximal portion of the PBB lying close to the spline's anterior margin converges rather abruptly toward the spline's midline (Fig. 16). A comparable convergence of the ABB has been reported for *Hypnum* (Duckett & Renzaglia 1986). Only two major configurations of BB orientation relative to the spline axis are represented among the six mosses listed in Table 2: C) ABB divergent and PBB parallel, viz., in *Sphagnum* (Duckett & Carothers 1979: Fig. 67), and 2) ABB parallel and PBB divergent as seen in the other five. The short, proximal convergences of the PBBs of *Archidium*, *Thuidium*, and *Hypnum* are considered here only as minor variations of the two major types.

Information (Table 2) regarding stellate patterns of the mosses is, unfortunately, incomplete. Available data show that the SPs of *Funaria* are of unequal length, the ASP averaging 110 nm and the PSP 150 nm. Those of *Archidium* are only slightly dissimilar (ASP, 140 nm; PSP, 130 nm), while both SPs of *Hypnum* measure 80 nm (Duckett & Renzaglia 1986). Those of *Thuidium* are also equal, but at about 45 nm, they are the shortest recorded for bryophytes. Comparable data for *Polytrichum* are unavailable. Structural studies of developing (Duckett & Carothers 1979) and mature (Duckett et al. 1983) spermatozoids of *Sphagnum palustre* revealed no evidence of stellate patterns in the transition zones.

Some Systematic Considerations

The discovery in liverworts (Heitz 1959) and mosses (Heitz 1960) of the stratified, intricate component we now term the multilayered structure (MLS) was followed by detailed studies elucidating its ultrastructure and morphology in *Polytrichum* (Paolillo 1965, Paolillo et al. 1968a, b) and *Marchantia* (Carothers & Kreitner 1967, 1968,

Kreitner 1970). These and similar investigations of *Pellia* (Suire 1970) and the horn-wort *Phaeoceros* (Moser 1970) were largely developmental, with their emphasis on the graphic description of spermatozoid differentiation. The first broadly comparative study of the MLS (Carothers & Duckett 1978) was limited to just two characters. The first concerned the number of horizontal strata constituting the bryophyte MLS – an inquiry prompted, in part, by an earlier implication (Paolillo 1974) that mosses and liverworts might differ by the former having three strata and the latter four. That com-parative study, which examined representatives of 19 liverwort, hornwort, and moss genera, revealed the 4-layered MLS in all but those in three moss genera. Since then, the number of bryophyte genera sampled has increased to nearly 40, of which only four have exhibited a tristratose MLS (see part 4 of the preceding section). It is note-worthy also that both forms of MLS stratification may occur in the same genus: *Poly-trichum juniperinum* (Paolillo et al. 1968a: Fig. 20) is apparently tristratose, but *P. commune* Hedw. (Paolillo 1965: Fig. 15) has a 4-layered MLS. In our view the pre-sumably tristratose MLSs are exceptions and warrant further investigation; however, their systematic significance, if any, is presently unknown.

The second MLS character investigated by Carothers and Duckett (1978) was the number of parallel microtubules at the spline's widest region. The feature was chosen because available information suggested that it might exhibit some variation according to the taxon and thus be systematically useful – even though the long, slender shape of the spline would impose fairly narrow limits. The species investigated then and since represent a taxonomically broad sampling of bryophytes, and the data acquired have given some idea of ranges within and among selected taxa, pertinent examples of which have been included in the preceding section. A few instances in which spline tubule number has special significance are cited below. In general, however, the sys-tematic utility of this character appears quite limited.

It is the ensemble of a dozen or so characters derived from structural analysis of the whole blepharoplast that makes this line of inquiry useful, particularly so when the micrographic evidence is interpreted diagrammatically and in comparable form. Such studies, together with those of more limited scope, have shown that blepharoplast morphology is a valuable supplement to the more familiar sources of information. Work to date has yielded a large amount of new information from a small number of species, and the need to expand the selection of species for study is evident. There are abundant grounds to suggest that further research in this area will give rise to new in-sights into bryophyte systematics and evolution. The observations that follow are of-fered primarily to illustrate the blepharoplast's contribution towards the characteriza-tion of selected taxa; however, it is beyond the scope of this paper to present cross-comparisons of each structural feature of the organelle for each of the taxa cited.

Hornworts

The sole hornwort order, Anthocerotales, comprises two families: Anthocerotaceae with five genera and Notothylaceae with one (Renzaglia 1978, Renzaglia & Carothers

1986). Although blepharoplast morphology has been analyzed in only two hornwort species, viz., *Phaeoceros laevis* (Moser et al. 1977, Carothers et al. 1977, Duckett et al. 1980) and *Notothylas orbicularis* (Renzaglia & Carothers 1986), both families are represented in the sample. Those studies reveal a remarkable degree of similarity between the blepharoplasts. The major distinguishing features summarized for *Notothylas* (see preceding section on hornworts) are virtually the same as those found in *Phaeoceros*. To reiterate some of the characters: the spline is inaperaturate and bilaterally symmetrical, the LS is wider than long with the greater extension on the spline's right-hand side, the basal bodies are short and aligned evenly with the spline's transverse leading edge, and the transition zones are devoid of stellate patterns.

The last-mentioned character has special significance because the presence of a stellate pattern in the transition region between flagellar shaft and basal body is characteristic of the chlorophyll *a*- and *b*-containing eukaryotes (Moestrup 1978). The absence of the stellate pattern in hornworts is currently interpreted as an evolutionary loss (Renzaglia & Carothers 1986). Another noteworthy aspect of the hornwort blepharoplast is its small size with the length of the short lamellar strip essentially equalled by that of each basal body (e.g., see *Notothylas*, Fig. 15). It appears likely that this feature is related to the overall length of the spermatozoid body, a dimension shown to be smaller in hornworts than in liverworts and smallest of all in *Notothylas* (Mehra & Sokhi 1976).

The close agreement in blepharoplast character-expression for *Phaeoceros* and *Notothylas* emphasizes the unity of the hornworts as a group and constitutes important evidence supporting its homogeneity as determined by comparative and developmental study of the plant body (Renzaglia 1978). These same unifying characters differ sharply from those exhibited by liverworts and mosses, and thereby emphasize the group's isolated position. Indeed, the uniqueness of the hornwort blepharoplast is completely consistent with taxonomic treatments placing the hornworts in their own division, the Anthocerotophyta (Stotler & Crandall-Stotler 1977, Renzaglia 1978, Schuster 1984b).

Liverworts

The major characters common (though not necessarily exclusive) to liverwort blepharoplasts include bilateral asymmetry, a bluntly rounded MLS-anterior, a relatively long-tapered lamellar strip, and a pair of staggered and subapical dimorphic basal bodies. In addition there is an assortment of qualitative and quantitative characters that, in various combinations, give hepatic blepharoplast morphology greater variability than found in either hornworts or mosses. The most conspicuous variations are those that collectively set the *Haplomitrium* blepharoplast apart from those of other liverworts. These features include the spline's distal widening in two stages (Fig. 14), its large number of constituent microtubules and consequent great width, its possession of an open aperture, the left-of-center location of aperture and basal bodies, and the elongated lamellar strip with its curved, drop-shape.

Calobryales are a small order including only two genera, *Takakia* and *Haplomitrium*, of which the former is known only from female plants. The dozen or so species of *Haplomitrium* are classified in two subgenera (Engel 1981), both of which are represented by the studies of *H. gibbsiae* (Carothers & Rushing, this paper) and *H. hookeri* (Carothers and Duckett 1979, with emendations in this paper). Thus, the data reported in the present account provide a representative characterization of blepharoplast morphology for the genus. However, conclusions regarding the *Haplomitrium*-type blepharoplast depend upon the interpretation of phylogenetic relationships between the Calobryales and other hepatic taxa. In Schuster's classification (1984a), the Calobryales, Metzgeriales, Treubiales, and Jungermanniales constitute the subclass Jungermanniidae, which, along with the Marchantiidae, constitute the class Hepaticae. Given the primitive status accorded *Haplomitrium* (Schuster 1971, 1979, 1984a), it would appear that there has been a tendency toward reduced structural complexity in the more advanced taxa (Carothers & Duckett 1979). A contrasting classification by Stotler and Crandall-Stotler (1977) interprets the liverworts as a division comprising three classes: Haplomitriopsida, Jungermanniopsida, and Marchantiopsida. According to this systematic treatment, the Haplomitriopsida are more remote from the Jungermanniopsida; therefore, blepharoplasts of the putatively primitive Jungermanniopsida would more likely resemble those of, for example, the metzgerialean *Pallavicinia*(Renzaglia et al. 1985) or the jungermannialean *Marsupella* (Duckett et al. 1983) than those of *Haplomitrium*.

The relative amount of information on blepharoplast morphology in the Calobryales contrasts sharply with that available for other Jungermanniopsida. Of approximately 550 species in the Metzgeriales, only three – representing three of the 30 or so genera – have been analyzed to extract appropriate comparable data: *Pellia neesiana* (Duckett et al. 1983), *Blasia pusilla* (Renzaglia & Duckett 1987), and *Pallavicinia lyellii* (Renzaglia et al. 1985). A much more extreme disparity exists for the Jungermanniales, an assemblage of roughly 7,000 species in some 180 genera. Blepharoplasts of only four of its species have been studied in detail: *Marsupella emarginata* (Duckett et al. 1983), *Chiloscyphus pallescens* and *Cephalozia lunulifolia* (Rushing et al. 1984), and *Bazzania trilobata* (Rushing & Carothers 1986a). No comparable study has been reported for a member of the Treubiales. These great inequalities between the huge amount of variation intrinsic to so many species and the few species actually investigated emphasize the need for a broader taxonomic sampling of the two orders; they also emphasize the need for caution in postulating evolutionary trends and relationships.

Blepharoplasts of the subclass Jungermanniidae exhibit a morphological diversity much exceeding that of the Marchantiidae; consequently, there are relatively few features shared in common by its members. In general the representatives of this subclass have narrower spline apertures, longer lamellar strips, and longer posterior basal bodies, but each of these characters varies (sometimes widely) within the group. As noted earlier in this section, the assemblage of unusual features characterizing the calobryalean blepharoplast makes it the most unlike of those reported for the subclass.

Comparison of blepharoplasts in species representing the Jungermanniales and Metzgeriales reveals less variation in the former than in the latter but more than in the Marchantiideae. Shared features contributing to the lower level of variation in the Jungermanniales include the lamellar strip having a lateral offset that is gradually sloped rather than abrupt and notch-like, the LS's extension beyond the anterior end of the spline (and right-lateral extension as well in *Marsupella*), and the LSs having about the same length. Excluding *Lophocolea*, the jungermannialean species listed in Table 2 have lamellar strips averaging 2.6 µm long and ranging from 2.2 µm (*Cephalozia*) to 2.9 µm (*Bazzania*). The basal bodies also exhibit close similarities in length. In the same four species the ABB averages 0.7 µm with a range of 0.65 µm (*Marsupella*) to 0.8 µm (*Chiloscyphus*), while the PBB's average is 2.3 µm with a range of 2.1 µm (*Cephalozia*) to 2.6 µm (*Marsupella*). The PBB is thus about three times longer than the ABB and in all cases overlaps the ABB. In contrast a character displaying wide variation is the number of microtubules at the spline's widest region. Averaging 19, the range for species representing seven jungermannialean genera – the five listed in Table 2 plus *Southbya* (Carothers & Duckett 1978) and *Nowellia* (Simone, cited in Lee 1974) – is from 13 (*Nowellia*) to 25 (*Chiloscyphus*). This spread of 13 is just over twice that observed for the Metzgeriales.

The subject of the first detailed study of blepharoplast morphology in the Jungermanniales was *Marsupella emarginata* (Duckett et al. 1983). In part, that report concluded that the *Marsupella* blepharoplast had a much simpler structure than that observed for the Marchantiideae, i.e., for *Marchantia* and the three sphaerocarpalean genera. Further, it hypothesized that *Marsupella* might (on the grounds of its simplicity) be taken as a starting point from which trends in blepharoplast morphology could be recognized. Since then the number of jungermannialean species investigated has been increased to four, each representing a different one of the 15 constituent suborders. This newer information has shown that the jungermannialean blepharoplasts share much in common with each other, though *Marsupella* still stands apart owing to such features as its unilaterally spatulate spline and its spline aperture that closes anteriorly by convergence of two adjacent microtubules – a feature unknown in other liverworts. It is our opinion that, structural "simplicity" notwithstanding, the phylogenetic significance of the morphology of the *Marsupella* blepharoplast cannot be realistically assessed given the limited amount of intraordinal information currently available. A preferable data-base from which to identify possible trends of blepharoplast specialization should include more representatives of diverse taxa such as *Blepharostoma*, *Anthelia*, *Radula*, and *Lejeunea*.

It is evident from the detailed studies of only three species of the Metzgeriales – *Pellia neesiana* (Duckett et al. 1983), *Pallavicinia lyellii* (Renzaglia et al. 1985), and *Blasia pusilla* (Renzaglia & Duckett 1987) – that the broad range of diversity manifested by the order (Renzaglia 1982) pertains also to blepharoplast morphology. Each of these three species belongs to a different one of the group's seven suborders (Schuster 1984a), and each possesses major characters that, in combination, distinguishes it from the others. Accordingly, there are only a few common features of the blepharoplast, the principal ones being a bilaterally unequal, spatulate spline-anterior and a spline aperture that is closed anteriorly by one or more short microtubules. To illustrate the distinctions among them, a brief selection of blepharoplast characters for each of the three species will suffice for present purposes. Since some of their contrasting MLS features are already cited in Table 2, the emphasis here is mostly on aspects of the locomotory apparatus.

The *Pellia* blepharoplast has short (0,4 µm) anterior basal bodies of only one-third the length of those in the other two hepatics. These basal bodies are equaled in smallness only by those of *Phaeoceros* and *Notothylas*. In contrast it has extremely long (10.5 µm) posterior basal bodies – the longest recorded for bryophytes. The "stagger" or distance between the anterior and posterior stellate patterns (12.5 µm) is also the longest on record; however, it is noteworthy that these data were derived from specimens of a somewhat later developmental stage than those of *Pallavicinia* and *Blasia*. The extreme lengths of the PBB and stagger in *Pellia neesiana* are accompaniments of the mature spermatozoid's size; having a cell-body length averaging 123 µm (Showalter 1926), it is the largest spermatozoid recorded for the bryophytes.

Major characters denoting blepharoplast individuality for the other two species are fewer than for *Pellia*. Those principal features of *Pallavicinia lyellii* include its very long (7.5 µm) lamellar strip – the longest recorded for bryophytes – and the orientation of the basal bodies, i.e., the ABB lies parallel to the spline axis while the PBB is outwardly divergent (Renzaglia et al. 1985: Fig. 18). (Both basal bodies have a parallel orientation in *Pellia* and *Blasia*.) The distinguishing features for *Blasia pusilla* include its spline aperture width of three microtubule-diameters, the presence of a right-lateral notch in the lamellar strip, overlapping basal bodies, and a short (0.1 µm) stagger between the stellate patterns.

From these selected features, it is evident that the blepharoplasts of the three metzgerialean species are clearly distinct, the different character-sets reflecting the wide variation in the order. *Blasia*, unlike *Pellia* and *Pallavicinia*, exhibits several major characters common to the Marchantiidae. These include the three microtubule-diameter-wide spline aperture, the well-documented similarity in outline of its lamellar strip (Carothers 1973: Fig. 10–12) with that of the Sphaerocarpales (Carothers et al. 1983, Brown et al. 1983) and *Marchantia* (Carothers & Kreitner 1968), the coincidence of the LS outline relative to the spline anterior, and the overlapping basal bodies. It follows that *Blasia* warrants further investigation regarding its systematic affinities. Most importantly, and as is true also for the Jungermanniales, much more information must be acquired before intraordinal variation and characterization of the Metzgeriales can be realistically determined. The needed studies must be detailed, comparable analyses such as the examples cited here; fragmentary observations, and especially undocumented ones, are of least value in constructing credible characterizations of blepharoplast morphology at any taxonomic level. The only order in the Jungermanniidae that, at present, is adequately known for comparative purposes is the Calobryales.

It is unsurprising that, given the paucity of comparative data, potential trends in specialization have for the most part been unrecognized or, at best, unconfirmed – as in the conjecture that there has been a tendency toward reduced structural complexity in the more advanced taxa (Carothers & Duckett 1979). Nevertheless, the available evidence suggests that there has been a general tendency toward increased length of the stellate patterns in the more advanced hepatic taxa. The basis for this supposition lies in a comparison of the ranges of length measurements for those orders represented by species that have been studied in detail. The sequentially overlapping ranges do not, of course, imply a linear evolutionary relationship among the orders. At the shorter end of the scale is the Calobryales, its stellate patterns ranging from 100 nm (*Haplomitrium gibbsiae*) to 120 nm (*H. hookeri*). A wider range but also at the shorter end of the scale is that of the Metzgeriales with *Pellia* (100 nm) equal with *H. gibbsiae* and followed by *Pallavicinia* (110 nm) and *Blasia* (140 nm). The overlapping range of the Jungermanniales begins with *Cephalozia* (135 nm) and continues with *Bazzania* (ASP = 150 nm; PSP = 140 nm), *Marsupella* (150 nm), and *Chiloscyphus* (180 nm). The widest range is exhibited next by the Sphaerocarpales beginning with *Geothallus* (160 nm) and extending through *Riella* (185 nm) to *Sphaerocarpos* (230 nm). The stellate pattern of *Marchantia*, the sole representative of the Marchantiales, is also 230 nm long.

The subclass Marchantiidae comprises three orders: Sphaerocarpales with three genera and about a dozen species, the monogeric Monocleales having one or two species, and the dominant Marchantiales with about 95 genera and some 2,000 species (Schuster 1984a). Of these the Sphaerocarpales are best known from the standpoint of blepharoplast morphology. A representative species has been studied for each of its genera: *Riella americana* and *Sphaerocarpos donnellii* (Carothers et al.1983) and *Geothallus tuberosus* (Brown et al. 1983). A detailed analysis has been made for only *Marchantia polymorpha* (Carothers and Kreitner 1967, 1968; see Fig. 15), while no comparable study has been reported for *Monoclea*.

The blepharoplasts of these four species are much alike (e.g., Table 2). Their principal common features include a bilaterally spatulate spline anterior with the greater expansion on the left side, a closed aperture of three microtubule-diameters wide, a coincident LS outline relative to the spline anterior, and a notch-like offset on the LS's right-hand margin. Lamellar strip lengths for these four representatives range from 1.1 to 2 μm, all of them shorter than those recorded for the Jungermanniales and Metzgeriales, except *Blasia* in which the LS is also 2 μm. The posterior basal bodies also are shorter, ranging in length from 1.3 to 1.9 μm. The principal differences between the blepharoplasts of the sphaerocarpalean and marchantialean species concern two features of the spline. In the former group the spline decreases in width rather abruptly from 20 (*Sphaerocarpos*) or 24 (*Riella* and *Geothallus*) parallel microtubules to only two which extend posteriorly constituting a long, very slender shank – the narrowest recorded for bryophytes. In contrast the narrower spline of *Marchantia*(17–18 microtubules) decreases in width to form a 6-membered shank.

Earlier systematic treatments differed in their handling of *Geothallus*, a monotypic genus with several unusual features; it was variously included in (Stotler & Crandall-Stotler 1977) or, pending more information, excluded from (Sharp 1974) the Sphaerocarpales. The presence of this narrow, distinguishing shank in the three sphaerocarpalean genera provides both significant additional evidence of their morphological similarity and strong support for the inclusion of *Geothallus* in the order. Further, the presence of so few differences in blepharoplast morphology between the Sphaerocarpales and Marchantiales is important supplementary evidence helping to characterize the Marchantiidae as a natural group and to distinguish it from the Jungermanniidae, in which there is a high degree of variation.

Mosses

As a group, the mosses number approximately 14,000 species in about 700 genera (Crum & Anderson 1981). Yet, as noted earlier, knowledge of blepharoplast morphology in mosses is drawn mainly from detailed studies of only six species: *Polytrichum juniperinum* (Paolillo et al. 1968a, b), *Sphagnum palustre* (Duckett & Carothers 1979), *Funaria hygrometrica* (Carothers & Brown 1985), *Archidium tenerrimum* (Brown & Carothers 1986), *Thuidium delicatulum* (Rushing & Carothers 1986b), and *Hypnum mamillatum* (Duckett & Renzaglia 1986). These six, belonging to five of the 18 orders recognized by Crum and Anderson, constitute an exceptionally small and nonrepresentative sample. Nevertheless, serveral structural attributes appear to be common to most or all of them.

The moss lamellar strip is typically truncated at its posterior end, the level of its distal extent largely coinciding with that of the anterior basal body. The dimorphic basal bodies are oriented with the ABB parallel to the spline axis and the PBB divergent toward the left – except in *Sphagnum* in which this is reversed, its PBB parallel and its ABB divergent toward the right. The moss PBB is typically long and its apex lies relatively close to the spline's anterior margin. The separation between the anterior and posterior stellate patterns ("stagger") is also typically long.

Thuidium and *Hypnum*, the only pleurocarpous mosses studied in detail, belong to different families in the Hypnobryales and thus allow the first intraordinal comparison of moss blepharoplasts. They are remarkably similar in overall morphology, and comparisons with other bryophytes indicate that *Hypnum* shares the greatest number of attributes with *Thuidium* (Duckett & Renzaglia 1986). Tabular data given in the studies of these two mosses reveal that most of the variations are dimensional and relatively minor. The principal differences between them are the type of spline aperture (open in *Thuidium* and closed in *Hypnum*) and the lamellar strip outline relative to the spline anterior (anterior LS extension in *Thuidium* and both anterior and right-lateral extension in *Hypnum*).

The remaining four mosses (*Sphagnum, Archidium, Funaria, Polytrichum*) belong to separate and perhaps mutually remote orders. Their blepharoplasts exhibit many variant characters and each of the organelles may be recognized by even a single distinguishing character: the absence of a stellate pattern in *Sphagnum*, the substantially larger number (24) of spline microtubules in *Archidium*, the very short (0.8 μm) anterior basal body of *Funaria*, and the very long (8 μm) PBB of *Polytrichum*. Comparisons of the MLS from these four mosses (Table 2) also show a greater number of major features shared by *Sphagnum* and *Polytrichum* than by other pair-combinations. Such characters include symmetry of the spatulate spline's anterior, presence of a closed aperture, shape of the LS outline, and coincidence of the LS outline relevant to the spline anterior. Quantitative data pertaining to the basal bodies of *Polytrichum* are largely unavailable. Albeit meager, this comparative information suggests that moss blepharoplasts are less variable than those of liverworts.

In concluding this review of comparative blepharoplast morphology we would stress the need for more studies of systematically significant bryophytes. The blepharoplast is a scarcely tapped source of supplementary information useful in characterizing bryophytes and in providing additional insight regarding their interrelationships. There is much of value to be learned by continuing the survey of major taxa still unexamined and by determining the ranges of morphological expression within major taxa.

Acknowledgments

We wish to express our sincere appreciation and gratitude to Dr. David Ratkowsky (Tasmanian Regional Laboratory, CSIRO, Hobart) and his wife Ann for their patient, persistent efforts and indispensible help in collecting antheridial material of *Haplomitrium gibbsiae* at the required stage of spermatid development. We wish also to thank Mr. Ralph Cruickshank (Faculty of Agricultural Sciences, University of Tasmania, Hobart) for his skilled assistance with the fixation and dehydration of those tissue samples.

The diagrammatic representations of blepharoplast structure in *Bazzania* and *Notothylas* (Fig. 15) appeared previously in the *Journal of the Hattori Botanical Laboratory*, and those of *Funaria*, *Archidium*, and *Thuidium* (Fig. 16) appeared previously in *The Bryologist*. We gratefully acknowledge the permission given by the Hattori Botanical Laboratory and the American Bryological and Lichenological Society to republish these figures. We also thank Dr. Karen Renzaglia for providing prepublication manuscripts regarding *Blasia* and *Hypnum*.

This investigation was supported by the National Science Foundation (U.S.A.) under Grant No. BSR-8313023.

Literature Cited

Brown, R. C. & Carothers, Z. B. 1986. Comparative studies of spermatogenesis in the Bryopsida. II. Blepharoplast morphology in *Archidium tenerrimum* Mitt. The Bryolo-

Brown, R.C., Carothers, Z. B. & Duckett, J. G. 1983. Comparative spermatogenesis in the Sphaerocarpales. II. Blepharoplast structure in *Geothallus tuberosus* Campb. The Bryologist 86: 234–243.

Carothers, Z. B. 1973. Studies of spermatogenesis in the Hepaticae. IV. On the blepharoplast of *Blasia*. American Journal of Botany 60: 819–828.

— 1975. Comparative studies on spermatogenesis in bryophytes, pp. 71–84. In: Duckett, J. G. & Racey, P. A. (Eds.), The Biology of the Male Gamete. Biological Journal of the Linnean Society 7 (Supplement No. 1). London. Academic Press.

Carothers, Z. B. & Brown, R. C. 1985. Comparative studies of spermatogenesis in the Bryopsida. I. Blepharoplast morphology in *Funaria hygrometrica* Hedw. The Bryologist 88: 325–332.

Carothers, Z. B., Brown, R. C. & Duckett, J. G. 1983. Comparative spermatogenesis in the Sphaerocarpales. I. Blepharoplast structure in *Sphaerocarpos* and *Riella*. The Bryologist 86: 97–105.

Carothers, Z. B. & Duckett, J. G. 1978. A comparative study of the multilayered structure in developing spermatozoids. Bryophytorum Bibliotheca 13: 95–112.

— — 1979. Spermatogenesis in the systematics and phylogeny of the Hepaticae and Anthocerotae, pp. 425–445. In: Clarke, G. C. S. & Duckett, J. G. (Eds.), Bryophyte Systematics. Systematics Association, Special Volume 14. London. Academic Press.

— — 1980. The bryophyte spermatozoid: a source of new phylogenetic information. Bulletin of the Torrey Botanical Club 107: 281–297.

Carothers, Z. B. & Kreitner, G. L. 1967. Studies of spermatogenesis in the Hepaticae. I. Ultrastructure of the Vierergruppe in *Marchantia*. Journal of Cell Biology 33: 43–51.

— — 1968. Studies of spermatogenesis in the Hepaticae. II. Blepharoplast structure in the spermatid of *Marchantia*. Journal of Cell Biology 36: 603–616.

Carothers, Z. B., Moser, J. W. & Duckett, J. G. 1977. Ultrastructural studies of spermatogenesis in the Anthocerotales. II. The blepharoplast and anterior mitochondrion in *Phaeoceros laevis*: later development. American Journal of Botany 64: 1107–1116.

Crum, H. A. & Anderson, L. E. 1981. Mosses of Eastern North America. 2 Vols. 1328 pp. New York. Columbia University Press.

Duckett, J. G. & Carothers, Z. B. 1979. Spermatogenesis in the systematics and phylogeny of the Musci, pp. 385–423. In: Clarke, G. C. S. & Duckett, J. G. (Eds.), Bryophyte Systematics. Systematics Association, Special Volume 14. London. Academic Press.

Duckett, J. G., Carothers, Z. B. & Miller, C. C. J. 1982. Comparative spermatology and bryophyte phylogeny. Journal of the Hattori Botanical Laboratory 53: 107–125.

— — — 1983. Gametogenesis, pp. 232–275. In: Schuster, R. M. (Ed.), New Manual of Bryology, Vol. 1. Nichinan, Japan. Hattori Botanical Laboratory.

Duckett, J. G., Carothers, Z. B. & Moser, J. W. 1980. Ultrastructural studies of spermatogenesis in the Anthocerotales. III. Gamete morphogenesis: from spermatogenous cell through midstage spermatid. Gamete Research 3: 149–167.

Duckett, J. G. & Renzaglia, K. S. 1986. The blepharoplast of *Hypnum*. Journal of Bryology 14: 375–385.

Durand, E. J. 1908. The development of the sexual organs and sporogonium of *Marchantia polymorpha*. Bulletin of the Torrey Botanical Club 35: 321–335.

Engel, J. J. 1981. *Haplomitrium monoicum*, a remarkable new species of Calobryales (Hepaticae) from New Caledonia, together with a reclassification of subg. *Haplomitrium*. Annals of the Missouri Botanical Garden 68: 668–676.

Gambardella, R. & de Lucia Sposito, M. L. 1982. Some ultrastructural features of the spermatid of *Timmiella barbuloides* (Brid.) Moenk. (Pottiales, Musci). Caryologia 35: 376–377.

Heitz, E. 1959. Elektronenmikroskopische Untersuchungen über zwei auffallende Strukturen an der Geisselbasis der Spermatiden von *Marchantia polymorpha*, *Preissia quadrata*, *Sphaerocarpus Donnellii*, *Pellia Fabroniana* (Hepaticae). Zeitschrift für Naturforschung 14b: 399–401.

— 1960. Über die Geisselstruktur sowie die Dreiergruppe in den Spermatiden der Leber- und Laubmoose. Proceedings of the European Regional Conference on Electron Microscopy, Delft 2: 934–937.

Ikeno, S. 1903. Beiträge zur Kenntnis der pflanzlichen Spermatogenese: Die Spermatogenese von *Marchantia polymorpha*. Botanischen Centralblatt 15: 65–88.

Kreitner, G. L. 1970. The ultrastructure of spermatogenesis in the liverwort, *Marchantia polymorpha*. Ph. D. Dissertation. University of Illinois. Urbana.

Kreitner, G. L. & Carothers, Z. B. 1976. Studies of spermatogenesis in the Hepaticae. V. Blepharoplast development in *Marchantia polymorpha*. American Journal of Botany 63: 545–557.

Lal, M. & Bell, P. R. 1975. Spermatogenesis in mosses, pp. 85–95. In: Duckett, J. G. & Racey, P. A. (Eds.), The Biology of the Male Gamete. Biological Journal of the Linnean Society 7 (Supplement No. 1). London. Academic Press.

Lee, S. 1974. Certain aspects of cytology of *Lophocolea heterophylla* (Schrad.) Dum. M. A. Thesis. State University of New York. Potsdam.

Lepper, Jr., R. 1956. The plant centrosome and the centrosome-blepharoplast homology. The Botanical Review 22: 375–417.

Mehra, P. N. & Sokhi, J. 1976. Observations on the spermatozoids of some Hepaticae and Anthocerotae from India. Journal of the Hattori Botanical Laboratory 41: 359–376.

Moestrup, Ø. 1978. On the phylogenetic validity of the flagellar apparatus in green algae and other chlorophyll a and b containing plants. BioSystems 10: 117–144.

Moser, J. W. 1970. An ultrastructural study of spermatogenesis in *Phaeoceros laevis* subsp. *carolinianus*. Ph. D. Dissertation. University of Illinois. Urbana.

Moser, J. W., Duckett, J. G. & Carothers, Z. B. 1977. Ultrastructural studies of spermatogenesis in the Anthocerotales. I. The blepharoplast and anterior mitochondrion in *Phaeoceros laevis*: early development. American Journal of Botany 64: 1097–1106.

Paolillo, Jr., D. J. 1965. On the androcyte of *Polytrichum*, with special reference to the Dreiergruppe and the limosphere (Nebenkern). Canadian Journal of Botany 43: 669–676.

— 1974. Motile male gametes of plants, pp. 504–531. In: Robards, A. W. (Ed.), Dynamic Aspects of Plant Ultrastructure. London. McGraw-Hill.

Paolillo, Jr., D. J., Kreitner, G. L. & Reighard, J. A. 1968a. Spermatogenesis in *Polytrichum juniperinum*. I. The origin of the apical body and the elongation of the nucleus. Planta 78: 226–247.

— — — 1968b. Spermatogenesis in *Polytrichum juniperinum*. II. The mature sperm. Planta 78: 248–261.

Renzaglia, K. S. 1978. A comparative morphology and developmental anatomy of the Anthocerotophyta. Journal of the Hattori Botanical Laboratory 44: 31–90.

— 1982. A comparative developmental investigation of the gametophyte generation in the Metzgeriales (Hepatophyta). Bryophytorum Bibliotheca 24: 1–253.

Renzaglia, K. S. & Carothers, Z. B. 1986. Ultrastructural studies of spermatogenesis in the Anthocerotales. IV. The blepharoplast and mid-stage spermatid of *Notothylas*. Journal of the Hattori Botanical Laboratory 60: 97–104.

Renzaglia, K. S., Carothers, Z. B. & Duckett, J. G. 1985. Comparative ultrastructural studies of spermatogenesis in the Metzgeriales (Hepaticae). I. The blepharoplast of *Pallavicinia lyellii*. American Journal of Botany 72: 588–595.

Renzaglia, K. S. & Duckett, J. G. 1987. Comparative ultrastructural studies of spermatogenesis in the Metzgeriales (Hepatophyta). II. The blepharoplast of *Blasia pusilla*. American Journal of Botany 74: 53–64.

Rushing, A. E. & Carothers, Z. B. 1986a. Comparative spermatogenesis in the Jungermanniales (Hepaticae). II. The blepharoplast of *Bazzania trilobata*. Journal of the Hattori Botanical Laboratory 60: 85–95.

— — 1986b. Comparative studies of spermatogenesis in the Bryopsida. III. Blepharoplast morphology in *Thuidium delicatulum*. The Bryologist 89: 144–151.

Rushing, A. E., Carothers, Z. B. & Duckett, J. G. 1984. Comparative spermatogenesis in the Jungermanniales (Hepaticae). I. Observations on the structure of the young spermatids of *Cephalozia lunulifolia* and *Chiloscyphus pallescens*. Journal of Bryology 13: 247–256.

Schuster, R. M. 1967. Studies on Hepaticae XV. Calobryales. Nova Hedwigia 13: 1–63.

— 1971. Two new antipodal species of *Haplomitrium* (Calobryales). The Bryologist 74: 131–143.

— 1979. The phylogeny of the Hepaticae, pp. 41–82. In: Clarke, G. C. S. & Duckett, J. G. (Eds.), Bryophyte Systematics. Systematics Association, Special Volume 14. London. Academic Press.

— 1984a. Evolution, phylogeny and classification of the Hepaticae, pp. 892–1070. In: Schuster , R. M. (Ed.), New Manual of Bryology. Vol. 2. Nichinan, Japan. Hattori Botanical Laboratory.

— 1984b. Morphology, phylogeny and classification of the Anthocerotae, pp. 1071–1092. In: Schuster, R. M. (Ed.), New Manual of Bryology. Vol. 2. Nichinan, Japan. Hattori Botanical Laboratory.

Sharp, A. J. 1974. Hepatopsida, pp. 779–781. In: Encyclopedia Britannica, 15th ed., Vol. 8. Chicago. H. H. Benton.

Showalter, A. M. 1926. Studies in the cytology of the Anacrogynae. I. Antherozoids. Annals of Botany 40: 691–707.

Stotler, R. & Crandall-Stotler, B. 1977. A checklist of the liverworts and hornworts of North America. The Bryologist 80: 405–428.

Suire, C. 1970. Recherches cytologiques sur deux Hépatiques: *Pellia epiphylla* (L.) Corda (Metzgériale) et *Radula complanata* (L.) Dum. (Jungermanniale). Ergastome, sporogénèse et spermatogénèse. Le Botaniste 53: 125–392.

Treeful, L. 1978. Spermatogenesis in the moss *Leptodictyum trichopodium* (Schultz) Warnst. M. S. Thesis. University of Wisconsin-Milwaukee.

Turner, F. R. 1966. Changes in cellular organization during spermatogenesis. Ph. D. Dissertation. University of Texas. Austin.

Wilson, M. 1911. Spermatogenesis in the Bryophyta. Annals of Botany 25: 415–457.

Appendix A

Materials and Methods

Male plants of *Haplomitrium gibbsiae* (Steph.) Schust. were collected by A. V. and D. A. Ratkowsky on 10 December 1979 from a sandstone outcrop just below The Springs, Mt. Wellington, Tasmania, at ca. 700 m elevation. Small portions of antheridium-bearing tissue were fixed in 4 % glutaraldehyde in M/10 phosphate buffer at pH 6.9 for 2 h at room temperature. The specimens were then washed in buffer, postfixed for one hour in 2 % phosphate-buffered osmium tetroxide, and dehydrated in a graded ethanol series to 70 % for shipment to Urbana. Following their subsequent complete dehydration the specimens were embedded in Spurr's low-viscosity epoxy resin. Thin sections for electron microscopic examination were stained with uranyl acetate and lead citrate.

All micrographs showing the blepharoplast in transverse section are oriented as they would appear when viewed from the cell's anterior end; accordingly, the posterior basal body lies closer to the left margin of the subtending multilayered structure than does the anterior basal body. All references to left or right pertain to positions occupied in the illustrations.

The diagrammatic representations shown in Fig. 14–16 are composite reconstructions based on ultrastructural analyses of many individual spermatids. In order to facilitate their comparison, each reconstruction is depicted at the same scale (x48,100) and in the same style (showing only the lamellar strip and the anterior portions of the cytoskeletal spline and flagellar apparatus). The microtubules constituting the spline are shown as parallel lines spaced to represent the center-to-center distance between adjacent tubules rather than tubule outside-diameters. Since the cumulative difference between these two dimensions becomes significant only when relatively large numbers of parallel microtubules are concerned, we have used the more familiar term "microtubule-diameters" in stating the width of the spline aperture, a characteristically narrow, slot-like feature. In determining the symmetry or asymmetry of the spline anterior, we have used whenever possible the midline of the shank as the axial reference.

The seven species listed in Table 1 are the only ones for which we presently have complete sets of comparable data. The rounded values given there are virtually the same as the dimensions used in drafting the original blepharoplast reconstructions. The new data for *Haplomitrium hookeri* (Smith) Nees were acquired from recently sectioned antheridia belonging to the same collection used in an earlier study (Carothers & Duckett 1979).

The taxonomic treatments followed here are those of Schuster (1984a) for liverworts, Renzaglia (1978) for hornworts, and Crum and Anderson (1981) for mosses.

Key to labeling: ABB, anterior basal body; AF, anterior flagellum; AM, anterior mitochondrion; ASP, anterior stellate pattern; ATZ, anterior transition zone; ET, extended microtubular triplet; LS, lamellar strip; N, nucleus; PBB, posterior basal body; PSP, posterior stellate pattern; PTZ, posterior transition zone; S_1, spline; S_1–S_4, specific strata of the multilayered structure; SA, spline aperture; SP, stellate pattern; bracket ([), lamellar strip. Except as otherwise noted for Fig. 14–16, each scale bar represents 0.2 μm.

Advances in Bryology 3: 135 – 158 (1988).

Peristome Structure and the Regulation of Spore Release in Arthrodontous Mosses

by

Dale M. J. Mueller and Alan J. Neumann

Department of Biology, Texas A&M University
College Station, Texas 77843-3258, U.S.A.

With 12 Figures

Abstract: Bryologists have long utilized characteristics of the peristome as important criteria in the taxonomy and systematics of mosses. The advent of scanning electron microscopy has produced a renaissance in the utilization of peristomial features in such bryological studies. In most mosses the peristome is an integral part of the spore release mechanism and generally serves to regulate the release of spores from the urn. Arthrodontous peristomes are composed of cell wall remnants of specific tissue layers of the capsule that are exposed by dehiscence of the operculum.

Arthrodontous peristomes may consist of a single ring of subunits (teeth) or two concentric rings of subunits that surround the mouth of the capsule. In the latter the outer ring is collectively referred to as the exostome and the inner ring is referred to as the endostome. The endostome may be composed of segments and cilia. The endostome is generally considered to play a more passive role in spore release, whereas the exostome frequently plays an active role in either hindering or facilitating spore release.

The subunits of both the exostome and endostome are trilaminate structures consisting of the adjacent periclinal walls and the intervening middle lamellae of three concentric tissue layers of the capsule, referred to as the outer, primary, and inner peristome cell layers. Both light and electron microscopy reveal that the orientation of cellulosic microfibrils in the cell walls, along with hydrophilic or hydrophobic substances, result in characteristic hygroscopic movements in response to fluctuations in atmospheric moisture. The specific construction and composition of the teeth result in bending movements, which tend to open and close the mouth of the capsule under specific moisture conditions. The hygroscopic movements of the teeth, or portions of the teeth, facilitate spore release under environmental conditions that are favorable for spore dispersal and germination.

Zusammenfassung: Merkmale des Peristoms wurden lange Zeit von Bryologen als wichtige Kriterien in der Taxonomie und Systematik der Moose erachtet. Seit dem Aufkommen der Rasterelektronenmikroskopie wird den Peristomstrukturen wiederum vermehrt Bedeutung beigemessen. Bei den meisten Moosen stellt das Peristom einen integralen Bestandteil des Sporenausbreitungsmechanismus dar und dient im allgemeinen der Regulierung des Sporenausstreuens aus der Kapsel. Arthrodonte Peristome bestehen aus Zellwandresten besonderer Gewebeschichten der Kapsel, welche durch das Aufspringen des Deckels freigelegt werden.

Arthrodonte Peristome bestehen entweder aus einem Ring von Untereinheiten, Zähne genannt, oder aus zwei konzentrischen Kreisen solcher Untereinheiten, welche die Kapselöffnung umgeben. Im zweiten Fall wird der äußere Kreis allgemein Exostom genannt, der innere Endostom. Das Endostom kann aus Zähnen und Zilien zusammengesetzt sein. Das Endostom spielt eine eher passive Rolle bei der Ausbreitung der Sporen, während das Exostom aktiv das Ausstreuen fördert oder erschweren kann.

Die Untereinheiten sowohl des Exostoms wie des Endostoms sind dreischichtige Strukturen, die aus den periklinen Zellwänden und dazwischenliegenden Mittellamellen von drei konzentrischen Zellagen der Kapsel, der äußeren, der primären und der inneren Peristomschicht, zusammengesetzt sind. Sowohl im Licht- wie auch im Elektronenmikroskop kann gezeigt werden, daß die Anordnung der Cellulose-Mikrofibrillen der Zellwände, zusammen mit hydrophilen oder hydropholen Substanzen, für die charakteristischen hygroskopischen, feuchtigkeitsabhängigen Bewegungen verantwortlich ist. Durch die besondere Bauweise und Zusammensetzung der Zähne ergeben sich Bewegungsmechanismen, die je nach den Feuchtigkeitsbedingungen die Kapselmündung öffnen oder verschließen. Die hygroskopischen Bewegungen der Zähne oder Teile der Zähne erleichtern die Ausstreuung der Sporen, wenn die Umweltbedingungen für Sporenausbreitung und -keimung günstig sind.

Keywords: arthrodontous, diplolepidous, endostome, exostome, haplolepidous, hygrocastique, hygroscopic, moss, ontogeny, peristome, spore dispersal, xerocastique.

Contents

Introduction

The peristome is considered to be a conservative and, therefore, a taxonomically important feature of the sporophyte, specifically in mosses belonging to class Bryopsida (Crum & Anderson 1981). The peristome is a morphological feature at the apex of the capsule of most mosses that is exposed upon dehiscence of the operculum (lid). It consists of subunits (teeth, segments, cilia) that surround the open end (mouth or stoma) of the spore-filled, lower portion of the capsule (urn). Bryologists have long used the presence or absence and structure of the peristome to circumscribe taxa; there is little doubt concerning the significance of peristomial features as important diagnostic criteria.

Mosses with peristomes have traditionally been categorized into subgroups based on anatomy and function of the peristome. Peristomes having teeth composed of intact cells are called nematodontous (as in the subclasses Polytrichidae and Tetraphidae), and peristomes in which the subunits are composed of variously thickened cell wall remnants are called arthrodontous (as in the subclasses Bryidae and Buxbaumiidae). Bryologists have commonly examined peristomes for characteristics that contribute to moss taxonomy and phylogeny and have too frequently coined terms to describe these characteristics. Unfortunately, this had led to ambiguities and misconceptions in the literature that continue to plague contemporary research.

This article is intended to be a general review of early and contemporary works that are considered to be the more important contributions to understanding peristome development, structure, and function. Particular attention is directed towards the clarification of terminology. Emphasis is placed on the necessity for comprehensive investigations of peristome structure and ontogeny in order to increase knowledge of peristome function.

Peristome Structure

Arthrodontous peristomes are derived from localized cell wall thickenings and selective lysis of unthickened walls of cells comprising the innermost layers of the amphithecial tissue of the capsule. The concentric cell layers involved in peristome formation are most often three in number. These are referred to as the inner (IPL), primary (PPL), and outer (OPL) peristomial layers (Blomquist & Robertson 1941). Arthrodontous peristomes may consist of a single ring (Fig. 1) or two (Fig. 2) to several concentric rings of teeth surrounding the mouth of the urn. The teeth of a single peristome are composed of adjacent periclinal walls of the PPL and IPL. Although there may be two or more rings of teeth, the double peristome type is most common. The innermost is the endostome (endoperistome) and is derived from adjacent periclinal wall remnants of the IPL and PPL. The outer ring is the exostome (exoperistome) and is derived from adjacent periclinal wall remnants of the PPL and the OPL. The similarities in construction and position between the endostome and the single-ring peristome have resulted in the assumption that the two are homologous (Philibert 1884, English translation by Taylor 1962).

It is generally accepted that the components of any arthrodontous peristome are, as Edwards (1984) stated, "...locally thickened periclinal wall-pairs that remain standing after the breakdown of the rest of the cells...." However, we would like to emphasize that these structures frequently are trilaminate, consisting of wall depositions from two distinct cell layers *and* the intervening middle lamella. Steinbrinck (1897), Pfaehler (1904), Lorch (1931), and Mueller (1970, 1973) have referred to these wall components as lamellae. Schnepf et al. (1978), Edwards (1984), and Shaw (1985) have used the term wall plates in reference to these layers. Others refer to the lamellae as the outer and inner surfaces or dorsal and ventral plates, respectively.

The concept of a central lamella (Mueller 1973), composed of the integrated primary cell walls and the intervening middle lamellae, afford a basis, in part, for explaining peristome movement. Hygroscopic movements are the result of differential expansion of the wall thickenings of the opposing walls, i. e., inner vs. outer lamellae. (See section entitled "Peristome Function.") Anatomists might call this the compound middle lamella because of staining or distinctive structural properties.

The surfaces of the inner and outer lamellae often exhibit varied ornamentation patterns, and many terms have been used to describe these features, which are often used as taxonomically important criteria. The remnants of horizontal (transverse) and vertical (longitudinal) cell walls (i. e., the anticlinal walls) of the OPL, PPL, and IPL are often evident on the surfaces of the teeth. Terms such as cross-striae, ridges, lamellae, divisural zigzag-lines, plates, and trabeculae have been applied to these remnants of cell autolytic processes. We recommend that the ridge-like remains of horizontal anticlinal walls be consistently called trabeculae and that vertical anticlinal wall remnants be called median or vertical ridges. (Other ornamentation patterns created by additional wall deposits require descriptive phraseology not apropos to this discussion.) The remnants of anticlinal walls are common but not always obvious. It is presumed that when they consist of no more than primary wall material, they may be nearly completely dissolved during autolysis. More importantly, endostome segments may ultimately be single-ply due to dissolution of primary walls and middle lamellae. When all wall deposits are preserved, sectioning reveals that these walls are trilaminate, not "two-ply" as frequently thought (Shaw & Rohrer 1984, Allen et al. 1985). Unfortunately, we have been lulled into thinking of the peristome as being two-ply by merely studying surface structure.

Depending upon the number of cells contributing cell wall materials to the outer lamella of a peristome tooth, arthrodontous peristomes are classified as haplolepidous or diplolepidous. (A variation in spelling, i. e., haplolepideous and diplolepideous, has appeared in several recent works.) The haplolepidous condition is distinguished by the wall remnants of a single, vertical file or column of cells (decreasing in size towards the apex) that contributes wall thickenings to the outer lamella of each peristome tooth, whereas the exostome of diplolepidous peristomes exhibits an outer lamella composed of wall thickenings contributed by two laterally adjacent files or columns of cells (Philibert 1884, English translation by Taylor 1962). In haplolepidous mosses numerous tiers of sixteen cells each are involved in the formation of the outer lamellae. Furthermore, the inner lamella of each tooth is characteristically formed by depositions from two adjacent cells. The result is that for each pair of teeth in haplolepidous peristomes there are a total of five files or columns of cells involved in their formation.

In diplolepidous mosses the outer lamellae of the teeth are the result of thickenings from two files (not one file) of cells. Readily seen in transverse section, there are characteristically 32 cells involved in the formation of the outer lamellae, whereas 16 cells are involved in the formation of the inner lamellae of the exostome. Only a few exceptions to this construction are known (Stone 1961, Shaw 1985, 1986). The number of cells involved in each tier is usually clearly evident by the remains of vertical anticlinal walls that form a continuous median ridge on the outer surface of each tooth. Transverse, anticlinal wall remnants on both the inner and outer lamellae are also evident, and these are referred to as trabeculae (Fig. 3). The haplolepidous condition has often been wrongly equated with uniperistomate mosses, while doubly peristomate mosses were considered to be diplolepidous, when, in fact, the usage of terms haplolepidous and diplolepidous must be tied in with the number of cells involved in the formation of the outer lamella of the peristome (exostome). (See reviews by Edwards 1979, 1984.)

The endostome is composed of delicate tooth-like structures (Fig. 2) called segments. These, in some cases, represent undissolved portions of the inner periclinal walls of the PPL and the outer periclinal walls of the IPL, as well as the middle lamellae. Occasionally, additional wall remnants (cilia) may occur between the segments. The bases of the segments (and cilia, when present) are often united into a basal membrane (Fig. 4). As Anderson and Palmer (1982) pointed out, there has been confusion concerning the construction of the endostome. Although some mosses have endostomes that are clearly trilaminate in construction, it is possible that when the periclinal walls of the IPL or PPL are not thickened with wall materials resistant to autolysis that all primary walls (and middle lamellae) may undergo destruction (Shaw & Allen 1985). It is an interesting idea that the products of this autolysis not only contribute in part to the ornamentation of the peristome but also to the ornmentation of the spores (Kreulen 1972, Mueller 1974). The relative amount of wall thickening contributed by either the IPL or PPL evidently varies in different mosses. Shaw and Rohrer (1984) conclude, "we cannot state with certainty how many lamellae are present or comment on their relative thickness." The thickenings to which they refer, presumably, are wall materials added to the primary walls; they did not study endostomes developmentally or always make definitive transverse sections, so their uncertainty is understandable. The relative amount of thickening contributed by these layers is variable from moss to moss and is diagnostic. In *Funaria hygrometrica* Hedw. thickenings that compose the endostome and exostome occur mainly on the inner periclinal walls of the OPL and the outer periclinal walls of the PPL and the IPL (Proskauer 1958). Allen (1981) has shown the peristomes of *Sorapilla sprucei* Mitt. and *S. papuana*

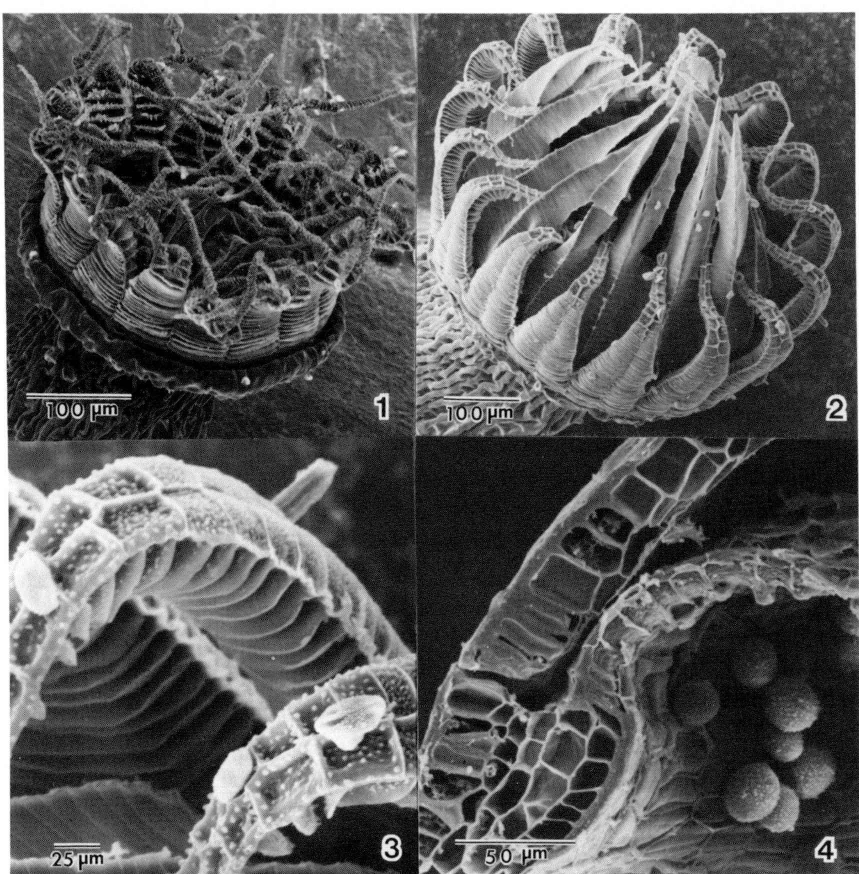

Figures 1–4. Scanning electron micrographs. **1**, *Fissidens ravenelii* Sull., haplolepidous, xerocastique peristome in dry, open (reflexed) position; **2**, *Sematophyllum adnatum* (Mx.) E. G. Britt., diplolepidous, xerocastique peristome, tips of exostome teeth lie between segments of endostome; **3**, *S. adnatum*, prominent trabeculae of inner lamina (facing downward) and vertical and transverse anticlinal walls on outer lamina of exostome tooth; **4**, *Forrstroemia trichomitria* (Hedw.) Lindb., spores and longitudinal section at level of annulus, showing attachment of mature exostome to urn.

Broth. & Geh. to be unusual. The endostome is more prominent than the exostome. Another exceptional case (Stone 1961) was noted in *Mittenia plumula* (Mitt.) Lindb. In this moss the OPL is apparently not involved in the formation of the exostome, and thus it gives the appearance of being haplolepidous.

Lantzius-Beninga (1847) was the first to illustrate in great detail the intricate and subtle differences in peristome structure between mosses. His epochal contribution was based on careful anatomical studies, and there have been few detailed investigations of this magnitude on peristome structure since then. Schofield (1985) and others before him recognized studies of structure and ontogeny of peristomes as valuable contributions in determining trends in phylogenetic relationships between major taxa of the Bryidae (while at the same time Schofield recognized that features of the gametophyte are indicators of trends within these taxa). Mueller (1970) in concluding a monograph on peristome structure stated, "it is possible that the peristome may indeed furnish clues to moss phylogeny." This has proven to be a greater understatement than originally intended. The need for investigations into the structure and function of the peristome in relation to spore disperal mechanisms is obvious, and the time is ripe for renewed efforts in this area. Much has yet to be learned about the biology of mosses through further investigations into peristome structure. How does structure and function relate to effectiveness in spore dispersal, and how does this affect the geographic distribution of the species? What impact has function had on evolution within this group of plants?

Detailed investigations on moss capsule apical development are necessary to elucidate peristome structure; however, such studies have been sporadic (Kienitz-Gerloff 1878, Evans & Hooker 1913, Blomquist & Robertson 1941, Saito & Shimoze 1955, Saito 1956, Mueller 1970, 1973, Kreulen 1972, Schulz & Schmidt 1974). Progress in taxonomic studies has been affected by misinterpretations of data on peristomes and inconsistency in descriptive terminology. The study of peristome development in *Aulacomnium heterostichum* (Hedw.) B.S.G. is particularly noteworthy because it helped standardize the terminology concerning the cell layers involved in the formation of arthrodontous peristomes (Blomquist & Robertson 1941). Emphasizing the specific cell layers involved, Kreulen (1972) compared the general developmental patterns in both single and double arthrodontous peristomes.

Peristome Function

The effectiveness of spore dissemination in mosses is influenced by features of the sporophyte such as length of the seta, attitude of the capsule (i. e., erect, inclined, pendent), as well as peristome structure (cf. Ingold 1965, Vitt 1981). Arthrodontous peristomes are considered to play an important role in spore release, but their actual role varies widely among mosses (Goebel 1895, Steinbrinck 1897, Lazarenko 1957, Ingold 1959, Pais 1964, Mueller 1973, Vitt 1981, Koponen 1982). Many arthrodontous peristomes exhibit hygroscopic movements that are interpreted as active mechanisms in spore release. Hygroscopic movements have been observed in the peristomes of many singly peristomate mosses. In exostomes of doubly peristomate mosses the teeth are frequently hygroscopic, whereas the segments and cilia of the endostome are generally thought not to react to changes in atmospheric moisture. In general the non-reduced endostome may serve as a passive regulator by restricting the area of the mouth of the urn, and may serve as an accessory to the exostome in closing the mouth of the urn to spore release.

Hygroscopic movements of the teeth presumably serve to permit spore release only during periods when environmental conditions are favorable to spore dissemination and germination (Lazarenko 1957, Ingold 1959, Pais 1964, Mueller 1973, Neumann 1983). In some cases no further movements of the teeth are observed after an initial reaction subsequent to dehiscence of the operculum (Vitt 1981, Koponen 1982). It is possible that dispersal mechanisms correlate with spore longevity. Species with meta-bolically active spores at the time of spore release (Mueller 1974) may require ade-quate moisture for immediate germination. Other mosses having spores with ample food reserves and poorly developed plastid membranes at the time of release (Seabury 1975) may not require active dispersal or a release mechanism that operates at a criti-cal time.

One spore release mechanism that has been customarily considered predominant in mosses, involves the inward bending of the teeth over the mouth of the urn, like a multi-sectional gate, in response to high atmospheric moisture (i. e., rain, fog, dew). Upon dehydration the teeth reflex, effectively opening the mouth of the urn and al-lowing spore release. This mechanism lessens the opportunity for spore release under "wet" conditions and is categorized as xerocastique (Ascherson 1892). Goebel (1895) presumed that the inflexing movement of the teeth was to keep moisture (water) from entering the spore-filled urn, thereby preventing the clumping of spores, which he thought would hinder dissemination. This mechanism allows spore release and dis-persal under relatively "dry" weather conditions, conditions viewed as favorable to the dissemination of spores of most mosses by wind currents (Smith 1938, Ingold 1959, Richardson 1981).

A second, sometimes overlooked, mechanism involves hygroscopic movements of the teeth that favor spore release during periods of high moisture availability and is categorized as hygrocastique (Ascherson 1892). In this case the hydrated teeth are re-flexed away from the mouth of the urn and upon dehydration inflex over the mouth of the capsule, effectively preventing spore release. This pattern of movement is the an-tithesis of the first, which is commonly described as characteristic of moss peri-stomes. Patterson (1953), Pais (1964, 1966) and Neumann (1983) consider the hy-grocastique mechanism to be correlated with the habit and habitat of these mosses.

There are several recognized variations of xerocastique and hygrocastique mechan-isms, as well as several mechanisms that may be considered as intermediate to these major categories. In several cases (Ingold 1959, Mueller 1973), the apices of the teeth are forced into the spore mass within the urn by the inflexing of teeth when hydrated. Dehydration causes the teeth to reflex and the tips of the teeth drag spores out of the urn through the interaction between the teeth and spores and, in some cases, through the rapid twisting of the tips of the teeth as these dry.

Pais (1964), Mueller (1973), Schnepf et al. (1978), and Neumann (1983) have all concluded that there is a correlation between the structure of the teeth and the hygroscopic movement they exhibit. The ultrastructure and chemical composition of wall remnants comprising the teeth are responsible for the widely varied hygroscopic movements. Subtle variation in structure and composition from species to species determines the specific response of the peristome to fluctuations in atmospheric moisture. A review of structure versus movement will be considered in later sections of this paper.

Historical Perspective

Hedwig (1782, 1787) was among the first to describe the structure of the moss sporophyte, including the peristome, and to consider the functions of these structures. He was aware of hygroscopic movements of certain peristomes and indicated (1787) that the peristome probably serves to close the mouth of the capsule to spore release under humid or moist conditions.

Lantzius-Beninga (1847) conducted extensive investigations into peristome structure and illustrated in intricate detail the subtle differences among moss peristomes. His contribution has served as a baseline and standard for all researchers. Only rarely are Lantzius-Beninga's illustrations marred by misinterpretation. Limitations of resolution, contrast, and depth of field in bright-field light microscopy resulted in a rare, but understandable, misinterpretation, such as attributing surface ornamentation patterns to the wrong lamella. [Such a problem need no longer exist with new developments in microscopy. The use of the scanning electron microscope has contributed greatly to the visualization and interpretation of the detail of peristome ultrastructure. Since the first application (Mueller 1970, Robinson 1971) of scanning electron microscopy, many bryologists have utilized this tool. Scanning electron microscopic observations have, understandably, grown to be almost obligatory in comparative studies of peristome morphology.]

Hutton (1874) was the first to elaborate upon the function of the peristome in spore dispersal. Hutton concluded that wind was the predominant means of disseminating spores and that the seta and capsule were both to be considered in spore dispersal. According to Hutton, the peristome was only one factor for regulating the ease with which spores were swept away from the mouth of the urn by the wind. (It has more recently been observed in *Fissidens limbatus* Sullivant that in the absence of any peristome movement the spores are squeezed out of the capsule, much like toothpaste from a tube; see Mueller 1970.) Hutton also observed that in mosses with double peristomes the endostome was not hygroscopic. Instead, as the exostome teeth become dehydrated and reflexed, their apices and trabeculae become entangled in the segments and ultimately spring free with a sudden jerk, which tends to propell spores from the capsule. Hutton coined the term trabeculae, referring to the horizontally oriented ridges on the teeth which are formed by remnants of transverse (anticlinal) cell walls (Fig. 3).

Early interest in spore release mechanisms reached a peak around the turn of the century with several noteworthy studies. Ascherson (1892) used the terms xerocastique and hygrocastique to describe the hygroscopic responses of various peristomes to changes in humidity or moisture. The term xerocastique was applied to mosses in which the peristome allowed spore release under relatively dry conditions and hygrocastique was applied to peristomes that allowed spore release under moist conditons. Goebel (1895) extended the study of peristome function to recognize the biological significance of the seta as well. Goebel attempted to create a logical arrangement for his observations and to categorize peristomes according to their structure and function in spore dispersal. He did recognize the limitations of such a scheme.

Steinbrinck (1897), like Goebel, studied the hygroscopic movements exhibited by moss peristomes and attempted to correlate these movements with peristome structure. He was aware that arthrodontous peristome teeth were, in essence, triangular, trilaminate remnants of thickend periclinal cell walls of specific cell layers. Steinbrinck employed polarized light microscopy to determine the plane of orientation of the cellulosic microfibrils comprising each thickened cell wall. His observations clearly indicated a correlation between peristome structure and hygroscopic movements. First, he showed a correlation between the lack of hygroscopic movement in endostome segments and their isotropic response to polarized microscopy. Second, Steinbrinck found a correlation between the structure of the outer teeth and the particular hygroscopic movements exhibited by individual teeth. He also noted that the relative rates of hydration/dehydration in the different lamellae contribute to the movements of portions of the individual teeth (i. e., basal region vs. apex).

Steinbrinck also attempted to categorize peristomes on the basis of their observed hygroscopic movements, but he warned of pitfalls in applying these characteristics to the systematic arrangement of mosses. To exemplify his point, Steinbrinck referred to the peristome of *Pylaisiella polyantha* (Hedw.) Grout (Hypnaceae). It was known that hypnaceous peristomes, such as that of *Amblystegium serpens* (Hedw.) B.S.G., inflex over the mouth of the urn when moistened. Steinbrinck concluded that this movement was due to the transverse orientation of microfibrils of the outer lamellae of the teeth and the swelling of the outer lamellae upon hydration. In contrast to most hypnaceous peristomes, the peristome teeth of *P. polyantha* when moistened reflex away from the opening of the capsule. The apices of the teeth do not exhibit the oscillating movements observed in other hypnaceous peristomes. Steinbrinck found that the microfibrils of the lamellae were parallel to the long axis of the teeth and that the inner lamellae exhibited a noticeable swelling upon hydration. He recognized that the composition of the additional wall thickening of the inner lamellae was responsible for the observed hygroscopic movement, but he did not attempt to identify the chemical constituents of the wall. In his closing comment on the peristome of *P. polyantha*, Steinbrinck noted that polarized light microscopy had a limit in its applicability to investigations into peristome function.

Derschau (1900) examined peristome structure by following peristome development. He employed a number of staining procedures to detect the presence of specific compounds in the developing cell walls, and he found that those cell walls that eventually remained as the peristome are laid down by apposition of wall materials. Cellulose was found to be the major wall constituent. Further wall deposition was believed to include pectin, cutin, and sphagnol, but no lignin or suberin. These wall materials were presumed to favor hygroscopicity and to protect the wall thickenings from decomposition. Students of peristome structure should further be aware of the noteworthy contributions of Schimper (1848), Limpricht (1890), Czapek (1899), Goebel (1930), Lorch (1931), and van der Wijk (1951).

The work of Pfaehler (1904) is often overlooked. He examined the peristomes of mosses in several families from both the structural and functional points of view. Apparently Pfaehler had been inspired by Hutton's work (1874) and continued the use of the term trabeculae in reference to the transverse wall remnants on the surfaces of the lamellae, which previous researchers had termed cross-striae. He elaborated on Hutton's observation that the trabeculae of the inner lamellae served dual functions. First, trabeculae assist in the dragging of spores from the urn. Second, trabeculae tend to snag neighboring segments and produce stress in both the segments and the ecumbered teeth. This causes the violent propulsion of spores from the capsule as the trabeculae disengage from the segments, thereby releasing the mechanical stress. Pfaehler concluded that controlled release of spores by the peristome was important in spore dissemination and depends on the hygroscopic movement of the peristome. It was further concluded that atmospheric conditions (wind and humidity), but not rain per se, were the most influential factors contributing to the hygroscopic movement of peristomes and release of spores for dissemination in some mosses.

Interest in spore dispersal mechanisms and peristome function waned over the next half century. However, the works of Evans and Hooker (1913) and Bloomquist and Robertson (1941) are noteworthy exceptions. Patterson (1953) reported observations of hygroscopic movements in peristomes of *Forrstroemia ohioensis* (Sull.) Lindb. He observed that, when dry, the exostome teeth inflexed along the interior wall of the urn and reflexed through an arc of 270 degrees when moistened. Patterson noted that Steinbrinck (1897) had observed a similar hygroscopic movement in *Pylaisiella polyantha*. He concluded that most researchers apparently had overlooked this mechanism and customarily emphasized the role of the peristome in preventing spore release under wet conditions (see e. g., Smith 1938). Patterson initially considered this to be an anomaly but found the movement characteristic of many mosses he examined.

In a brief review of peristome function, Lazarenko (1957) reiterated the conclusions drawn by Pfaehler (1904) on the relationship between environmental conditions and peristome movement. Lazarenko emphasized that the hygroscopic movement of peristomes is most noticeable at the onset and end of rain and during a heavy fog or dew. The differences in peristome structure account for the variability in hygroscopic movements, but all peristome movements in epiphytic mosses are adapted to the release of spores under favorable moisture conditions, rather than complete air dryness. The underlying premise is that spores require a small amount of moisture to adhere to a substrate and for germination.

Ingold (1959, 1965) concurred with Lazarenko (1957) and Pfaehler (1904) that atmospheric conditions were important influences on the hygroscopic movements of moss peristomes, but he was convinced that the fundamental purpose of hygroscopic peristomes was the closure of the mouth of the urn under damp conditions. Ingold elaborated on the structure of the peristome of *Eurhynchium confertum* (Dicks.) Milde, a doubly peristomate moss. The outer lamellae of the exostome are apparently composed of numerous closely spaced transverse plates of unknown composition that swell upon hydration. The composition of the inner lamellae was not described but was considered non-hydrophilic or at least much less hydrophilic than the outer lamellae. The different rates and magnitude of expansion exhibited by the lamellae supposedly account for the straightening and inflexing of the exostome when hydrated, thereby closing the gaps between endostome segments and effectively preventing spore release.

Extensive studies of the relationship between peristome structure and movement were conducted by Pais (1964). She examined representative mosses in both categories –hygrocastique (having a peristome that allows spore release in wet weather) and xerocastique (having a peristome that allows spore release under dry conditions). Pais found that valid observations of hygroscopic movement could be made using old herbarium specimens, as well as with fresh material. She established a correlation between structure of peristomes and the characteristic movements that these exhibit. Her observations corroborated those of Lazarenko (1957) and provided support for the hypothesis that the mechanisms of spore dispersal in operculate (stegocarpous) mosses depend on relative atmospheric humidity and air currents. Pais (1964, 1966) demonstrated that mosses with hygrocastique peristomes are common in tropical areas of Africa with high annual precipitation (i. e., humid atmospheric conditions), while mosses with xerocastique peristomes dominate arid regions.

Structure and Composition of Xerocastique Peristomes

Investigations into the development of the peristome began prior to the twentieth century and continued sporadically afterward (Goebel 1887, Strasburger 1902, Evans & Hooker 1913, Blomquist & Robertson 1941, Proskauer 1958). Mueller (1973) studied peristome development in *Fissidens limbatus*, including the structure and hygroscopic movements of its xerocastique peristome. The peristome of *F. limbatus* is single and haplolepidous. The inner lamellae of the teeth result from wall deposition to the exterior periclinal walls in cells of the IPL (Fig. 5), while the outer lamellae represent depositions to the interior periclinal walls of what he called the outer peristomial layer. The teeth exhibit heavy wall deposition on the outer lamellae in the basal one-

third of the teeth; there is much less wall deposition in the formation of the inner lamellae. The upper portions of each tooth attenuate into two delicate, slender filaments
that are uniformly thickened, but much less than the basal parts (Fig. 1). Thick trabeculae are found along the entire length of a tooth, although those of the inner lamella are much more prominent at tooth bases. The outer lamella of each tooth base was
revealed by both transmission and scanning electron micrographs as dense stacks of
transverse ridges and clefts (plates) between the trabeculae (Fig. 6). These ridges
were determined to consist of a matrix of hemicellulose and carbohydrates. Mueller
concluded that the inflexing of the basal region was due to hydration and resulting
swelling of the hemicellulosic matrix of the outer lamellae.

The bases of teeth in *Fissidens limbatus* are very sensitive to changes in humidity.
In contrast the filaments twist and untwist when wetted or dried but do not respond to
gaseous water as do the teeth bases. The orientation of the cellulosic microfibrils in
each lamella of the filaments was determined to be diagonal to the longitudinal axis of
the tooth and perpendicular to the plane of orientation in the adjacent lamella. Therefore, the mechanical stress created by these opposing forces cause these filaments to
twist upon drying. Water causes the teeth to inflex over the mouth of the urn, thereby
effectively preventing spore release (Mueller 1970, 1973). The filiform apices of the
teeth are forced into the mass of spores at the mouth of the capsule due to a bend in
each tooth above the heavily thickened base. Shrinkage of the basal region of the outer lamellae due to dehydration causes a rapid reflexing of the teeth, which leaves the
mouth of the urn open, thereby allowing spore release. This rapid movement jerks the
slender filaments out of the spore mass, dragging along spores caught on or between
the filaments. The rapid twisting of the filamants dislodges spores from the teeth.
Only the basal portions of the teeth are sensitive to changes in humidity, while whole
teeth are reactive to wetting and drying.

In a study of the ontogeny of plant cell walls, Schnepf et al. (1978) followed the
development of a moss peristome because the peristome is essentially composed of
secondarily thickened cell walls. They chose *Racopilum tomentosum* (Hedw.) Brid.,
a moss with a double peristome, and employed polarized light microscopy, histochemical techniques, and tranmission electron microscopy. Although their descriptions and terms are not in keeping with generally accepted terminology, the study produced a wealth of information on peristome structure and movement.

In *Racopilum tomentosum* the endostome exhibits no hygroscopic movement and
appears isotropic in polarized light microscopy (Schnepf et al. 1978). Only the inner
lamellae of the segments, deposited by IPL cells, are noticeably thickened and appeared to be weakly suberized (when compared to the outer lamellae of the segments).
The outer lamellae of the exostome are composed mainly of cellulose, deposited in
transverse plates between trabeculae. Each plate appears to contain small deposits of
pectinaceous material as a matrix for the cellulosic microfibrils in both early and late
wall deposition. The middle lamellae appear to become suberized by the time the teeth
reach maturity. The inner lamellae consist of wall thickenings of cellulosic microfibrils in a pectic matrix. The wall components are deposited in successive layers with

the microfibrils oriented diagonally from the transverse and radial anticlinal walls to the center of the periclinal wall. The surfaces of the lamellae become suberized during the latter stages of deposition.

The exostome of *Racopilum tomentosum* inflexes when moistened, closing the openings between endostome segments over the mouth of the urn. Schnepf et al. (1978) determined that this movement results from rapid hydration and swelling of the outer lamellae. The inner lamellae hydrate and swell at a much slower rate, resulting in a slight straightening of the teeth. Upon drying, the outer lamellae dehydrate and shrink rapidly, causing the exostome to reflex from the mouth of the urn. Again, the change in the inner lamellae is much slower, and as the inner lamellae dehydrate and shrink, the teeth flex inward, placing the teeth apices between the segments and effectively hindering spore release. Many of their observations on composition, structure, and movement corroborate an earlier study (Mueller 1970, 1973) that also utilized, light and electron microscopy, as well as histochemistry.

As part of a continuing study of the Splachnaceae, Koponen (1982) examined the structure of peristomes of this moss family. All splachnaceous peristomes are xerocastique but differ in structure and in the degree of hygroscopic response to atmospheric conditions. Koponen observed a positive correlation between the relative thickness of OPL versus PPL depositions and the degree of hygroscopic movement of different peristomes. For example in *Splachnum*, the peristome reflexes upon drying until the teeth are appressed to the outer wall of the urn. Koponen determined that wall deposits in the OPL are much thicker than in the PPL and IPL and declared, without citing supporting histochemical analyses, that the outer lamella is composed of non-suberized cellulose.

Studies by Mueller (1973) and Schnepf et al. (1978) focused on two different types of peristomes. Mueller worked with a haplolepidous, single peristome, whereas Schnepf et al. studied a diplolepidous peristome. The hygroscopic movements in each of these peristomes, however, were shown to be similar; thus both are categorized as xerocastique (i. e., the wet teeth arch over the mouth of the urn). The basic structure of the peristomes in these mosses was also shown to be similar, substantiating the many correlations between structure and hygroscopic movement.

There is a distinct intercellular layer between cells of adjacent peristomial layers that comprise each peristome tooth. Mueller (1970, 1973) called this the central lamella (Fig. 7, 8). It consists of the true middle lamella and primary walls of adjacent cells.

Figures 5–8. Micrographs of sectioned peristomes. **5**, *Fissidens limbatus* Sull., transverse section near base of peristome teeth, showing the groups of five cells involved in formation of each pair of teeth; **6**, *F. limbatus*, longitudinal section of peristome tooth near base, showing large trabeculae of inner lamina and transverse ridges and clefts between trabeculae on the outer lamina; **7**, *Leucodon julaceus* Sull., transmission electron micrograph of longitudinal section, showing upper part of exostome tooth with central lamina and wall thickenings, electron dense coating is forming over tooth; **8**, *L. julaceus*, transmission electron micrograph of central lamina and wall thickenings in transverse view.

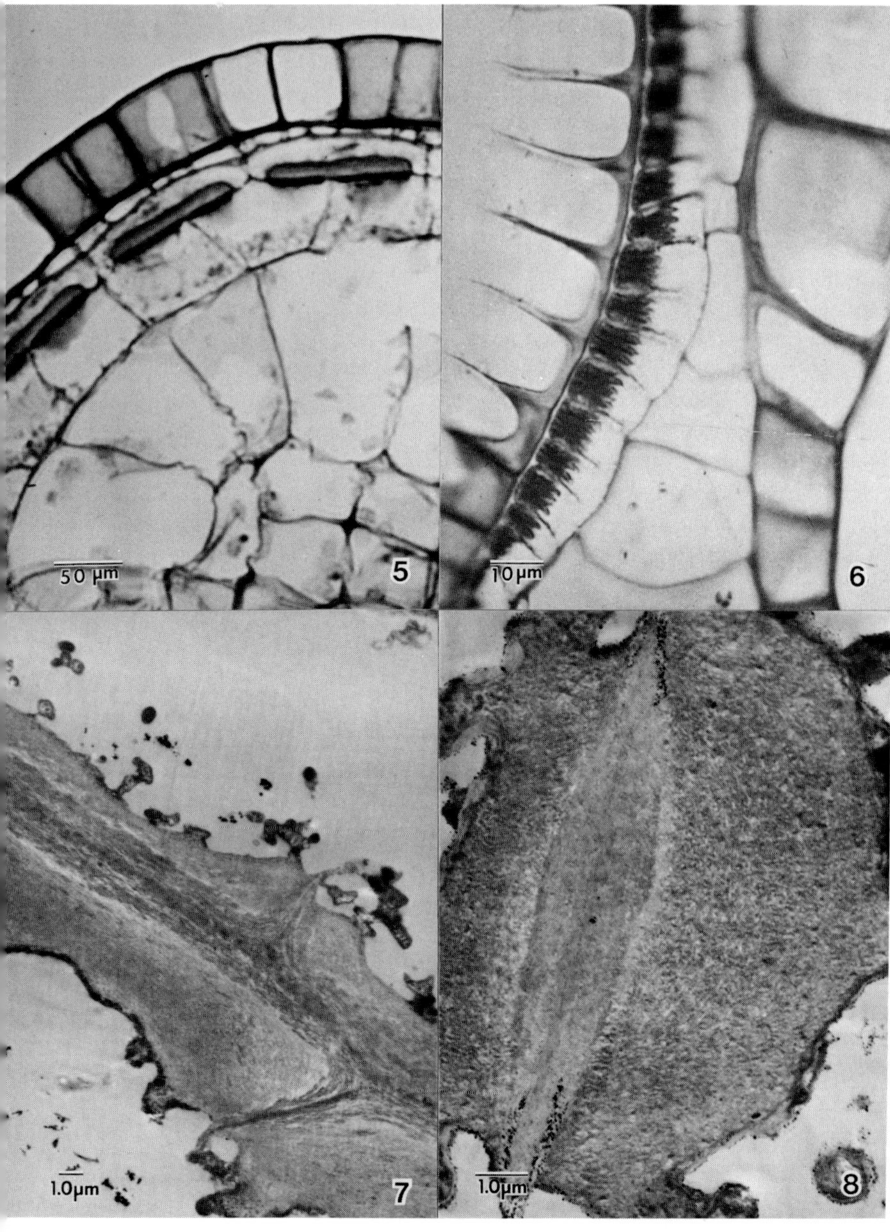

The undissolved cell walls that ultimately form each peristome tooth are, therefore, composed of thickenings formed by adjacent files of cells, between which is a common central lamella. The central lamella appears to be amorphous in structure and is composed of pectic compounds, as indicated by staining and by polarization microscopy. Extractions of pectins from the peristome with hot ammonium oxalate cause a noticeable reduction in stainability, which gives supporting evidence for the presence of pectin in the middle lamella. The central lamella gives positive stain reactions with nile blue, sudan black, Schiff's reagent, and the periodic acid-Schiff's (PAS) reagent (Mueller 1973). This indicates that lipoidal compounds are present. Although the PAS test is generally regarded as a test for carbohydrates, especially polysaccharides, Pearse (1968) noted that phospholipids also gave positive results with PAS. Derschau (1900) reported the presence of cutin in the middle lamellae of peristomes, while Mueller only indicated the possible presence of cutin in the central lamellae of the peristome of *Fissidens limbatus*. Schnepf et al. (1978) reported the presence of suberin in the middle lamellae of the exostome of *Racopilum tomentosum*. Positive staining by Schiff's reagent (Mueller 1973) indicates the presence of natural aldehyde groups, which may be indicative of either suberin or cutin (Kolattukudy et al. 1981). It is likely that the central lamella allows movement of teeth due to the uneven expansion and shrinkage of the outer and inner lamellae.

The middle lamellae are flanked by the cell walls of the contiguous cells, which become much thickened in localized areas. These walls appear strongly birefringent when viewed with polarization microscopy (Mueller 1973, Schnepf et al. 1978). This is a positive indication of the presence of cellulose, which is supported by the absence of positive staining with PAS (Feder & O'Brien 1968). The lack of staining by iodine-sulfuric acid and zinc-chlor-iodide (Mueller 1973) cannot be interpreted as evidence for or against the presence of cellulose because the reactions may be obscured by other substances in the matrix (Roelofsen 1959, Jensen 1962). The thickened walls are considered to be predominantly composed of cellulosic microfibrils with small amounts of pectic compounds as a matrix. From the results of various tests it is obvious that there is a predominance of densely packed cellulosic microfibrils in these localized thickenings and that the microfibrils are primarily oriented parallel to the longitudinal plane of the wall in at least parts of the teeth. Schiff's reagent indicates the presence of natural aldehydes but does not indicate the specific group of compounds responsible for the reaction to this stain.

The basal regions of the teeth are thicker than the apical regions, and it is the basal regions that exhibit the greatest degree of hygroscopic movement. At the base of the teeth the inner lamellae are generally homogeneous and anisotropic. The principal direction of cellulosic microfibril orientation is parallel to the longitudinal axes of the teeth. It was noted by Mueller (1973) and Schnepf et al. (1978) that the orientation pattern of microfibrils in the trabeculae was perpendicular to the longitudinal axis of the teeth. In both peristomes the outer lamellae in the basal region have horizontal ridges. Mueller found these plates in *Fissidens* to be isotropic and composed of cellulose in a matrix of hemicellulose. Schnepf et al. determined that the outer lamellae of the peristome teeth of *Racopilum* were composed of cellulose and were anisotropic. The outer lamellae were found to be hydrophilic in both studies. The outer lamellae swell upon hydration or shrink with dehydration, whereas the inner lamellae change very little, if at all, in response to hydration or dehydration. Therefore, the expansion of the outer lamellae upon hydration causes it to override the inner lamellae, producing an inward curvature that bends the teeth over the mouth of the urn. The reflexive movement observed during dehydration is caused by the shrinking of the outer lamellae.

Structure and Composition of Hygrocastique Peristomes

Neumann (1983) studied mosses in the Cryphaeaceae and Leucodontaceae with exostomes and reduced endostomes of the hygrocastique type. The OPL consists of a cylinder of 32 cells, hence the outer lamella of each tooth is formed by two adjacent files of cells (diplolepidous). Portions of transverse anticlinal walls remain as trabeculae and portions of the vertical anticlinal walls remain as the vertical median ridges. Wall thickenings of the OPL generally do not exceed four micrometers. The PPL consists of a conical cylinder of 16 cells (Fig. 9) immediately interior to the OPL, hence the inner lamella of each tooth is formed from the wall thickenings from individual PPL cells. Wall thickenings along the inner lamellae may exceed 18 μm in the basal region (Fig. 10) but are only about four micrometers above middle of the tooth. These observations are similar to those of Pais (1964).

It can be deduced from polarized light microscopy that the cellulosic microfibrils of the inner lamellae are parallel to the periclinal walls of the PPL at the mid-point of the wall. Microfibrils radiate outward in broad arcs to become parallel to the transverse anticlinal walls in these mosses with large trabeculae (Neumann 1983). Thus microfibrils parallel to the vertical plane of the periclinal wall at mid-cell arc at a 90 degree angle to become parallel to the horizontal plane (Fig. 11) of the transverse anticlinal walls forming the trabeculae. Observations of transverse sections of the basal region of the exostome also indicate that the last microfibrils to be deposited in the periclinal wall thickenings are parallel to the horizontal plane of the trabeculae instead of the periclinal wall. It has been observed with the aid of transmission electron microscopy that a pectin-rich substance appears to coat the surfaces (Fig. 7, 8) of the wall thickenings (Neumann 1983). The pectinaceous composition of this substance was indicated from ruthenium and iron stains. Both Pais (1964) and Schnepf et al. (1978) reported similar deposits. The composition and the structure of the cell walls of both the exostome and endostome (Fig. 12) are identical. The degree of wall thickening, ornamentation, and autolysis differs somewhat among lamellae of the same structure, regions of the same structure, and different structures.

Observations of the hygroscopic movements exhibited by the exostomes of the mosses studied by Neumann coincide with those reported by Patterson (1953) and Pais (1964). There is a definite correlation between the observed movements and the structure of the teeth. Neumann found in peristomes of the Cryphaeaceae and Leucodontaceae that the wall remnants of the exostome were primarily composed of densely packed cellulosic microfibrils with a definite pattern of orientation. The majority of microfibrils were in linear arrays parallel to the longitudinal axes of the periclinal walls in both the OPL and the PPL. The microfibrils tended to bend 90 degrees in the wall angles to become parallel to the planes of the transverse anticlinal walls. The walls appear to contain only small quantities of matrix materials. Histohemical analyses (Neumann 1983) indicated that the matrix materials contained very little, if any, pectin, although the middle lamella was rich in pectinaceous compounds. The matrix of the walls probably contained suberin- or cutin-like wax, as suggested by results of the Schiff's reagent, silver methenamine, and iron stains. The thickenings provide mechanical support and tensile strength in the exostome. The orientation of the microfibrils allows, rather than hinders, the bending movements exhibited by the teeth.

Wall thickenings of the inner and outer lamellae of the exostome are composed of a loose framework of cellulosic microfibrils embedded in a dense matrix rich in polysaccharides (Neumann 1983). Histochemical analyses also indicated the presence of pectin and lipids but did not indicate whether these are free compounds or complexes bound to the sugar moieties as pectic polysaccharides or lipoidal side-chains. The matrix polysaccharides are hydrophilic, and imbibition of water molecules causes the volume of the wall thickenings to increase. Conversely, loss of water through evaporation causes the wall thickenings to decrease in volume. Both reactions were extremely rapid. Density of microfibrils within this matrix decreased with distance from the primary wall, and the orientation complemented that of the primary wall. The microfibrils provide mechanical support and strength to the wall thickenings and the teeth in general.

Hygrocastique peristomes such as in *Leucodon* serve as a regulatory gate that effectively prevents spore release during periods of low moisture availability. Patterson (1953) and Pais (1966) both noted that the majority of mosses in the hygrocastique category are corticolous. Indeed, the mosses selected for Neumann's study (1983) are all corticolous and are commonly found on the bark of hardwood trees in mesic woods. The gametophytes of these mosses are adapted for survival in mesic forests, necessitating a spore dispersal mechanism that supports spore dissemination at appropriate times in these habitats.

Figures 9–12. Micrographs of sectioned peristomes. **9,** *Forrstroemia trichomitria* (Hedw.) Lindb., light micrograph of diplolepidous exostome in transverse view just above level of annulus; **10,** *F. trichomitria*, light micrograph of basal portion of exostome in longitudinal section, clearly showing trilaminate structure of tooth; **11,** *Leucodon julaceus* Sull., transmission electron micrograph of exostome (inner lamella) in longitudinal section at junction of trabeculum and central lamina, electron dense striations indicate orientation of iron "stained" microfibrils; **12,** *L. julaceus*, light micrograph of trilaminate exostome and reduced endostome in longitudinal section, the latter clearly layered.

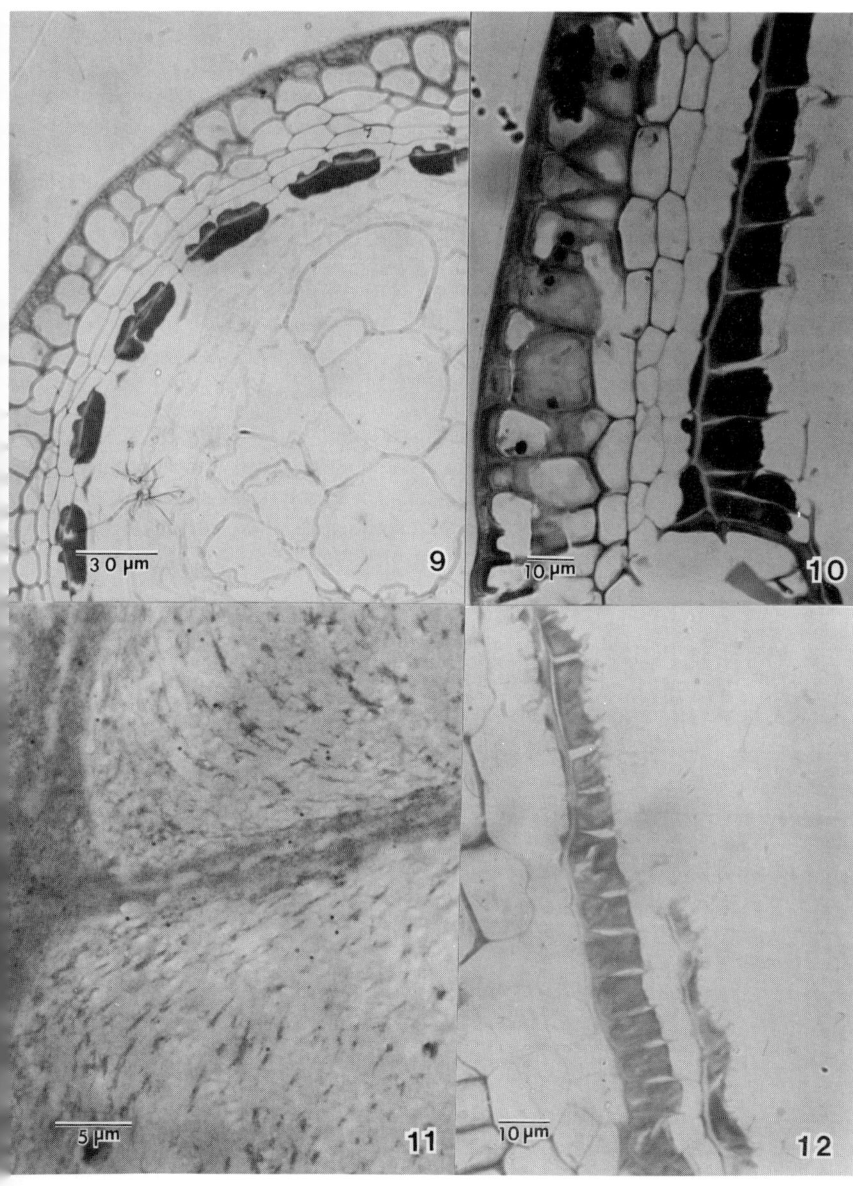

The exostome teeth of hygrocastique mosses serve as a moveable gate to regulate spore release. Only during periods of high moisture (e. g., fog, dew, rain) are the teeth reflexed outward so that the barrier to spore release is effectively eliminated. Spores may be washed out of the urn by rain. Spores probably do not drift very far and may land on the bark of the same tree on which the parent plant is located or on a neighboring tree branch or trunk. Most likely, the majority of spores are washed out of the urn and travel along the water run-off patterns of the tree bark where they have the greatest probability of lodging in a place suitable for germination and growth of a new gametophyte.

Conclusions

There has been a trend in bryological investigations over the past decade toward renewed emphasis on peristome structure and function. Most bryologists apparently accept the concept that the peristome is a conservative morphological characteristic in class Bryopsida and utilize peristome characteristics in studies of phylogeny. Peristome structure is recognized as an important taxonomic criterion above the rank of family.

Edwards (1979) concluded that "peristomes have largely been ignored since Philibert, and those studies which have been made have generally been concerned with precise investigations of one species." We feel that this situation is rapidly changing. There are, however, several pitfalls that investigators must be wary of as they interpret and communicate their observations. There is need for standardization of terminology. As a result of inconsistencies in the application of terms, the literature is sometimes confusing and, worse still, often subject to mis-interpretation. Secondly, there is a tendency to make inferences based on too few observations and to apply what is known for a few species too widely. Data from one type of peristome may or may not be applicable to the peristomes of other taxa. Concepts of phylogeny, as well as ecology, have been obfuscated by generalizations based on too few studies of peristome ontogeny and structure.

Comparisons of data from developmental studies provide a basis for re-evaluating contemporary concepts of peristome structure, as well as for making correlations between structure and function. Although the peristomes of *Fissidens*, *Racopilum*, and *Leucodon* may belong in different structural and functional categories, there are similarities that deserve reiteration. In each case the teeth are trilaminate structures largely made up of periclinal wall depositions contributed by two adjacent layers of cells. The initial wall thickenings of cellulosic microfibrils provide the tensile strength for mechanical support and, eventually, the movement of the tooth. For the most part, these microfibrils are aligned parallel to the long axis of the teeth. Additional wall materials are deposited that consist of fewer cellulosic microfibrils and larger amounts of matrix materials of what appear, in some cases, to be hemicellulose, pectins, and suberin. The key to hygroscopic movement appears to be whether or not, and to what degree, these materials are hydrophilic or hydrophobic. One lamella consists of matrix materials that are more hydrophilic or hydrophobic than materials comprising the other, or there may be more of this material within one lamella than the other. As the wall matrix material hydrates it swells. The increase in volume places mechanical stress against the trabeculae or adjacent lamella so that when lamellae have differences in deposition or hydration the mechanical stress is relieved as the main axis of the tooth bends in response to the differential swelling of the lamellae. In *Leucodon* the additional wall depositions are part of the inner lamellae (PPL side) and cause the teeth to reflex when hydrated and inflex over the mouth of the urn as the inner lamellae dehydrate – the definitive reaction of hygrocastique peristomes. The converse is true in *Fissidens* and *Racopilum*, which represent xerocastique peristomes. More studies of peristomes are necessary to support or modify these conclusions.

There is critical need for additional detailed investigations into developmental patterns and structure of peristomes, including moss capsules that are presently classified as gymnostomous. Edwards (1979) stated that studies of selected species "may hide valuable information that can be revealed by cruder, wider surveys." His conviction was reaffirmed in 1985 when he asserted that "sections. . .do provide valuable insight" but that "simply viewing mature peristome patterns face-on. . .gives. . .a far better assessment." We feel instead that it is the paucity of thorough studies of individual taxa that has been detrimental. As Anderson and Palmer (1982) noted, the type of detailed study of peristomes that is required at this time involves the more traditional techniques of microtomy and histochemistry for both light and electron microscopy. Observations of whole or dissected peristomes under bright-field or scanning electron microscopy alone are inadequate and may be tenuous at best, unless fortified with studies of ontogeny and histochemistry at both the light and ultrastructural levels. "Before they can be securely incorporated into bryophyte systematics all features seen under the SEM also require rigorous developmental studies using TEM" (Duckett 1986). We concur! In addition, ecological studies are necessary to confirm or modify the notion that spore release in mosses with hygrocastique peristomes is correlated with the dispersal mechanism of washing spores from the urn and along drainage pathways where the spores might lodge and germinate.

Literature Cited

Allen, B. H. 1981. A reevalution of the Sorapillaceae. The Bryologist 84: 335–338.

Allen, B. H., Magill, R. E. & Crosby, M. R. 1985. Observations on the peristome and systematic position of *Tetrastichium fontanum* (Musci). Journal of Bryology 13: 515–522.

Anderson, L. E. & Palmer, P. G. 1982. The peristome of *Anacamptodon splachnoides*. The Bryologist 85: 193–203.

Ascherson, P. 1892. Hygrochasie und zwei neue Fälle dieser Erscheinung. Berichte der Deutschen Botanischen Gesellschaft (Berlin) 10: 94–114.

Blomquist, H. L. & Robertson, L. L. 1941. The development of the peristome in *Aulacomnium heterostichum*. Bulletin of the Torrey Botanical Club 68: 569–584.

Crum, H. A. & Anderson, L. E. 1981. The Mosses of Eastern North America. 2 vols. New York. Columbia University Press.

Czapek, F. 1899. Zur Chemie der Zellmembranen bei Laub- und Lebermoosen. Flora 86: 361–381.

Derschau, M. von 1900. Die Entwicklung der Peristomzähne des Laubmoossporogoniums. Botanisches Centralblatt 82: 161–168, 193–200.

Duckett, J. G. 1986. Ultrastructure in bryophyte systematics and evolution: an evaluation. Journal of Bryology 14: 25–42.

Edwards, S. R. 1979. Taxonomic implications of cell patterns in haplolepideous moss peristomes, pp. 317–346. In: Clarke, G. C. S. & Duckett, J. G. (Eds.), Bryophyte Systematics. London. Academic Press.

— 1984. Homologies and inter-relationships of moss peristomes, pp. 658–695. In: Schuster, R. (Ed.), New Manual of Bryology. Vol. 2. Nichinan, Japan. Hattori Botanical Laboratory.

Evans, A. W. & Hooker, H. D., Jr. 1913. Development of the peristome in *Ceratodon purpureus*. Bulletin of the Torrey Botanical Club 40: 97–109.

Feder, N. & O'Brien, T. P. 1968. Plant microtechnique: Some principles and new methods. American Journal of Botany 55: 123–142.

Goebel, K. 1887. Outlines of Classification and Special Morphology of Plants. (English Translation). Oxford. Clarendon Press.

— 1895. Archegoniaten Studien. VII. Über die Sporenausstreuung bei den Laubmoosen. Flora 80: 459–486.

— 1930. Organographie der Pflanzen. 3: 643–1378. Jena.

Hedwig, J. 1782. Fundamentum Historiae Naturalis Muscorum Frondosorum. Leipzig.

— 1787. Descriptio et Adumbratio Microscopico-Analytica Muscorum Frondosorum. Teil I. Leipzig.

Hutton, F. W. 1874. Observations on the different modifications in the capsules of mosses, with reference to the dispersion of their spores. Transactions of the New Zealand Institute, Botany, 3: 342–347.

Ingold, C. T. 1959. Peristome teeth and spore discharge in mosses. Transactions of the Botanical Society of Edinburgh 38: 76–88.

— 1965. Spore Liberation. Oxford. Clarendon Press.

Jensen, W. A. 1962. Botanical Histochemistry. San Francisco. Freeman.

Kienitz-Gerloff, F. 1878. Untersuchungen über die Entwicklungsgeschichte der Laubmooskapsel und die Embryoentwicklung einiger Polypodiaceen. Botanische Zeitung (Berlin) 36: 33–64.

Kolattukudy, P. E., Espelie, K. E. & Soliday, C. L. 1981. Hydrophobic layers attached to cell walls. Cutin, suberin and associated waxes. Encyclopedia of Plant Physiology 13B: 225–254.

Koponen, A. 1982. On the structure and function of the peristome in Splachnaceae. Journal of the Hattori Botanical Laboratory 53: 73–98.

Kreulen, D. J. W. 1972. Features of single- and double-peristomate capsules. Homology of layers and ontogeny of outer spore sac. Lindbergia 1: 153–160.

Lantzius-Beninga, S. 1847. Beiträge zur Kenntnis des innern Baues der ausgewachsenen Laubmooskapsel, insbesondere des Peristoms. Nova Acta Academiae Caesarae Leopodino-Carolinae Germanicae Naturae Curiosorum. 22: 559–604.

Lazarenko, A. S. 1957. One some cases of singular behavior of the moss peristome. The Bryologist 60: 14–17.

Limpricht, K. G. 1890. Die Laubmoose Deutschlands, Oesterreichs und der Schweiz. In: Rabenhorst, L., Kryptogamen Flora. 3 vols. Leipzig.

Lorch, W. 1931. Anatomie der Laubmoose. In: Linsbauer, K. (Ed.), Handbuch der Pflanzenanatomie 7. 358 pp. Berlin.

Mueller, D. M. J. 1970. Structure and development of the moss *Fissidens limbatus* Sullivant. Ph. D. Dissertation. University of California. Berkeley.

— 1973. The peristome of *Fissidens limbatus* Sullivant. University of California Publications in Botany 63: 1–34.

— 1974. Spore wall formation and chloroplast development during sporogenesis in the moss *Fissidens limbatus*. American Journal of Botany 61: 525–534.

Neumann, A. J. 1983. Peristome structure and movement of selected species of the Leucodontaceae and Cryphaeaceae (Musci). Ph. D. Dissertation. Texas A&M University. College Station.

Pais, M. S. 1964. Sur le mouvement du péristome chez quelques espèces de mousses et son rapport écologique. Portugaliae Acta Biologica, Seriae A, 8: 301–317.

— 1966. Les types de mouvement du péristome des mousses et les conditions climatiques. Revista Biologia (Lisboa) 5: 239–250.

Patterson, P. M. 1953. The aberrant behavior of the peristome teeth of certain mosses. The Bryologist 56: 157–159.

Pearse, A. G. E. 1968. Histochemistry. Theoretical and Applied. Edinburgh. Churchill Livingstone.

Pfaehler, A. 1904. Étude biologique et morphologigue sur la dissemination des spores chez les mousses. Bulletin de la Société Vaudoise des Sciences Naturelles 40: 41–132.

Proskauer, J. 1958. On the peristome of *Funaria hygrometrica*. American Journal of Botany 45: 560–563.

Richardson, D. H. S. 1981. The Biology of Mosses. New York. Halsted.

Robinson, H. 1971. Scanning electron microscope studies on moss leaves and peristomes. The Bryologist 74: 473–483.

Roelofson, P. A. 1959. The plant cell wall. In: K. Lindsbauer (Ed.), The Encyclopedia of Plant Anatomy. Berlin. Gebrüder Borntraeger.

Saito, S. 1956. Studies on the development of the peristome in Musci. II. On the peristome in *Dicranum japonicum* Mitt. Botanical Magazine (Tokyo) 69: 53–58.

Saito, S. & Shimoze, S. 1955. Studies on the development of the peristome in Musci. I. On the peristome in *Bartramia crispata* Schimp. Botanical Magazine (Tokyo) 68: 55–60.

Schimper, W. P. 1848. Recherches anatomiques et morphologiques sur les mousses. 92 pp. Strasbourg.

Schnepf, E., Stein, U. & Deichgraber, G. 1978. Structure, function and development of the peristome of the moss, *Rhacopilum tomentosum*, with special reference to the problem of microfibril orientation by microtubules. Protoplasma 97: 221–240.

Schofield, W. B. 1985. Introduction to Bryology. New York. MacMillan.

Schulz, D. & Schmidt, W. 1974. Entwicklung der Peristoms von *Funaria hygrometrica*. Flora 163: 451–465.

Seabury, F. 1975. Sporogenesis in selected genera of the Musci. Ph. D. Dissertation. Texas A&M University. College Station.

Shaw, J. 1985. Peristome structure in the Mitteniales (ord. nov.: Musci), a neglected novelty. Systematic Botany 10: 224–233.

— 1986. Peristome structure in the Orthotrichaceae. Journal of the Hattori Botanical Laboratory 60: 119–136.

Shaw, J. & Allen, B. H. 1985. Anatomy and morphology of the peristome in *Discelium nudum* (Musci: Disceliaceae). The Bryologist 88: 263–267.

Shaw, J. & Rohrer, J. R. 1984. Endostomial architecture in diplolepidous mosses. Journal of the Hattori Botanical Laboratory 57: 41–61.

Smith, G. M. 1938. Cryptogamic Botany. Vol. 2, Bryophytes and Pteridophytes. New York. McGraw-Hill.

Steinbrinck, C. 1897. Der hygroskopishe mechanismus der Laubmoosperistoms. Flora 84: 131–158.

Stone, I. G. 1961. The gametophyte and sporophyte of *Mittenia plumula* (Mitt.) Lindb. Australian Journal of Botany 9: 124–151.

Strasburger, E. 1902. Das botanische Practicum. 4th Ed. Jena.

Taylor, E. C. 1959. Peristome teeth in polarized light. The Bryologist 62: 149–155.

— 1962. The Philibert peristome articles, an abridged translation. The Bryologist 65: 175–212.

Vitt, D. H. 1981. Adaptive modes of the moss sporophyte. The Bryologist 84: 166–186.

Wijk, R. van der 1951. Vergelikende ontogenie van het peristoom der Bryales. Buxbaumia 5: 10–17.

Advances in Bryology 3: 159 – 223 (1988).

Sporogenesis in Bryophytes

by

Roy C. Brown and Betty E. Lemmon

Department of Biology, University of Southwestern Louisiana
Lafayette, LA 70504-2451, U.S.A.

With 70 Figures

Abstract: Sporogenesis in bryophytes is a complex developmental process characterized by precise nuclear and cytoplasmic divisions resulting in free-living spores with distinctive walls. The following structural evidence from various bryophytes indicates that the genetically controlled cytomorphogenesis is programmed in the prophase I sporocyte: cytoplasmic infurrowing in the future division planes, plastid division and migration, development of patterned wall precursors in meiotic prophase, and microtubule systems developed during wall ontogeny that reflect polarity established in meiotic prophase. The ontogeny and configuration of the metaphase I spindle in mosses, and possibly other bryophytes, is unusual and relates to polarity established in prophase I. A tetrahedral system of microtubules encages the nucleus in prophase I; this system is transformed into a functionally bipolar spindle that retains evidence of quadripolarity throughout metaphase I. Two principal layers occur in the spore walls of bryophytes, the inner intine (endospore) and outer exine (exospore). An additional layer, the perine, is deposited outside the exine of the spore walls of mosses and perhaps of hornworts. The spore wall is typically initiated by deposition of sporopollenin on tripartite lamellae at the plasma membrane. Notable exceptions are the Anthocerotae and Andreaeopsida. Organized systems of microtubules may be associated with initiation of exine, intine, and the aperture. Ultrastructural studies of sporogenesis have revealed several characters important to the study of bryophyte phylogeny. Three distinct types of walls correlate with the major taxa of Musci: the Andreaeopsida, Sphagnopsida, and Bryopsida. Spore wall ontogeny in Marchantiidae is distinct from Jungermanniidae. Several patterns of spore wall development occur in Metzgeriales. Anthocerotae share features with Musci and Hepaticae.

Zusammenfassung: Sporogenese bei Bryophyten ist ein komplexer Entwicklungsprozeß mit präzisen Kern- und Cytoplasmateilungen, aus denen freilebende Sporen mit Zellwänden entstehen. Folgende strukturelle Gegebenheiten bei verschiedenen Bryophyten machen deutlich, daß die genetisch kontrollierte Cytomorphogenese im Prophase-I-Sporocyten programmiert wird: Cytoplasmaeinfurchung in der späteren Teilungsebene, Plastidenteilung und -wanderung, Entstehung der Vorläufer der Wandstruktur in der Meioseprophase und ein bei der Wandontogenie entstandenes Mikrotubulisystem, das die Polarität der meiotischen Prophase aufzeigt. Die Ontogenie und Konfiguration der Metaphase-I-Spindel bei Laubmoosen, und möglicherweise auch anderen Moosen, ist ungewöhnlich und steht in Zusammenhang mit der in Prophase I entstandenen Polarität. Ein tetraedrisches Mikrotubulisystem umgibt den Kern in Prophase I. Dieses System wird in eine funktionell bipolare Spindel umgewandelt, deren Quadripolarität während der Metaphase I deutlich bleibt. Die Sporenwand der Bryophyten besteht aus zwei Hauptschichten, der inneren Intine (Endospor) und der äußeren Exine (Exospor). Eine zusätzliche Schicht, die Perine, wird auf der Außenseite der Exine der Laubmoose und vielleicht auch der Hornmoose abgelagert. Die Sporenwandbildung wird im allgemeinen mit der Ablagerung von Sporopollenin auf den dreiteiligen Lamellen der Plasmamembran eingeleitet. Bemerkenswerte Ausnahmen sind die Anthocerotae und die Andreaeopsida. Die Organisation der Mikrotubulisysteme steht vermutlich in Zusammenhang mit der Bildung von Exine, Intine und Öffnung verbunden. Ultrastrukturuntersuchungen der Sporogenese zeigten, daß verschiedene Merkmale für phylogenetische Interpretationen wichtig sind. Drei deutlich unterscheidbare Wandtypen entsprechen den drei Klassen der Musci, den Andreaeopsida, Sphagnopsida und Bryopsida. Die Sporenwandontogenie bei den Marchantiidae ist verschieden von der der Jungermanniidae. Mehrere Möglichkeiten der Sporenwandentwicklung finden sich bei den Metzgeriales. Die Anthocerotae zeigen Merkmale sowohl der Musci wie der Hepaticae.

Keywords: aperture, bryophytes, division site, exine, intine, meiosis, microtubules, MTOC, perine, plastid migration, polarity, primexine, spindle, sporogenesis, spore wall.

Contents

Introduction

Sporogenesis in bryophytes is a complex cytomorphogenetic process by which sporocytes are transformed into free-living spores. It is characterized by precise nuclear and cytoplasmic divisions resulting in tetrads of haploid spores and includes the development of distinctive, resistant walls around the spores. Sporogenesis is a reproductive event of fundamental importance in the life cycle of cryptogams. Spores are the dispersal units of sexual reproduction, and the evolutionary impact of genetic recombination depends upon the success of spores to establish gametophytic growth. Spores are covered by a specialized cell wall known as the sporoderm, the most important component of which is sporopollenin. The function of the sporoderm is protection of cellular contents during spore dispersal in terrestrial plants. Not only do spores provide a means of dispersal, but more importantly, they may increase gene flow potential by introducing gamete-producing plants in the vicinity of those with which they can hybridize. Gene flow via the male gamete is extremely limited, fertilization being accomplished only within a few centimeters of the antheridia (Wyatt 1977).

Bryophytes comprise pivotal groups for investigations into the origin of land plants, and comparative morphological studies of sporogenesis have provided a wealth of new characters. While the taxonomic utility of surface ornamentation of spores and pollen has long been appreciated, it is is now recognized that a large number of developmental features of sporogenesis are of phylogenetic significance. These additional characters of sporogenesis are particularly useful in determining homologies and will be valuable in the effort to understand relationships of the three major taxa of bryophytes and the relationships of the bryophytes to various groups of vascular plants and algae. The classification of major taxa of bryophytes used in this paper follows that of Crum and Anderson (1981) who recognized the mosses, liverworts, and hornworts as subdivisions of Division Bryophyta. Classification and nomenclature of mosses follows that of Crum and Anderson (1981). Classification of hepatics follows that of Schuster (1966). Names and authorities of hepatics follows usage of Stotler and Crandall-Stotler (1977).

Knowledge of bryophyte sporogenesis begins with the studies of Nageli (1844) and von Mohl (1839), which are followed by a series of wonderfully detailed descriptive studies conducted around the turn of the century by Farmer (1894, 1895) Davis (1899, 1901), Moore (1905), and Allen (1916). Frequent reference will be made to the observations of these early workers because modern ultrastructural studies address many of the same fundamental questions framed by cytologists using the light microscope. Since the 1930s when the squash technique for staining chromosomes with aceto-orcein and aceto-carmine replaced paraffin sectioning and differential staining of cells for the study of cytology, knowledge of chromosome structure and chromosomal behavior has increased enormously, but information about the cytoplasmic events of sporogenesis has remained nearly static. The advent of the transmission electron microscope (TEM) has allowed more detailed investigations of the carefully integrated series of cytological systems in sporogenesis that were recognized by the early cytologists. We are challenged to derive as much benefit from the TEM as the keen observers and astute thinkers at the turn of the century did from the light microscope.

The extremely complex cytomorphogenetic process of bryophyte sporogenesis comprises many separate developmental sequences that are played out in a compressed time frame. Comparative studies of ontogeny have allowed us to recognize component systems of sporogenesis that, for the purpose of study and discussion, are considered distinct one from the other. The major stages are the two successive nuclear divisions of meiosis, cytoplasmic cleavage into a tetrad of spores, development of a spore wall around each spore, and maturation of the spore into an independent dispersal unit. Ultrastructural studies of sporogenesis have provided insight into characters of fundamental importance in the life cycles of bryophytes.

It is becoming increasingly apparent that the sporocyte, once isolated from its vegetative neighbors, is genetically programmed for a carefully orchestrated developmental pathway that results in independent dispersal units with distinctively ornamented spore walls. The entire process seems to be programmed in the early sporocyte. This conclusion is based on the following phenomena that occur in various combinations during sporogenesis of the different taxa of bryophytes.

1. Precocious marking during meiotic prophase of the eventual cleavage planes of the spore tetrad, either by lobing of the cytoplasm or discrete infurrowing of the sporocyte wall.

2. Two divisions of the single plastid (in mosses and hornworts) and migration during meiotic prophase of the resultant four plastids into the cytoplasmic domains that will eventually be cleaved into spores.

3. Development of structural wall precursors during meiotic prophase that predict the pattern of spore exine.

4. Development of microtubule systems associated with initiation of the spore wall and the development of an aperture, both relating directly to polarity established during meiotic prophase.

While no one bryophyte exhibits all four of the phenomena listed above, it is clear that bryophyte sporogenesis, when viewed as a whole, provides clear evidence that polarity for division and subsequent development in young spores is established in meiotic prophase.

The question posed by Bradley Moore Davis in 1899 (p. 95), "Can it be supposed that cytoplasm would be intrusted with so important a task as the preparation of a chloroplast for each of the four nuclei that are later to preside over the spores before there is any indication that such nuclear division is to take place?" touches on issues central to modern cell biology – cell polarity, intracellular movement, and the regulation of cytomorphogenetic events. Since the discovery of microtubules in plant cells by Ledbetter and Porter (1963), techniques of TEM, and more recently techniques of indirect immunofluorescence microscopy, have been used to study the different systems of microtubules that develop during the cell cycle. Of outstanding importance to an understanding of cell division has been the discovery by Pickett-Heaps and Northcote (1966) of the preprophase band of microtubules (PPB) that predicts the future plane of division before the nucleus enters prophase. In the parenchymatous tissue of higher plants, the PPB girdles a cell in the plane that the new cell wall will occupy, and thereby provides structural evidence that the cytoplasm is entrusted with important roles and that cells are indeed programmed for future events. Subsequent studies have shown unequivocally that the PPB predicts the plane of division in all classes of division (formative, proliferative, symmetrical, and asymmetrical) in higher plants (Gunning et al. 1978, Eleftheriou 1985). The actual function of the PPB remains obscure in that the PPB is disassembled prior to actual formation of the phragmoplast. The highly ordered process of sporogenesis in bryophytes provides an alternate system for studying the determination of division polarity in plant cells. Comparative studies of sporogenesis are contributing to the basic understanding of plant cell division.

The goals of this review are to present recent advances in knowledge of the ultrastructure of bryophyte sporogenesis, to synthesize current information about the process, and to pose questions for further investigation. As the topic of sporogenesis is enormously complex, it is not our intention to rework material covered in previous reviews but to emphasize the newly discovered, rediscovered, and yet to be discovered. This treatment will reflect our interest in developmental aspects of sporogenesis. We have chosen the last comprehensive review of sporogenesis in bryophytes by Neidhart (1979) as a convenient starting point. Since Neidhart's review, certain aspects of sporogenesis have been reviewed in the *New Manual of Bryology* (Schuster 1983). The pertinent topics are: the cytology of Hepaticae and Anthocerotae (Newton 1983), the cytology of mosses (Ramsay 1983), and the spore (Mogensen 1983). Additionally, a recent evaluation of ultrastructure in bryophyte systematics by Duckett (1986) contains thoughtful coverage of sporogenesis and the emerging role of ultrastructure in the study of bryophyte phylogeny.

Cytoplasmic Lobing and the Division Site

Bryophyte sporogenesis provides superb examples of determination and precocious marking of the future division plane before the spindle is organized. As is the case in all other plant meiocytes (Buchen & Sievers 1981, Hogan 1985), no PPBs are present in the sporocytes of bryophytes. However, a different manifestation of cell polarity clearly indicates that the cytoplasmic division sites for both divisions are established early in meiotic prophase. The sporocyte cytoplasm of mosses, hornworts, and hepatics (except the Marchantiidae) becomes clearly lobed (Fig. 1–4) with cleavage furrows predicting the future planes of division that will occur following second nuclear division. In *Sphagnum* (Brown et al. 1982a) the division sites are precisely marked by delicate thickened infurrows of the sporocyte wall. Early wall deposition at the sites of the future partitioning septa is a manifestation of the establishment and functioning of the division site early in meiotic prophase. The mechanisms of pro-

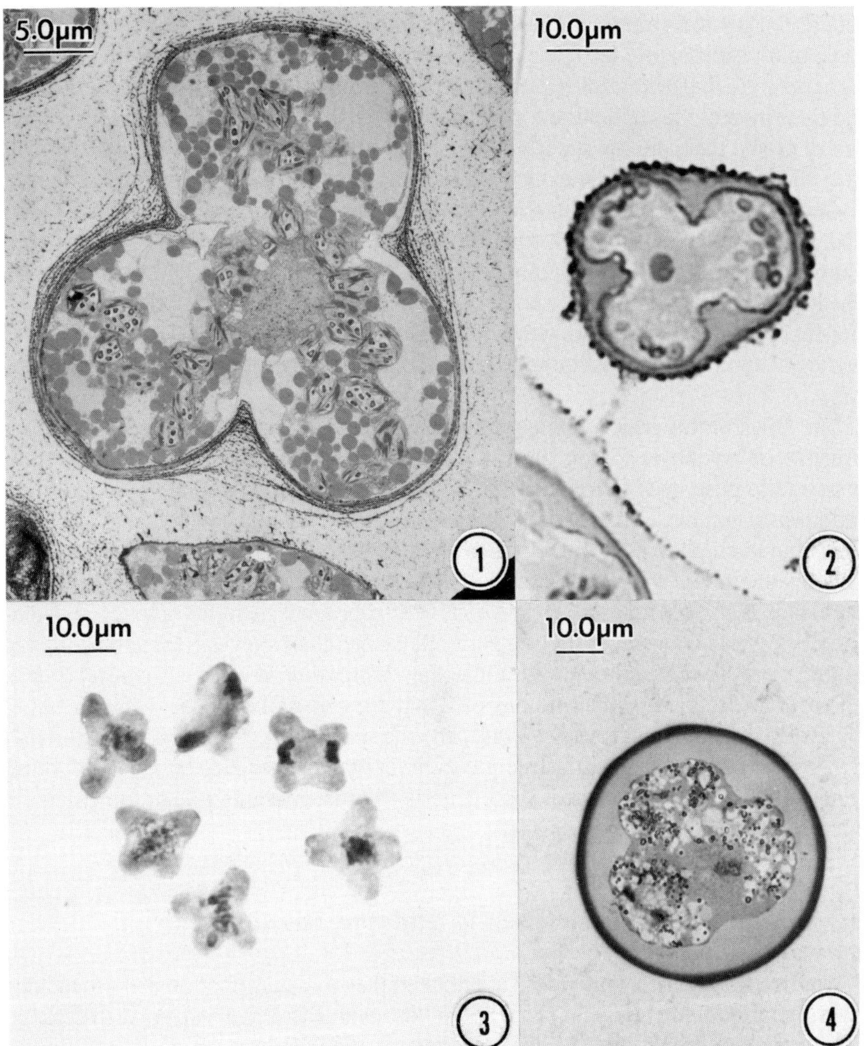

Figures 1–4. Cytoplasmic lobing in sporocytes of bryophytes. **1,** *Herbertus aduncus* (Dicks.) S. Gray (Hepaticae), TEM of prophase I sporocyte, the nucleus in central area is surrounded by numerous plastids, x1,900; **2,** *Notothylas orbicularis* (Schwein.) Sull. (Hepaticae), LM of plastic section showing prophase I sporocyte, the nucleus contains a nucleolus, starch grains mark positions of plastids in cytoplasmic lobes, x900; **3,** *Atrichum undulatum* (Hedw.) P.-Beauv. (Musci), sporocytes in prophase I to telophase I stained with aceto-orcein, lobing is achieved during prophase I, plastids in the cytoplasmic lobes are not visible at the low magnification of this light micrograph, x1,000; **4,** *Archidium tenerrimum* Mitt. (Musci), LM of plastic section showing sporocyte in prophase I, nucleus is still in acentric position, a plastid is located in each cytoplasmic lobe, x650.

nounced cytoplasmic lobing in other bryophyte sporocytes are unknown. It may be that an extreme form of activity at the division site results in continued wall ingrowth with consequent deep infurrowing of the cytoplasm. Microtubules have not been reported to be associated with the initiation of cytoplasmic lobing in any bryophyte (Brown & Lemmon 1982, 1987a). It is unlikely that the sporocyte is shaped by microtubules, although other contractile proteins of the cytoskeleton (e.g., actin) may be involved.

The division sites that are established in meiotic prophase mark the six planes along which cleavage of the cytoplasm into a tetrad of spores will occur following meiosis. See Figure 5 for a series of interpretive drawings that summarize polarity in moss sporogenesis and Figure 6 for a series of indirect immunofluorescent images that summarize changes in microtubule patterns that occur during moss meiosis. In a majority of bryophytes, the spores are tetrahedrally arranged within the sporocyte wall, exceptions being isobilateral tetrads in the hepatics *Conocephalum conicum* (L.) Lindb. (Meyer 1929) and *Riccia perssonii* Khan (Parihar 1965). Each young tetrahedral spore immediately after cleavage has a rounded outer (distal) surface and three flattened faces abutting the other three spores of the tetrad. Each of the four spores has a more or less pointed proximal pole positioned at the original center of the sporocyte cytoplasm (Fig. 5l). Thus, none of the six division planes lies in a straight line through the sporocyte cytoplasm. Even in mosses that undergo successive cleavage, such as *Amblystegium riparium* (Hedw.) B.S.G. (Brown & Lemmon 1982b), the eventual cleavage pattern reflects establishment of six division planes that converge in the center of the sporocyte.

Lobing of the cytoplasm in meiotic prophase is most extreme in hepatics of the Jungermanniidae (Fig. 1) and in mosses of the Polytrichidae (Fig. 3) in which the cleavage furrows reach nearly to the nucleus, lying in the small central portion of the sporocyte. Lobing begins early in prophase and results in pronounced definition of the future spores. In *Pellia*, Davis (1901) illustrated a nucleus returning to the central area from an early prophasic position in one of the cytoplasmic lobes. The nucleus is shown as being distorted as it squeezes through the restricted neck of the lobe. The meiotic spindles are contained in the common central portion of the sporocyte. A septum is reported to form after first meiosis, resulting in a peculiar dyad, each half of which consists of a pair of cytoplasmic lobes. Small, second-division spindles that form on either side of the septum barely separate chromosomes. Presumably, the nuclei migrate from their telophasic positions in the narrow necks of the lobes to a more central position in each tetrad member. The only portion of the spore circumference not covered by the original sporocyte wall is the extreme proximal pole. The surface ornamentation of hepatic spores with this type of development may be uniform except for the localized proximal poles (Brown & Lemmon 1986).

Rather surprisingly, members of the Marchantiidae undergo meiosis in sporocytes that are not lobed or marked in any way that predicts the eventual cleavage planes. In *Conocephalum conicum*, the only member of this group studied by TEM, the cleavage pattern following meiosis is variable, resulting in spore arrangements that range from tetrahedral to linear. Modern studies are needed to elucidate the nature of the division site and arrangement of phragmoplasts in the alternate pattern of cleavage seen in the Marchantiidae.

Plastid Division and Migration

In the monoplastidic sporocytes of mosses and hornworts, the plastid divides twice and the four resultant plastids achieve equidistant positions in the peripheral cytoplasm of the prophase I sporocyte. Although modern studies have centered on mosses, similar polarized behavior of plastids was reported in hornworts by von Mohl in 1839. In the early sporocyte, the single plastid is a shallow bowl-like structure lying close to the nucleus.The first plastid division typically occurs while the nucleus is in an acentric position during the synaptic stage of prophase I (Fig. 5b, 7).

Microtubules have been observed associated with the two long, slender plastids as the plastids move from the nucleus and rotate to lie at right angles to each other (Fig. 5c). Each divides by constriction at its midpoint, and the four small plastids move to equidistant positions in the cortical cytoplasm. The plastids are arranged at the apices of a tetrahedron (Fig. 5d). This positioning of the four plastids establishes the polarity of sporogenesis, with each plastid marking one of the four polar regions of second division spindles (Fig. 5i). The four nuclei of the newly formed spore tetrad will lie adjacent to plastids at the tetrad poles (Fig. 5 *l*). Thus, the behavior of the single plastid in meiotic prophase of monoplastidic sporocytes anticipates the second nuclear division rather than the first. The precocious behavior provides an outstanding example of organelle migration as a presumably genetically programmed component of cytomorphogenesis. Such programmed migration appears to be independent of influences from surrounding cells and has been termed "morphogenetic organelle migration" (Brown & Lemmon 1985c).

Figure 5. Interpretive drawings summarizing meiosis in mosses. All figures are in the same orientation as those in Fig. 6. **a–d**, Early to late prophase I. **a**, The single plastid cups the nucleus; **b**, The nucleus migrates to an acentric position and the single plastid divides; **c**, The nucleus returns to a centric position and the two plastids rotate so as to lie at right angles to each other, each plastid divides; **d**, The four plastids are tetrahedrally arranged and interconnected by six bands of microtubules (compare Fig. 6a); **e**, Metaphase I, four principal bands of microtubules from the prophasic tetrahedral system converge in pairs toward the spindle axis and contribute to the metaphase spindle, the spindle axis terminates in opposite, perpendicularly oriented cleavage furrows between pairs of plastids, each half-spindle comprises a pair of microtubule focal points straddling a cleavage furrow (compare Fig. 6b, 6c); **f**, Anaphase I, focal points of microtubules continue to converge toward the spindle axis as chromosomes migrate toward poles (compare Fig. 6c); **g**, Telophase I, bean-shaped nuclei curving around opposite cleavage furrows are oriented at right angles to each other, phragmoplast microtubules extend between nuclei (compare Fig. 6d); **h**, Prophase II,

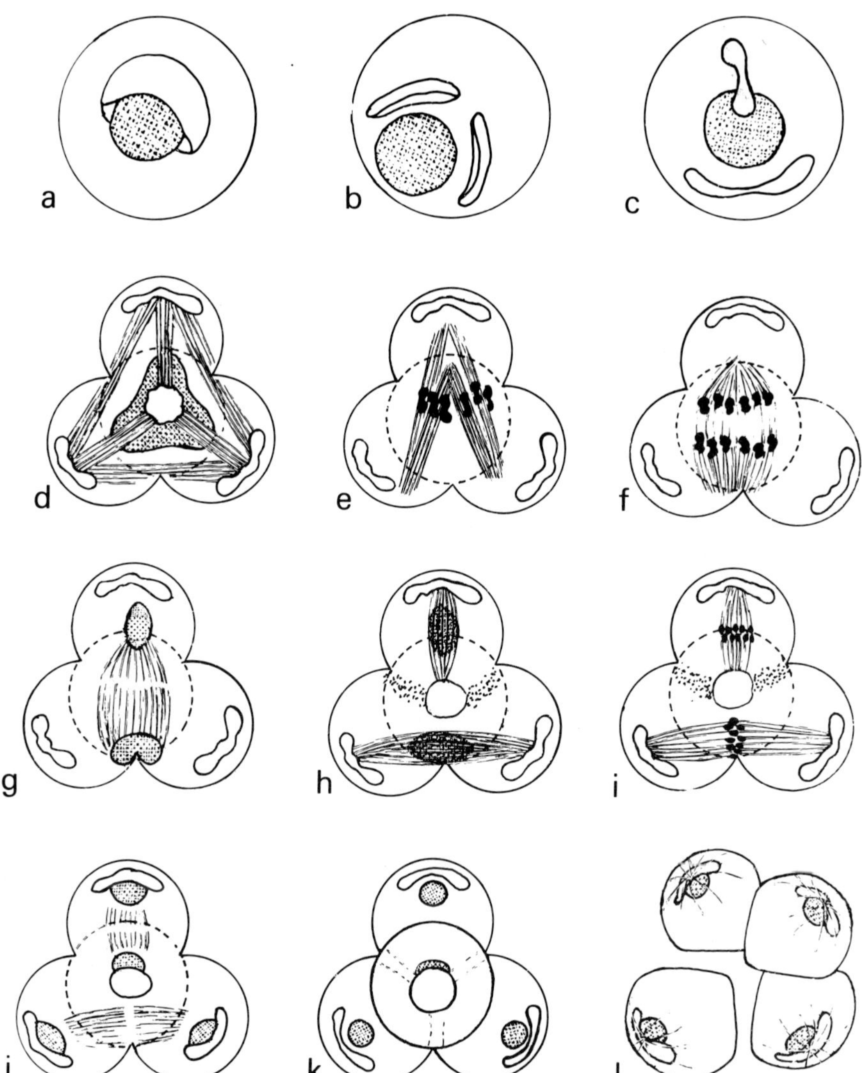

microtubules ensheath spindle-shaped nuclei (compare Fig. 6e); **i**, Metaphase II, chromosomes lie at equators of spindles that extend between pairs of plastids (compare Fig. 6f); **j**, Telophase II, nuclei are adjacent to plastids at the tetrad poles, phragmoplast microtubules extend between pairs of daughter nuclei but are absent in the first division site (compare Fig. 6g); **k**, Cytokinesis, cleavage occurs simultaneously along all six predetermined division planes (compare Fig. 6i); **l**, Tetrad of spores, each spore contains a plastid and a nucleus, the outer (distal) surface is rounded and the inner (proximal) face consists of three flattened surfaces meeting at a common point in the original center of the sporocyte (compare Fig. 6i).

Nuclear Events of Meiosis

Meiosis in bryophytes is characterized by a distinct succession of nuclear phases that is generally typical of plant meiosis. Our observations on moss meiosis are in agreement with recent interpretations of meiotic prophase in plants by Klasterska and Ramel (1979). This scheme includes three prophase stages recognized by Wilson (1937) but generally ignored in later literature: synizesis, diffuse, and second contraction (chromatin condensation) leading to discrete chromosomes in metaphase I. The duration of meiosis in *Ceratodon purpureus* (Hedw.) Brid. under a constant temperature of 12.8°C has been estimated at about 47 hours with meiosis I lasting about 38 hours or approximately 80 % of the total time (Luomajoki 1985). Condensation of the chromatin begins slightly before migration of the nucleus to an acentric position in the cytoplasm. It is while the nucleus is in acentric position, referred to as the "bouquet" stage (Fig. 5b, 7, 8) by light microscopists, that synaptinemal complexes may be observed. For reasons unknown to us, synaptinemal complexes are most easily recognized in Marchantiales and Funariales (Fig. 9–11). If lobing is precocious and extreme, as in mosses of the Polytrichales and liverworts of the Jungermanniidae, the nucleus is located in one of the lobes during this stage. Heterochromatin is conspicuous in most bryophytes throughout prophase, and the nucleoli remain prominent (Fig. 2, 4, 7–9, 11). The nucleus returns to a central position in the cell, and the gradual condensation of chromatin culminates in a typical pachynema stage. A nuclear stage characterized by decondensation of chromatin is intercalated between pachynemata and metaphase I (Fig. 12–14). In mosses, hornworts, and hepatics with deeply lobed sporocytes the nucleus becomes tetrahedrally shaped with lobes extending into the cytoplasmic lobes. In mosses, a tetrahedral system of microtubules encages the four-angled nucleus at this stage (Fig. 13, 14). Chromatin at this "diffuse" stage of meiotic prophase resembles the chromatin of an interphase nucleus. However, the tetrahedral shape of the nucleus, the microtubules of the tetrahedral system encaging the nucleus, and the cytoplasmic lobing between cleavage furrows are reliable markers of an advanced stage of prophase. It is not known how long the nucleus remains in the diffuse stage in different mosses under various environmental situations. Dill

Figure 6. Immunofluorescent images of sporocytes stained by FITC conjugated antibodies to anti-tubulin showing changes in microtubule patterns during meiosis, x2,000. **a**, Prophase I, somewhat flattened sporocyte showing tetrapolar system of microtubules interconnecting the four plastids in cytoplasmic lobes (compare Fig. 5d); **b**, Prometaphase I, convergence of microtubule focal points from tetrad poles toward the spindle axis results in spindle with an arrowhead configuration, two focal points straddle the lower cleavage furrow and the two focal points at the upper pole are superposed (compare Fig. 5e); **c**, Early anaphase I, focal points continue to converge toward spindle axis, dark areas at the equator are chromosomes, kinetochore bundles extend between chromosomes and polar regions (compare Fig. 5f); **d**, Telophase I, phragmoplast microtubules extend between unstained nuclei (compare Fig. 5g); **e**, Prophase II, microtubules ensheath the perpendicularly oriented nuclei, the

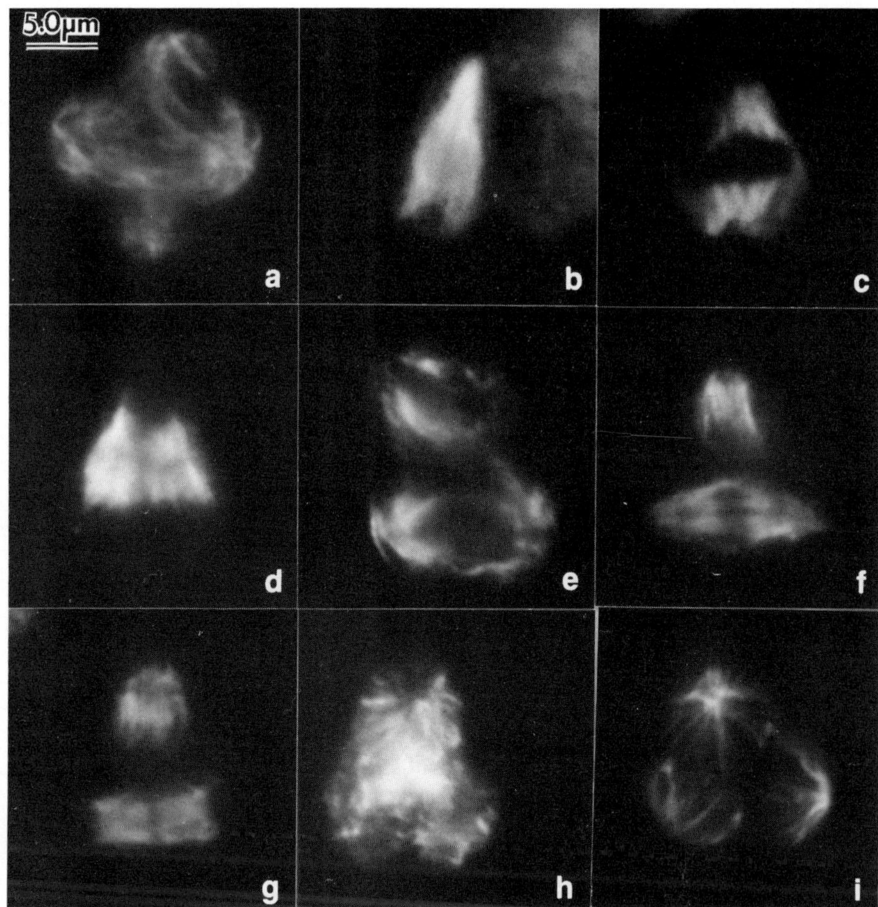

two nuclei each lie between a pair of plastids and an organelle band lies in the first division site between the two pairs of lobes (compare Fig. 5h); **f**, Metaphase II, spindles extend between pairs of plastids, chromosomes are aligned at equators located in cleavage planes (compare Fig. 5i); **g**, Telophase II, phragmoplast microtubules extend between pairs of daughter nuclei located at plastids, microtubules are absent in the first division site occupied by the organelle band (compare Fig. 5j); **h**, Post-meiotic stage showing transition of microtubules to the distal systems typical of young spores of the tetrad (compare Fig. 5h); **i**, Spore tetrad with organized distal systems of microtubules associated with exine initiation (compare Fig. 5 *l*).

(1964) estimated that sporocytes of the moss *Hypnum circinale* Hook. remained in the diffuse stage for 2–3 days. Dill suggested that meiosis in many mosses may extend over a period of 2–3 weeks. Klasterska (1976) has compiled data on the diffuse stage, termed dictyonema, in vascular plants and reports that it is not uncommon for the nuclei of sporocytes to remain in this stage of chromatin decondensation for long periods, with some reports of overwintering. A similar situation may exist among mosses, and an interrupted meiotic prophase may provide a plausible explanation for the speed with which meiosis is completed in the early spring.

Chromosomes of mosses are attached to microtubules of the spindle by highly structured kinetochores, whereas hepatics and hornworts appear to have diffuse kinetochores of simpler structure that are more typical of vascular plants. The moss kinetochore (Fig. 16, 17) typically exhibits layers of electron dense material to which a number of microtubules are attached (Lambert 1977, Brown & Lemmon 1982a, 1987a). The metaphase spindle of first meiotic division (Fig. 18) is extremely complex in ontogeny and configuration. It will be discussed in a separate section of this paper.

Following telophase I, the nuclei of mosses undergo a complete, albeit short, infra-meiotic interphase. The shortness of this phase in hepatics, is apparently what misled Farmer (1894) and contributed to his erroneous conclusion that chromosomes were distributed to four nuclei in a single meiotic division. He recognized typical second division spindles following quadripolar first division spindles in most hepatics but defended his interpretation of a single meiotic division in *Pallavicinia* (Farmer 1895, 1904). From our observations using TEM and immunocytochemistry, it appears that the ontogeny of the second division spindle is quite typical of that of other plants. The bean-shaped telophase I nuclei (Fig. 5g) become spindle-shaped in prophase II (Fig. 5h, 6e, 19). Each nucleus is oriented between a pair of plastids and lies at right angles to the other. Microtubules converging at opposite spindle poles ensheath the nuclei (Fig. 6e, 19–21) and develop into the mature spindles of metaphase II (Fig. 5i, 6f). We have not observed discrete microtubule organizing centers at the poles, but metaphase II spindle microtubules seem to terminate in the narrow region of cytoplasm immediately adjacent to the plastid, if not actually in association with the plastid envelope.

Figures 7, 8. Synizesis stage of prophase I in the moss *Rhynchostegium serrulatum* (Hedw.) Jaeg. & Sauerb. (Reproduced with permission from Brown & Lemmon 1982d.) 7, The single plastid (P) has divided once, the acentric nucleus contains synapsed chromatin (C), x9,300; **8**, Detail of an acentric nucleus (N) with heterochromatic center (H), M mitochondrion, D dictyosome, x13,700.

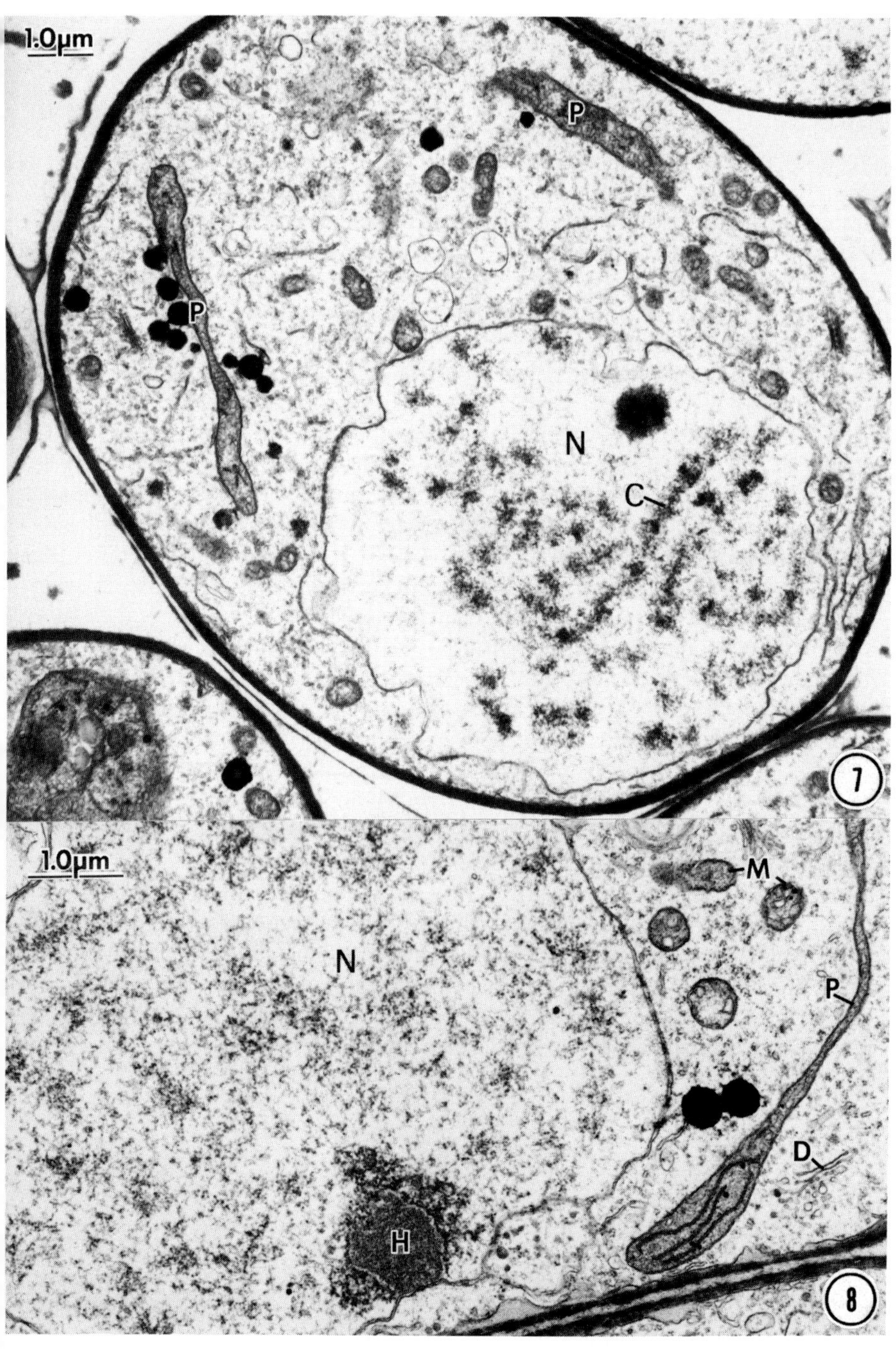

Prophasic Tetrahedral System of Microtubules and the Meiotic Spindle

Once the four plastids are located at the future tetrad poles (the second meiotic division poles), they mark the focal centers for four cones of microtubules that converge to form a tetrahedron of microtubules surrounding the nucleus (Fig. 5d, 6a, 13, 14). This unique system of microtubules attains its maximum development during the diffuse stage of late meiotic prophase just before prometaphase. Developmental studies of the tetrahedral system point to microtubule nucleation in the immediate vicinity of the plastids, if not in association with the plastid envelope (Fig. 13, 15). It could be supposed that this microtubule system, which is peculiar to the sporocytes of bryophytes, might function in final plastid positioning (by forcing the plastids to equidistant positions at the tetrad poles) and in cytoplasmic lobing. While it is true that these phenomena are usually coincidental events of prophase I, important exceptions are known. In *Atrichum undulatum* (Brid.) B.S.G., for instance, cytoplasmic lobing begins before plastid division and migration and before the tetrahedral system of microtubules develops (Brown & Lemmon 1987a) and in liverworts of the Jungermanniidae pronounced lobing occurs in polyplastidic cells. We must consider, therefore, that plastid positioning and cytoplasmic lobing could be independent responses to the establishment of division polarity in the young sporocyte.

We believe that the prophasic tetrahedral system of microtubules in mosses is a direct precursor of the meiotic spindle (Brown & Lemmon 1987a). The tetrahedral system is transformed into a unique quadripolar spindle that is functionally bipolar. Chromosomal movement results in two well-defined telophase I nuclei lying at opposite ends of a single division axis. At metaphase I, each half-spindle consists of a pair of poles that straddles a cleavage furrow on opposite sides of the sporocyte (Fig. 5e, 6b, c, 18). The polar cleavage furrows and therefore the two half-spindles are oriented at right angles. Thus, a whole spindle in any side view will have the appearance of an arrowhead (Fig. 5e, 6b, c). One pair of poles will be seen straddling a cleavage furrow and the other pair will be seen superposed with one directly behind the other. These configurations are commonly seen in immunofluorescence preparations (Fig. 6b, c) but are not visible in a single thin section. A median section through the spindle could pass through one pair of poles, but both poles of the pair in the other half-spindle (at right angles to the first) would be out of the plane of

Figures 9–11. Synaptinemal complexes during synizesis in bryophytes. **9,** *Corsinia coriandrina* (Spreng.) Lindb. (Marchantiales), portion of nucleus showing synaptinemal complexes and heterochromatin, x6,800; **10,** Detail of the synaptinemal complex boxed in Fig. 9, x37,000; **11,** *Funaria hygrometrica* Hedw. (Funiarales), portion of nucleus showing synaptinemal complexes and heterochromatin, x13,500.

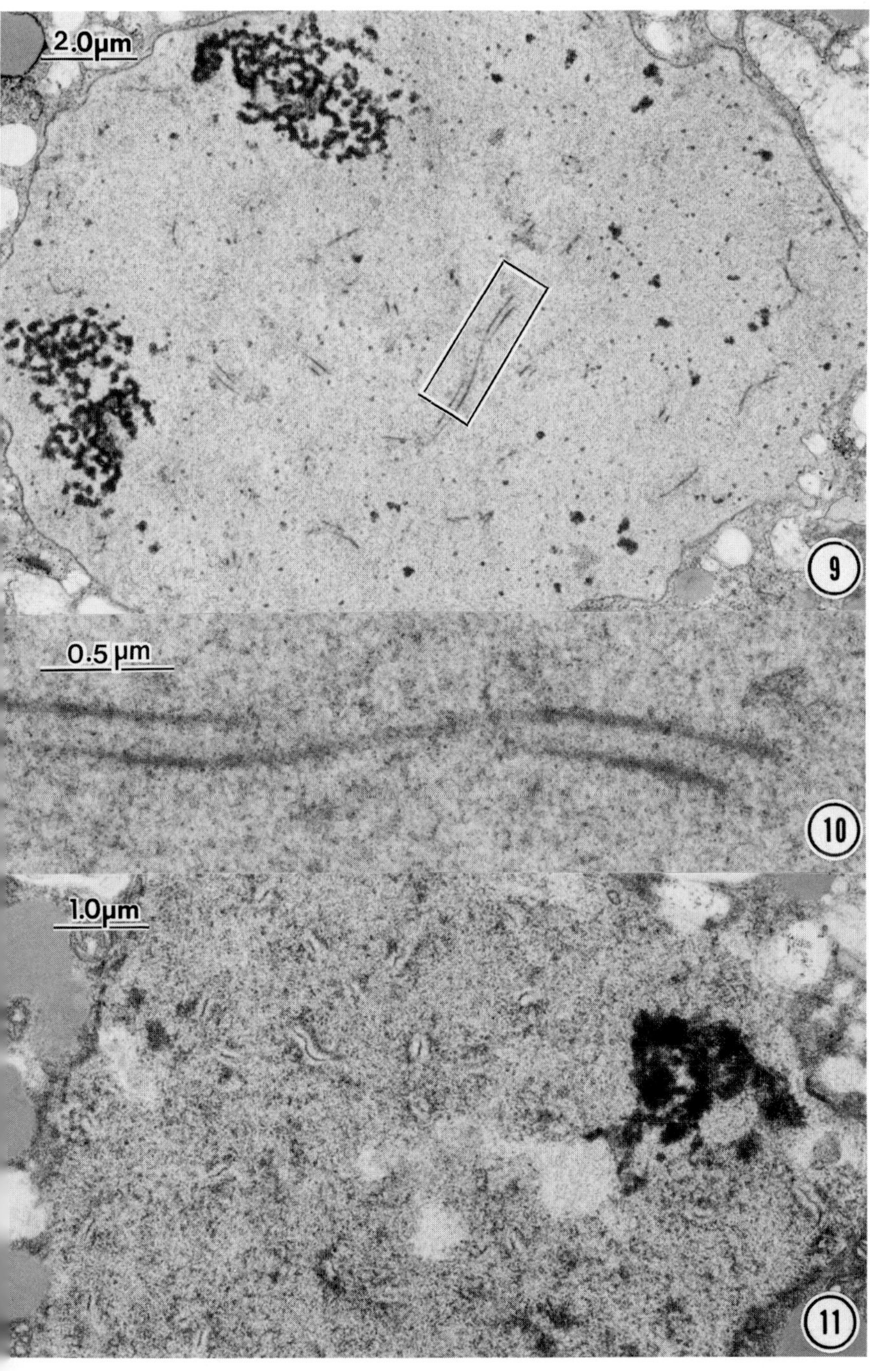
2.0μm
9
0.5 μm
10
1.0μm
11

section. A tangential section of the spindle could include as many as three of the four poles, i.e., both of a pair in one half-spindle and one of the pair in the other half-spindle. Occasionally, a spindle stained by immunofluorescence will be seen flattened or from an angle so that all four of the spindle poles are visible.

As anaphase I progresses, the pole-pairs converge toward the mid-line of the spindle axis (Fig. 5f, 6c). The chromosomes congress at the broad polar regions. Reforming nuclei are shaped like kidney beans, each being curved over cleavage furrows and lying at right angles to each other (Fig. 5g). Microtubules, as revealed by TEM studies, are no longer associated with the plastids, and in fact, the plastids in many mosses are ensheathed by endoplasmic reticulum (Brown & Lemmon 1982a, 1987a).

The origin and poleward movement of kinetochore microtubules is an intriguing problem of spindle dynamics and chromosomal movement in mosses. Is there a substantial capture of microtubules from the tetrahedral system by the kinetochores, or are kinetochore microtubules newly synthesized in prometaphase? Are some of the kinetochore microtubules directed to one or the other of a pair of poles? Studies of the origin of kinetochore bundles and determination of the course of kinetochore movement are needed. Such information would help answer fundamental questions concerning the mechanism of nucleation and of attachment of kinetochore microtubules involved in chromosomal movement.

Earlier reports on spindle dynamics in the moss *Mnium hornum* Hedw. by Lambert (1974, 1977, 1980) suggested that the nuclear envelope serves to nucleate microtubules of the developing spindle. While we have frequently observed intimate association of microtubules and nuclear envelope, we feel that evidence from ultrastructural studies, and particularly from recent immunofluorescence studies, is more convincingly in favor of nucleation in the cytoplasm surrounding the plastids or possibly in association with the plastid envelope itself. In any case development of the tetrahedral system of microtubules begins early in prophase whereas the prophase spindles of other plants typically develop during prometaphase. Our interpretation of the meiosis I spindle differs from Lambert's interpretation principally in that we recognize the spindle as a direct modification of a tetrahedral prophasic system, while Lambert concluded that the spindle arises at the nuclear envelope and consists of two opposite cones of microtubules with the narrow polar ends located near plastids.

It is remarkable to find that the quadripolar nature of the meiosis I spindle was accurately described by early light microscopists. The microtubules of the tetrahedral sys-

Figures 12, 13. Diffuse stage of prophase I in the moss *Rhynochostegium serrulatum*. (Reproduced with permission from Brown & Lemmon 1982d.) **12,** Chromatin in the diffuse nucleus (N) is decondensed after condensation in pachynemata, three of the four tetrahedrally arranged plastids (P) are in this section of a whole sporocyte, x10,600; **13,** Detail showing diffuse nucleus distorted toward plastid (P), microtubules (arrows) of the tetrahedral system that encage the nucleus are focused at a plastid (P), NE nuclear envelope, x27,100.

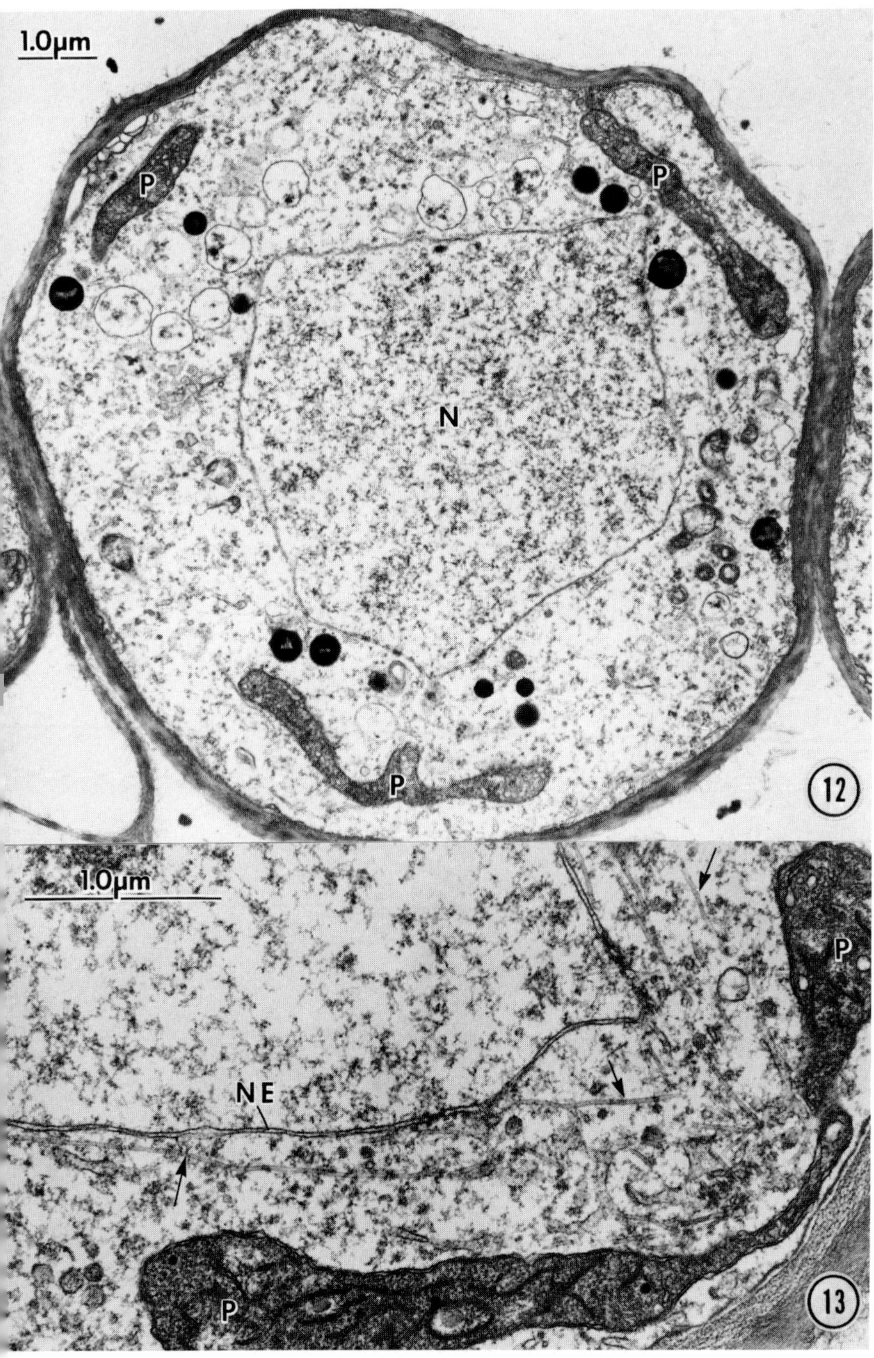

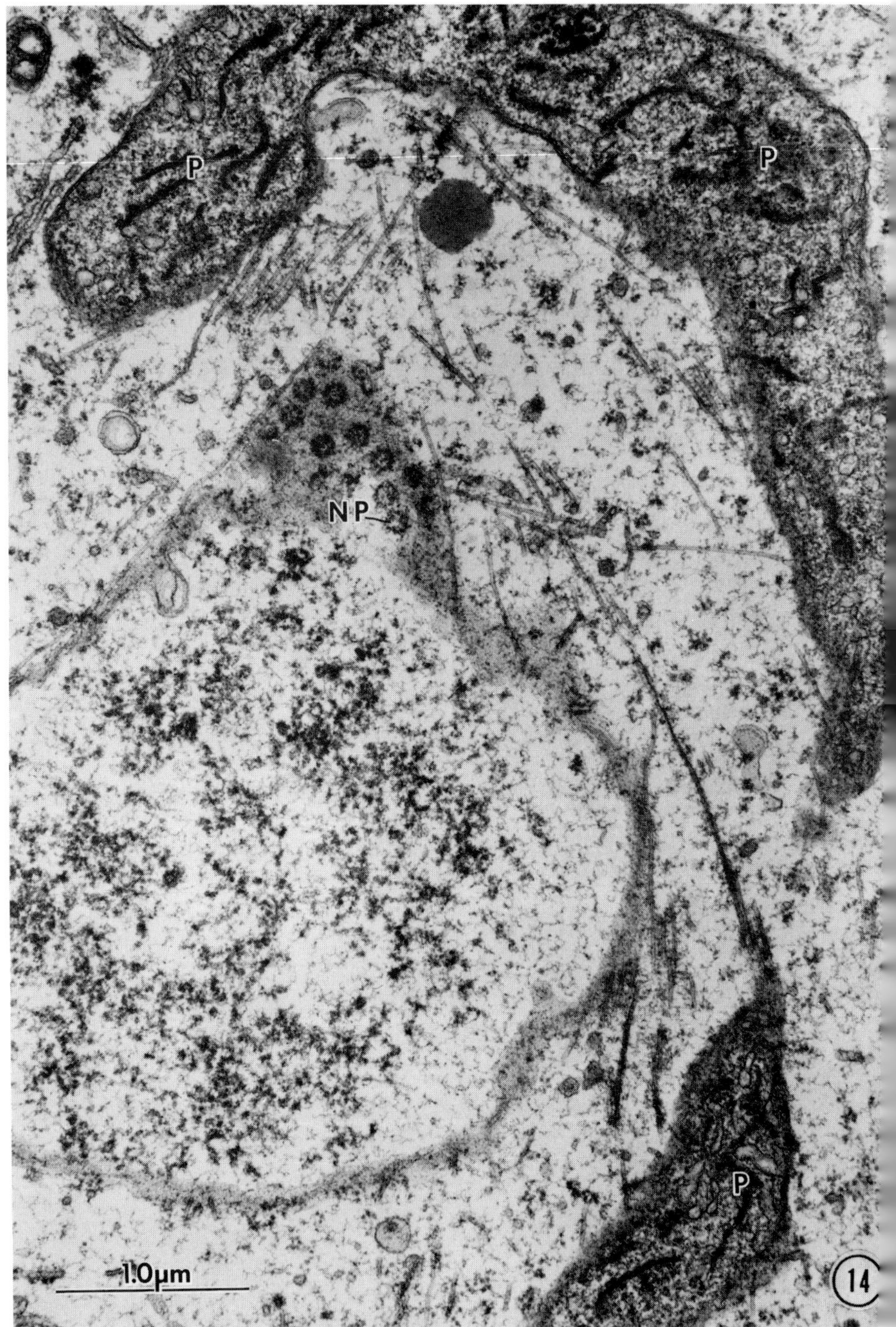

Figure 14. Portion of the tetrahedral system of microtubules interconnecting plastids (P) and encaging the diffuse nucleus in the moss *Rhynchostegium serrulatum*, NP nuclear pores, x27,200. (Reproduced with permission from Brown & Lemmon 1982d.)

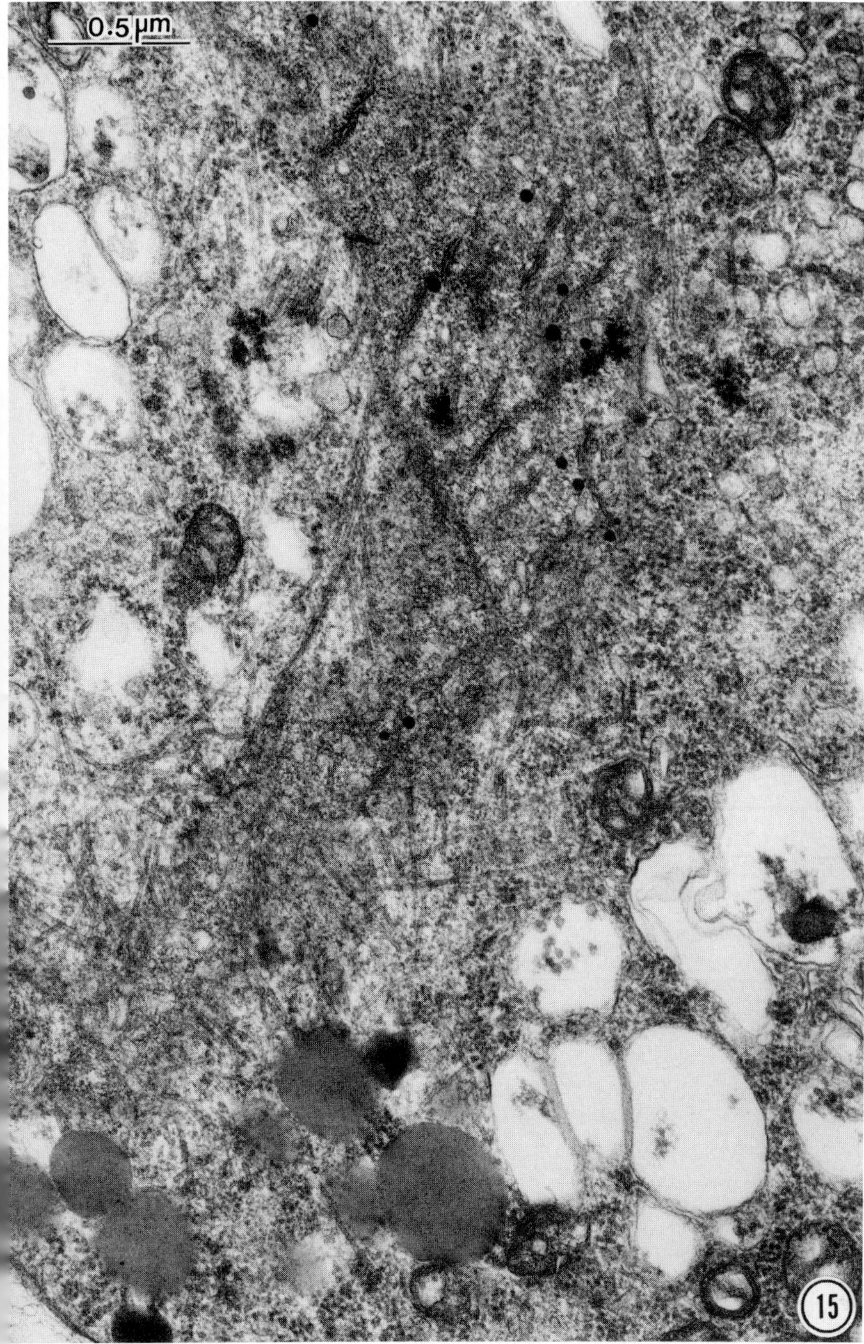

Figure 15. Microtubules associated with plastid during meiotic prophase in the moss *Entodon seductrix* (Hedw.) C.M. The close association of microtubules with the plastid surface is visible in this grazing section of a plastid, x39,600.

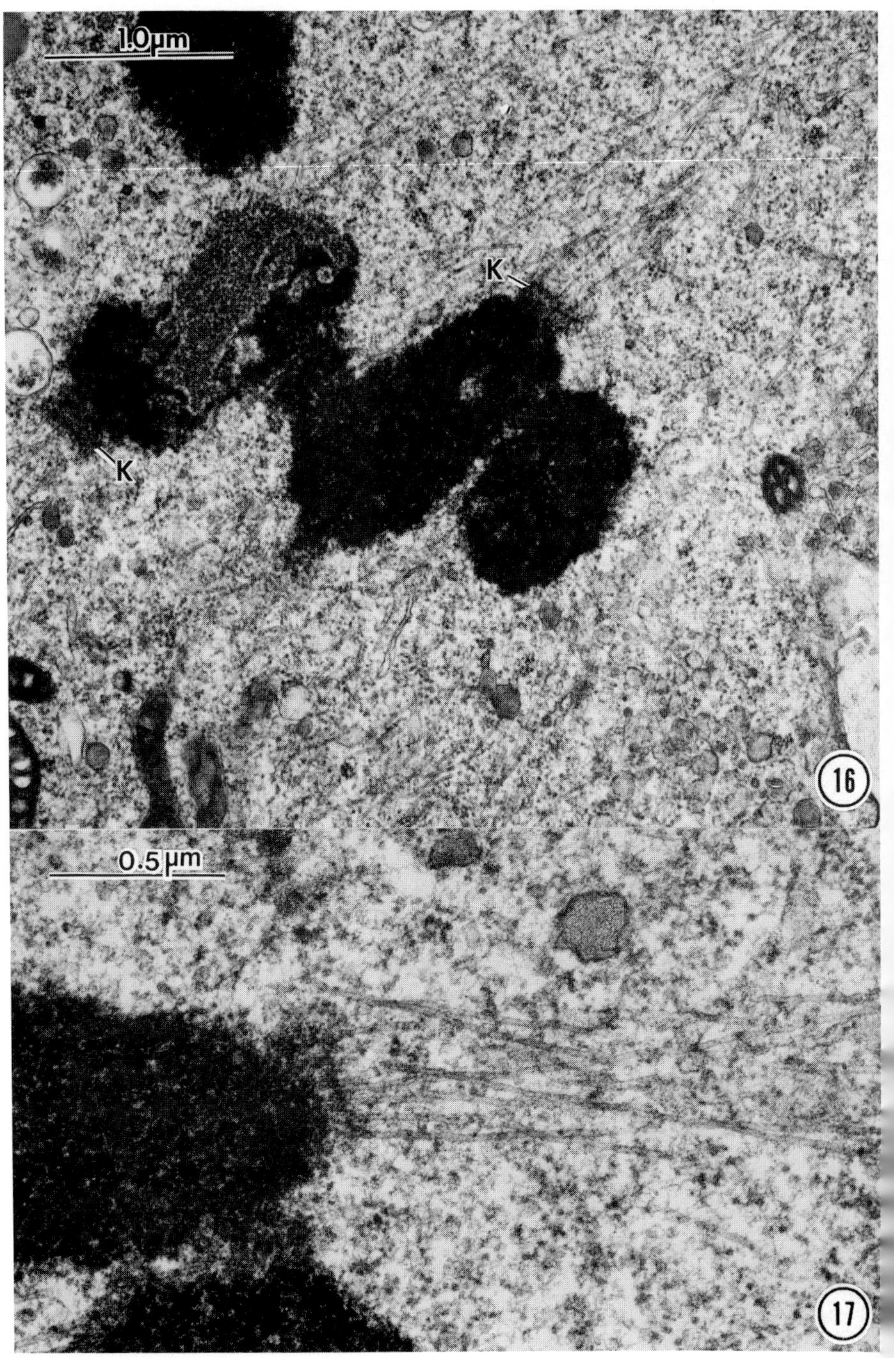

Figures 16, 17. Kinetochores in the moss *Entodon seductrix*. **16,** Metaphase I bivalent with kinetochores (K), x25,700; **17,** Detail of a layered kinetochore with attached microtubules, x49,700.

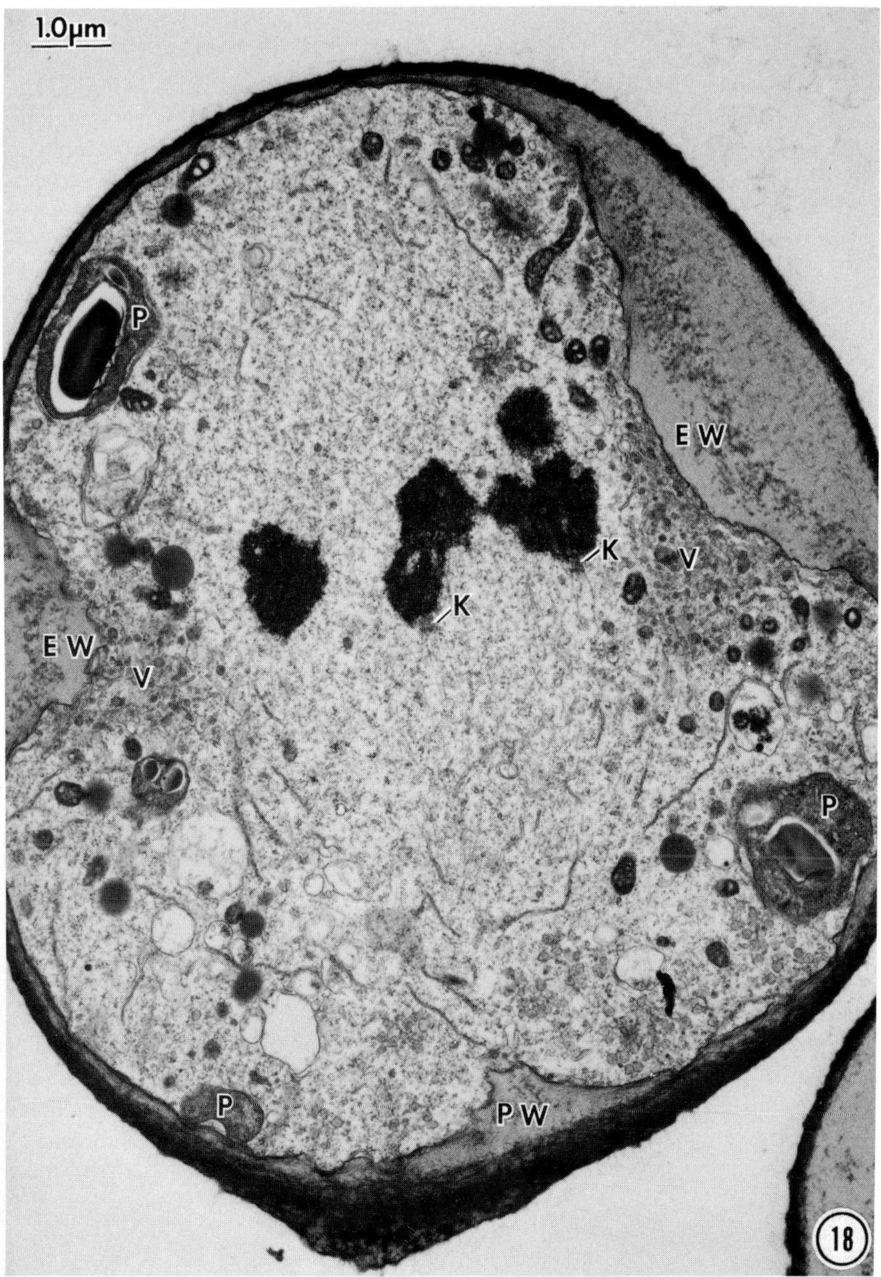

Figure 18. Metaphase I in the moss *Entodon seductrix*. Chromosomes are aligned with the equatorial cleavage furrows (EW). Poles terminate at perpendicularly oriented opposite polar cleavage furrows (PW). Girdling plate of vesicles (V) is associated with the equatorial wall ingrowth. A plastid (P) is located in each cytoplasmic lobe, K kinetochore, x10,700.

tem are so numerous that they apparently were observed and described as fibrils. The prophasic tetrahedral system of microtubules can be added to the short list of microtubule arrays (e.g., metaphase spindles and phragmoplasts) that were recognized through light microscopy. We are indebted to these pioneer cytologists for the accuracy of their observations. Unfortunately, the observations were sometimes lost when an erroneous interpretation was discredited. Rejection of the concept of a quadripolar spindle in bryophytes can be traced to Farmer's misinterpretation (1894) that the four nuclei were distributed to the four poles in a single nuclear division. When the two nuclear divisions of meiosis and the transitory nature of the quadripolar spindle were correctly described in later studies on hepatics (Farmer 1904, Farmer & Moore 1905), reports of the quadripolar spindle were forgotten or ignored. Even though Allen (1916) correctly reported convergence of pairs of poles toward the spindle axis in a moss, the phenomenon of a quadripolar origin for the meiotic spindle remained obscure until it was recently documented in studies using indirect immunofluorescence and TEM (Brown & Lemmon 1987a).

Not all bryophytes have a meiotic spindle that can be traced to a tetrahedral system of microtubules developed in meiotic prophase. The hepatic *Conocephalum conicum* (Marchantiales) produces a small, flat meiotic spindle with blunt poles that is located acentrically in the sporocyte (Meyer 1929). From immunocytochemical images of sporocytes in all developmental stages, we conclude that the *Conocephalum* spindle is organized on the outer surface of the nucleus, which apparently remains in an acentric position throughout meiosis I. Whether or not other members of the Marchantiidae produce the same type of spindle is eagerly awaited information. This large group of hepatics is isolated from other bryophytes by the fact that the sporocytes show no structural evidence of the determination of division plane, which is conspicuously marked by lobing and/or sporocyte wall infurrowing in all other groups of bryophytes.

The monoplastidic hornworts and polyplastidic hepatics of the Jungermanniidae are both expected to have meiotic spindles that are similar to the quadripolar spindle of mosses. Comparative data are lacking as neither group has been studied with TEM and immunocytochemical methods. However, based on early reports, we would expect to find that the meiotic spindle is derived from a prophasic tetrapolar system of microtubules. Strasburger (1880) and Davis (1899) illustrated fibrils connecting the four plastids following plastid partitioning in *Anthoceros*, and quadripolar spindles were reported in *Pallavicinia* by Farmer (1894, 1895). Assuming that spindle ontogeny in both hornworts and hepatics is similar to spindle ontogeny in mosses, there

Figure 19. Prophase II in the moss *Rhynchostegium serrulatum*. Nucleus (N) is elongated in the plane of the spindle axis between a pair of plastids (P). Microtubules (arrows) of the forming spindle ensheath the nucleus. An organelle band (B) marks the first division site, x22,000. (Reproduced with permission from Brown & Lemmon 1982c.)

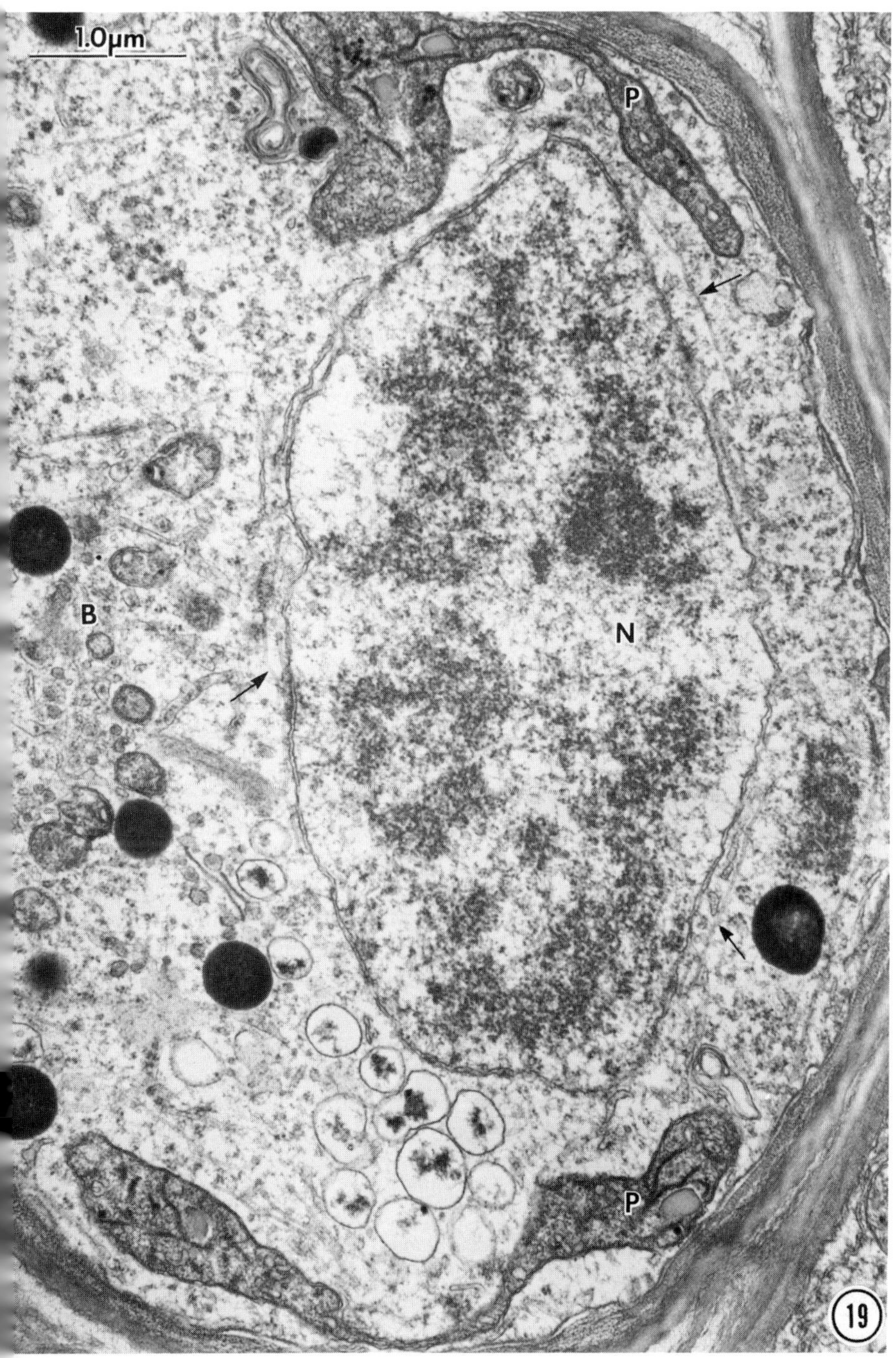

1.0µm
P
N
B
P
19

remains the question of microtubule nucleation. Certainly, we would expect that microtubule nucleation centers (MTOCs) in polyplastidic hepatics would differ from those in monoplastidic sporocytes. It will be interesting to determine the relationship of these nucleation centers to those intimately associated with plastids.

As in all land plants, the designation of a MTOC is ambiguous. In typical plant cells, the cytoplasmic centers of microtubule organization are either undetectable with the TEM or detected as vague areas of electron-dense material associated with a convergence of microtubules. Discrete MTOCs have been reported in plant cells undergoing specialized morphogenesis such as flagellated sperm cells (Hepler 1976, Kreitner & Carothers 1976, Robbins 1984), guard cells (Galatis et al. 1983), and spores (Brown & Lemmon 1983). Among land plants, well-defined MTOCs have been observed most frequently in cells of bryophytes. Vegetative cells of *Marchantia* are unusual in having MTOCs at the mitotic poles where the spindle microtubules converge (Fowke & Pickett-Heaps 1978, Steer 1984, Apostolakos & Galatis 1985). Sporocytes of *Rhynchostegium serrulatum* (Hedw.) Jaeg. & Sauerb. (Brown & Lemmon 1982a) exhibit MTOCs at the migrating poles of the tetrahedral microtubule system as the poles converge during transformation to the functionally bipolar spindle. Young spores of *Ditrichum pallidum* (Hedw.) Hampe (Brown & Lemmon 1980) and *Rhynchostegium serrulatum* (Fig. 59) exhibit MTOCs at the distal surface during exine initiation. In *Tetraphis pellucida* MTOCs occur at the proximal pole during aperture development (Fig. 63) and at the distal surface during exine initiation (Brown & Lemmon 1983).

Meiotic Cytokinesis

Meiotic cytokinesis in bryophytes is generally delayed until afer second division when cleavage occurs simultaneously along all six predetermined cleavage planes (Fig. 22). There are exceptions, however. In certain mosses (Genevès 1974, Brown & Lemmon 1982b) and in many hepatics (Blair 1926), cytokinesis occurs successively, a septum being formed after first nuclear division and two additional septa formed after second nuclear division. Nevertheless, the four spores assume a tetrahedral arrangement with six planes intersecting at the center of the former sporocyte. Thus, the tetrahedral spore tetrad reflects the polarity established during meiotic prophase, even in cases where successive cleavage occurs. In *Amblystegium riparium* (Hedw.) B.S.G., the only moss in which successive cytokinesis has been studied in detail (Brown & Lemmon 1982b), the sporocytes are not released from the archesporial walls during meiotic prophase as they are in most mosses, and the tetrahedral ar-

Figures 20, 21. Prophase II in the moss *Entodon seductrix*. **20,** A grazing section of one nucleus (N) showing microtubules of the forming spindle that ensheath the nucleus, B organelle band, P plastid, x10,400; **21,** Detail of the perinuclear array of microtubules, N nucleus, x37,500.

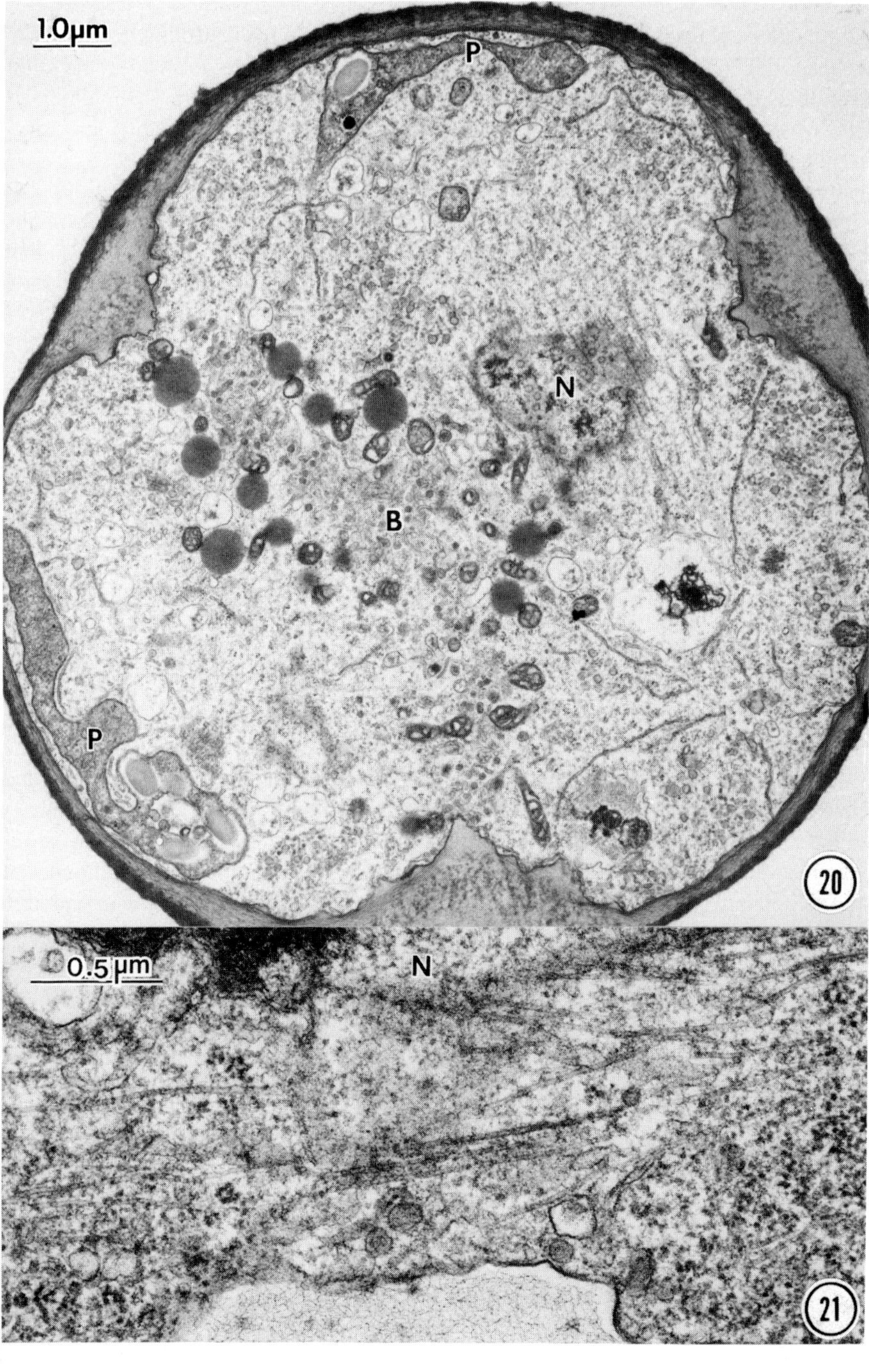
1.0μm
P
N
B
P
20
0.5μm
N
21

rangement of plastids is often distorted or flattened due to the confining nature of the archesporial cell walls. It seems that the tetrahedral polarity is established in the same fashion as in sporocytes released into the spore sac. However, polarity is manifested later when the spore tetrad is finally released and space permits the tetrahedral configuration.

In many mosses with simultaneous spore cleavage, the plane of the first division phragmoplast is marked by a distinct organelle band comprising mitochondria, endoplasmic reticulum (ER), and several populations of vesicles (Fig. 19, 20, 22). The organelle band persists until after second nuclear division when phragmoplasts are formed in the equatorial regions of the two second division spindles (Brown & Lemmon 1982c, 1987b). Intersporal septa, formed by coalesence of vesicles containing material visually similar to that of the sporocyte wall, develop simultaneously along all six predetermined planes. In some cases septa form in the second division sites in advance of septa in first division sites occupied by the organelle band (Fig. 22). Septa intersect at the center of the sporocyte and fuse with the sporocyte wall material in cleavage furrows.

The organelle band is of evolutionary interest. This distinctive structure appears as a component of meiosis in a variety of plants. A polarization of organelles results in an approximately equal distribution of organelles to the tetrad members in bryophytes (Brown & Lemmon 1982d, 1987b), pteridophytes (Marengo 1977, Sheffield & Bell 1979, Lee 1982, Bednara & Rodkiewicz 1985, Bednara et al. 1986), and gymnosperms (Vasil & Aldrich 1970, Wolniak 1976, Rodkiewicz et al. 1984). In certain angiosperms with simultaneous cleavage, a type of organelle band is reported to form after first division (Rodkiewicz et al. 1984). In pteridophytes and gymnosperms the organelle band is made up of dictyosome vesicles, ER fragments, mitochondria, and plastids. An important aspect of meiosis in mosses is the selective response to polarity at the end of first nuclear division. The plastids remain locked in their positions at future spindle poles and do not respond to the apparently strong polarizing influence at the cell plate site. In mosses phragmoplast microtubules appear at the time of band initiation and are associated with migration of vesicles and other organelles to the equatorial region. Vesicles and ER fragments are usually thought to be somehow guided by phragmoplast microtubules (Gunning 1982) and in this manner contribute to formation of the cell plate. The phragmoplast microtubule array appears to be disassembled in prophase II. The organelle band is maintained throughout second division in spite of the lack of an interlinking system of microtubules, and the cell plate is eventually formed in the absence of microtubules.

It is notable that phragmoplast microtubules form after first division (Fig. 6d), but the septum does not develop until after second division. The first division septum, when it belatedly forms after second division, develops in the absence of microtubules (Fig. 6g). The configuration of the first division septum is complex. It does not occupy a simple plane, but rather has the shape of a hyperbolic paraboloid, comprising four tilted planes that meet at a common point in the center of the sporocyte. It is usually possible to observe only one curve of the complex configuration in any one

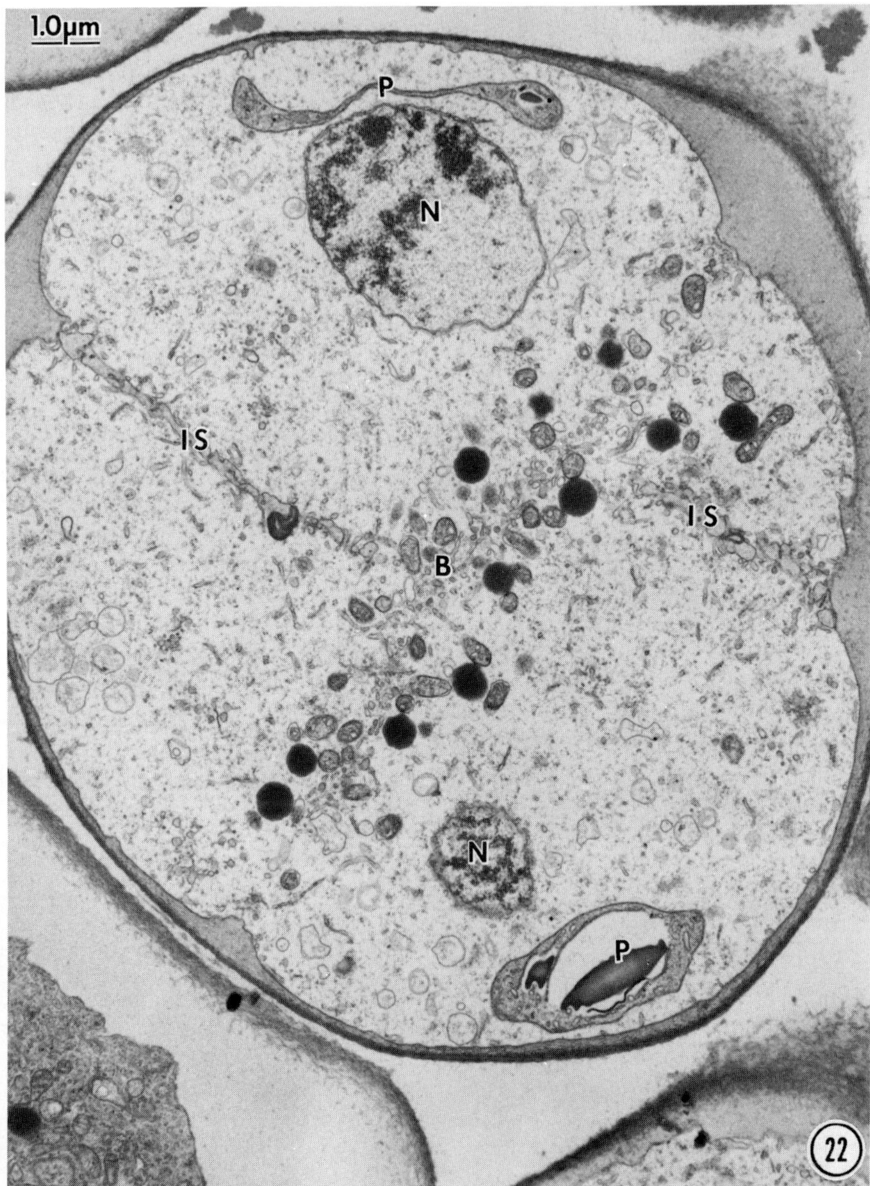

Figure 22. Meiotic cytokinesis in the moss *Entodon seductrix*. Intersporal septa (IS) form by coalescence of vesicles in the second division sites while the first division site is still occupied by the organelle band (B) that formed immediately after first nuclear division. Intersporal septa fuse with peripheral wall ingrowths formed during prophase I. Each spore will contain a plastid (P) and a nucleus (N), x9,300.

view of an organelle band. The planes of the first division intersporal septum, when joined by the two flat septa of second division, cleave the sporocyte cytoplasm into four tetrahedrally arranged spores (Fig. 5k). Immediately after cleavage each spore has a rounded distal surface and three flattened faces that abut those of the other three spores (Fig. 5 *l*).

Since tetrahedral arrangement of spores is the most economical use of the spherical space within the confines of a sporocyte wall, it could be supposed that the trilete spore marking is mechanically induced. Only in *Andreaea* spores do we find evidence for this. In other trilete spores such as *Sphagnum* and in spores with a hilate proximal aperture, highly polar microtubule systems are associated with the initiation of the exine on the distal surface and aperture on the proximal surface. In *Sphagnum* the aperture is trilaesurate, extending for some distance along ridges on the edges of the three proximal spore faces (Brown et al. 1982a, 1982b). A system of microtubules is associated with development of the distinctive proximal wall of *Sphagnum* spores. Although *Andreaea* also has trilete spores, the swellings along the ridges, which do not surround a trilaesurate aperture, are perhaps mechanically induced as no microtubule system was observed (Brown & Lemmon 1984a). Thus, whereas *Sphagnum* and *Andreaea* may seem to share a common feature (that of trilete spores), ultrastructural evidence indicates that the resemblance is only superficial. Moss spores that have a hilate proximal aperture and those that lack an aperture tend to lose the trilete appearance during maturation, the proximal faces becoming nearly as rounded as the distal faces. These "spherical" spores are typical of mosses in the Bryidae. Spores of the Jungermanniidae frequently exhibit a distinctly sculptured proximal face. The proximal face develops along cytokinetic planes in the small central area of cytoplasm that is common to the four lobes of the sporocyte. The exine on the proximal face is usually thinner than elsewhere along the perimeter of the spore and appears to function as a germinal aperture.

Spore Wall Development

At maturity the bryophyte spore wall consists of a minimum to two principal layers, an innermost layer of intine (the endospore), and an acetolysis-resistant layer of exine (the exospore) consisting of sporopollenin. These two layers have been interpreted as being basically similar to intine and exine in higher plants (Buchen & Sievers 1981, Mogensen 1983, Brown et al. 1986, Blackmore & Barnes 1987), hence we use the same terminology in describing walls of spores and pollen. The outer exine surface is frequently sculptured in distinctive patterns and is sometimes referred to as the sculptoderm. Although spore walls of bryophytes may be quite elaborately ornamented, none is known to include the complex tectate sculpturing that characterizes the outer layers of exine in the walls of angiosperm pollen grains. An additional spore wall layer, the perine, is deposited over the exine of moss spores. Perine is an osmiophilic, non-acetolysis resistant outer spore wall layer that is of extrasporal origin. This additional outer wall layer is characteristic of moss and fern spores, while other wall layers, such as the pollenkitt, also occur as an outer coating on the surface of angiosperm pollen grains. The perine provides additional surface ornamentation, or in moss spores with thin homogeneous exines, the only surface ornamentation. In hepatics the patterned exine develops within a polysaccharide wall layer laid down around tetrad members following cytokinesis. This layer is known as the spore special wall and is thought to function much as does the primexine of developing pollen grains (Horner et al. 1966). This layer is not known to occur in mosses. Tripartite lamellae of near unit-membrane dimensions, which are characteristic of early stages of wall deposition, may occur throughout the exine of hepatic spore walls, whereas the exine of moss spore walls is mostly homogeneous. Certain of the Hepaticae, particularly the Metzgeriales, may develop multilayered walls of considerable complexity.

Spore Wall Development in Hepatics

Wall development and morphology in hepatics (Fig. 23–33) comprises a character set of considerable taxonomic importance and has been employed by Schuster (1966, 1984) to help define the two major subclasses. The sporocytes of the Jungermanniidae are deeply quadrilobed in meiotic phophase, whereas those of the Marchantiidae are unlobed. Spores of the Jungermanniidae are said to be inconspicuously trilete, whereas those of the Marchantiidae are strongly polar with triradial ridges (Schuster 1984). Large and highly ornate spores occur in the Marchantiidae, whereas most studies of spores of the Jungermanniidae report a relatively simple architecture (Horner et al. 1966, Rowley & Southworth 1967, Heckman 1970, 1972, Suire 1970, Oltmann 1974). Recent studies give evidence of spore walls of considerable complexity in the order Metzgeriales (Inoue & Hibino 1984, Steinkamp & Doyle 1984, Brown et al. 1986).

Spore wall development in hepatics typically proceeds in centripetal fashion, the outermost layer of exine being the first formed and the innermost intine the last formed. In many hepatics the first observable event associated with wall development is the appearance of a spore special wall (Fig. 23) around the spores in the tetrad (Horner et al. 1966, Heckmann 1972, Oltmann 1974). Following cytokinesis the poly saccharide wall is deposited immediately outside the plasma membrane of the newly cleaved spores. This wall appears to function as a primexine in the original sense of Heslop-Harrison (1963, 1968) and Horner et al. (1966), i.e., as a special wall layer that is laid down around the young spores of a tetrad that contains elements that predict the eventual sculpture pattern of the exine. Because the terminology regarding special walls in bryophytes and vascular cryptogams is inconsistent and confusing, we restrict use of the term "special wall" to the original sense (Beer 1906) to refer to the wall layer formed around each spore of the tetrad following cytokinesis and within which the exine develops.

An important variation in the time of production of extracellular exine precursors has been reported in *Cephalozia, Pallavicinia* (Heckman 1972, Brown et al. 1986), and *Haplomitrium* (Brown & Lemmon 1986, Brown et al. 1986), which belong to three different orders. In these hepatics the mature sculptured exine is prepatterned by exine precursors produced well in advance of cytokinesis while the sporocytes are still in meiotic prophase. The exine precursors have a function similar to that of primexine but differ significantly in that they are produced by the sporocyte rather than the newly cleaved spores. We were recently surprised to discover that Farmer reported this significant phenomenon in 1894, yet no mention has been made of this until its rediscovery with TEM. In *Haplomitrium* (Fig. 28), *Apotreubia* (Fig. 29), and *Pallavicinia* (Fig. 30, 31), the exine precursors are locally produced outside the plasma membrane as caps on projections of the cytoplasm. The exine precursors are distinctly less dense than the sporocyte wall and were simply called "low-dense" areas by Heckman (1972). We found that the exine precursors were not stained by the Thiéry (1967) reaction for neutral polysaccharides, whereas the investing sporocyte wall stained densely (Brown & Lemmon 1986). The aniline-blue test for callose was positive, indicating that the exine precursors contain callose. This is an unusual and unexpected finding. Callose encases and isolates developing microspores of seed plants (Currier 1957) and is said to be present in the sporocyte wall in the Marchantiidae (*Riccia*, Beer 1906; *Geothallus*, Doyle 1962). Callose had not previously been found in the sporocyte walls of the Jungermanniidae and is not known to occur in moss sporocytes (Neidhart 1979, Lal & Chauhan 1982). The callosic exine precursors thus give evidence of an entirely new form of exine precursor in members of the Jungermanniidae. The deposition of callose on the inner surface of the spore special wall has been reported in *Riccia* (Beer 1906) and *Geothallus* (Doyle 1962), but no specific relationship to exine patterning was shown. Recently, evidence that callose isolates developing spores of *Coleochaete* suggests that callosic isolation in sporogenesis is a plesiomorphic character among land plants (Graham & Taylor 1986). The sporadic record of exine precursors in four different orders, Calobryales, Jungermanniales, Metzgeriales, and Treubiales emphasizes the need for additional studies of the distribution of this character and assessment of its importance in deciphering phylogentic relationships within the Jungermanniidae.

The first activity leading to the permanent exine is the production of tripartite lamellae (TPL) external to the spore cytoplasm (Fig. 24, 30). Exine lamellae in liverwort spores were first observed in early studies using TEM when they were described as "platelets" in *Riccardia* (Horner et al. 1966) or simply as "lamellae" in *Scapania* (Rowley & Southworth 1967). Such TPL are associated with initiation of the exine in

Figures 23–25. Early wall development in spores of the hepatic *Fossombronia wondraczekii* (Corda) Dum. **23,** Young spores of a tetrad still encased by the sporocyte wall (SW), each spore is invested by a primexine (PE) after cytokinesis, N nucleus, x2,900; **24,** Tripartite lamellae form at the surface of the plasma membrane, x64,600; **25,** Accumulation of lamellae give rise to multilamellate layers, x100,000.

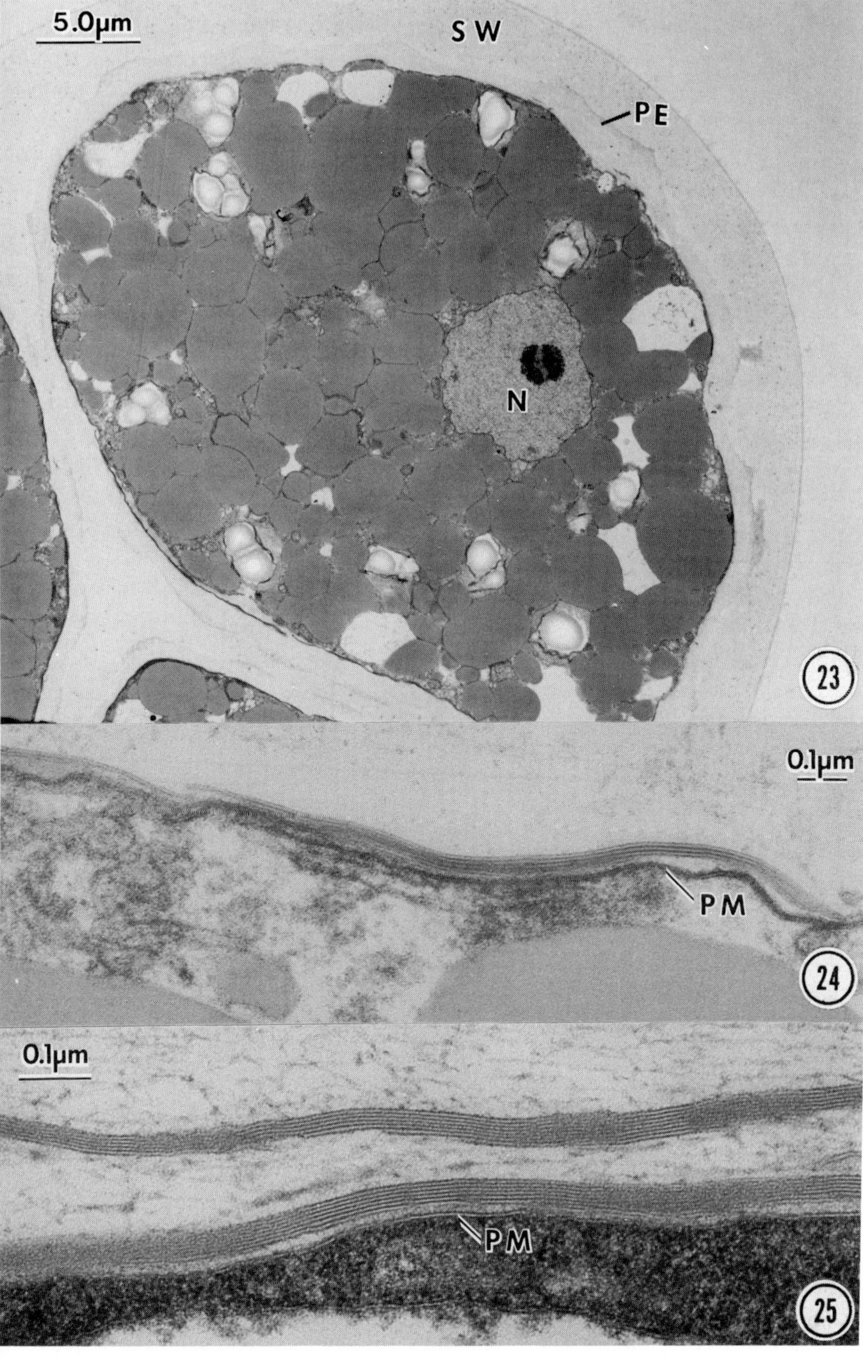

5.0μm
SW
PE
N
23
0.1μm
PM
24
0.1μm
PM
25

spores of most hepatics, vascular cryptogams, and in pollen grains of a number of seed plants. (For a review, see Brown & Lemmon 1986.) In seed plants lamellate exine is usually restricted to the inner layer of unsculptured nexine, but in cryptogams lamellae may be involved in the outer sculptured zone of exine as well (Fig. 32, 33). Of the free-sporing plants, only the hornwort *Phaeoceros* (Ridgway 1965), the hepatic *Corsinia* (Wiermann & Weinert 1969), and the moss *Andreaea* (Brown & Lemmon 1984a, Filina & Filin 1984) have been reported to lack tripartite lamellae in exine initiation. Variously named as white-line structures, platelets, sheet lamellae, slip lamellae, folia, feuillets or ensembles exiniques elementaires, these structures are similar in all groups. They are of unit-membrane dimension and consist of two osmiophilic layers separated by a translucent layer, the "white-line" that is often observed in mature spore walls (Fig. 32.).

Hepatic spores are unusual in that TPL, or layers compounded from TPL, may occur thoughout the entire exine. In a comparative study of spore walls in the Jungermanniales, Heckman (1970) recognized two types of lamellae based on relative dimensions, slip lamellae of definite rectangular dimension and sheet lamellae of indefinite expanse. The lamellae occur as single tripartite structures with a white-line core and are thickened on both faces by the addition of sporopollenin. In mature spores the outer lamellar face is often the most heavily coated. Single lamellae may be variously oriented and may occur throughout the spore wall, as in the simple spore walls of the Jungermanniales.

More recently, a second major category of lamellar organization has become apparent, that of a multilaminar band consisting of several to many TPL in a strict parallel array (Fig. 25–27, 30). This compound unit is thickened only on the two exposed outer faces. One or more of these multilaminar layers form an inner exine layer that is continuous around the spores of certain liverworts (Fig. 25, 26, 30). This feature of the spore wall has been reported in descriptions of mature walls in *Pallavicinia* (Heckman 1972), *Fossombronia* (Steinkamp & Doyle 1984), and *Athalamia* (Steinkamp & Doyle 1981). It seems quite likely that spore exines in the majority of the Marchantiidae are composed of several to many multilaminar layers (Steinkamp 1973, Denizot 1974, 1976, Steinkamp & Doyle 1979, 1981, Thaithong 1982).

The origin and chemical nature of tripartite structures in exine development are not understood (Rowley & Dunbar 1967, Buchen & Sievers 1981). TPL are intensely stained by the phosphotungstic-chromic acid (PTA-CA) post-stain. This stain is known to have a high affinity for carbohydrates and glycoproteins and preferentially stains the plasma membrane of plant cells (Roland et al. 1972). The staining of

Figures 26, 27. Late spore wall development in the hepatic *Fossombronia wondraczekii*. **26**, Outer exine (E2) accumulates outside the earlier formed multilamellate exine (E1), intine (I) forms adjacent to the cytoplasm, x27,000; **27**, Nearly mature wall comprising sculptured exine (E2), multilamellate inner exine (E1) and innermost intine (I), x17,500.

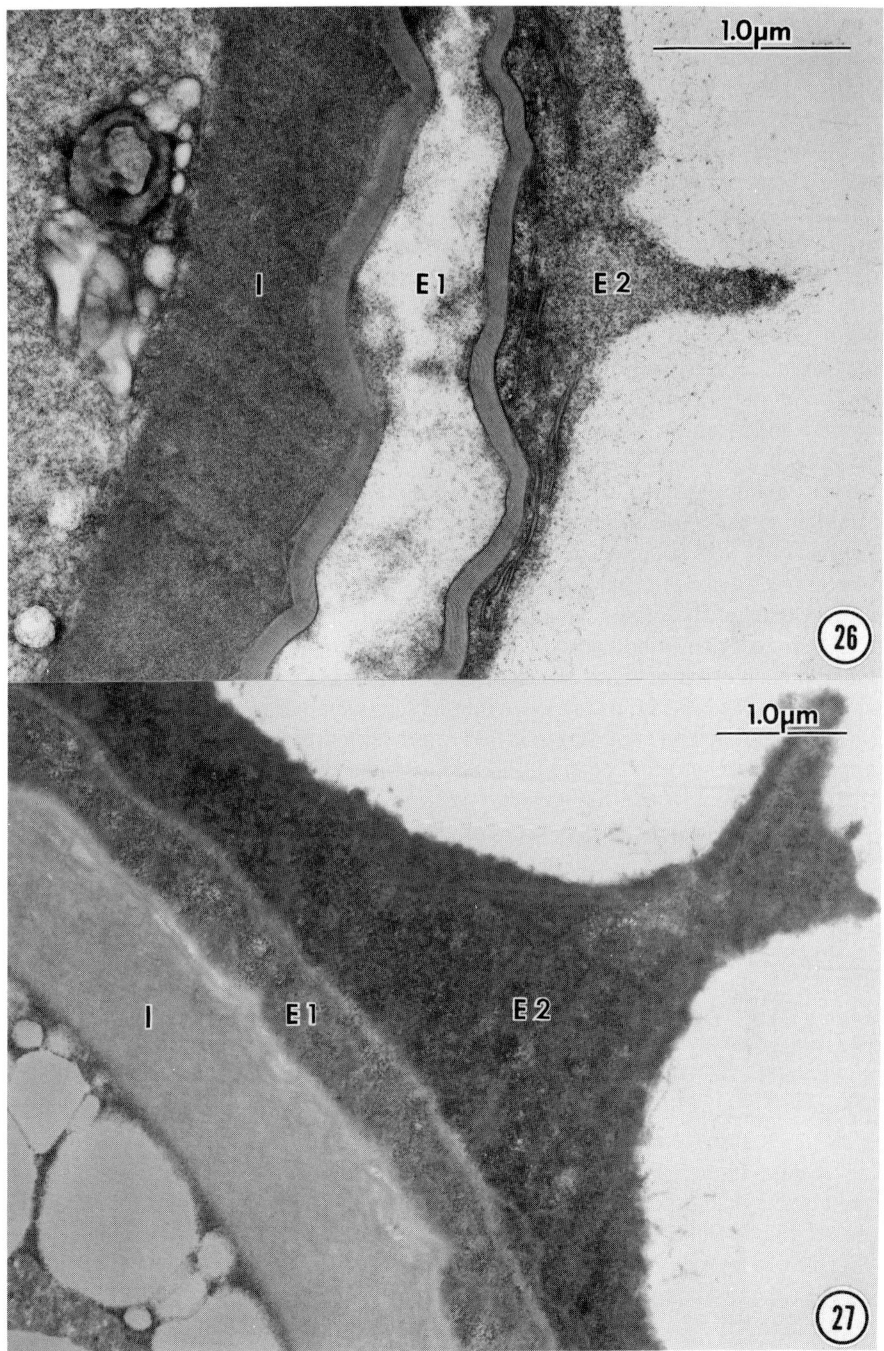

1.0μm
I
E 1
E 2
26
1.0μm
I
E 1
E 2
27

TPL is probably due to the presence of glycoproteins or lipoproteins, which are known to occur in plant exines (Rowley et al. 1981). There is a growing body of indirect evidence that the tripartite exine precursors polymerize outside the cell, as proposed by Horner et al. (1966). Studies of the prominent "A" layer in the spore wall of *Sphagnum* showed unconsolidated precursor material outside the cell membrane. This material may contribute to the construction of tripartite structures (Brown et al. 1982a). The development of tripartite slips in the exine template of *Haplomitrium* may occur at some distance from the plasma membrane (Brown & Lemmon 1986). In a recent review of pollen wall development, Dickinson (1982) concluded that tripartite structures involved in nexine development are formed directly at the surface of the plasma membrane. It is possible that certain wall materials of a complex lipid nature may assume a trilayered configuration.

The production of TPL in the large ornate spores of *Fossombronia* is a striking process and promises to provide an experimental system in which the process could be studied in depth. The TPL first appear as isolated bowl-shaped platelets. These overlap and fuse and a multilaminar layer is built up centripetally by the apposition of lamellae at the surface of the plasma membrane (Fig. 24, 25). As in *Sphagnum* (Brown et al. 1982a), there seems to be some compression of lamellae during fusion. In both *Fossombronia* and *Sphagnum* the lamellae of the fused layer are noticeably thinner than newly formed lamellae at the plasma membrane. As the first formed lamellae are similar in both structure and PTA-CA staining properties to the plasma membrane, it is not possible to state beyond a doubt that plasma membrane fragments are not involved in TPL formation. Microtubules were occasionally seen beneath the plasma membrane in early stages of TPL production. However, microtubules did not appear to figure as prominently in *Fossombronia* spore wall ontogeny as they did in the development of the "A" layer of exine in spores of *Sphagnum* (Brown et al. 1982a). The microtubules may serve to variously shape the plasma membrane, thus forming spaces in which wall materials may be deposited, as suggested by Schnepf (1974), or to direct enzyme complexes or wall precursor material to the site of localized formation of TPL.

Spore Wall Development in Mosses

Spore wall development in mosses (Fig. 34–66) also involves the production of tripartite lamellae in all groups except the Andreaeopsida. Wall initiation apparently occurs in the absence of the spore special wall described for liverworts. In mosses the developing intersporal septa fuse with the sporocyte wall to form a continuous wall around the young spores of the tetrad. No additional wall material is laid down within the sporocyte wall on the distal spore surfaces. The first visible sign of exine initiation is typically seen on the distal surface just outside the plasma membrane. Our studies suggest that exine initiation in bryopsid and sphagnopsid mosses is fundamentally like that in hepatics and pteridophytes, i.e., sporopollenin accumulation on tripartite

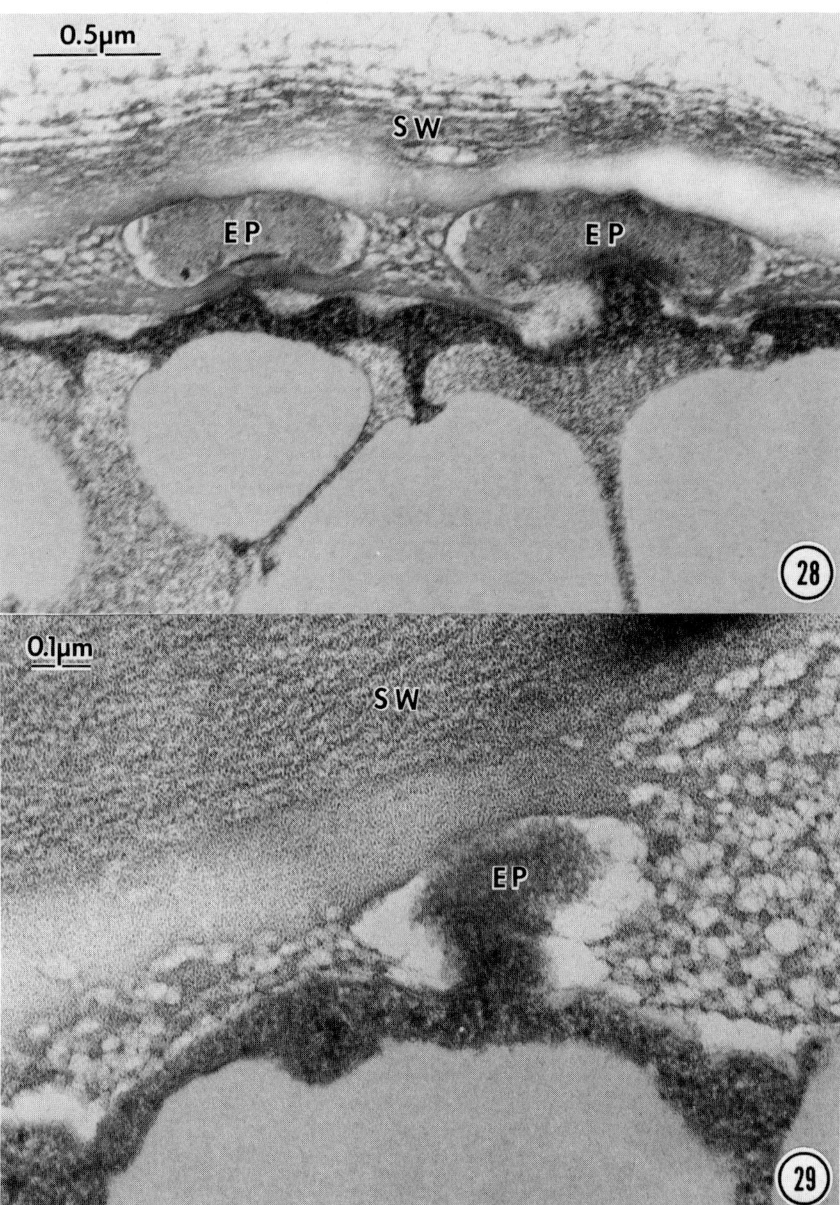

Figures 28, 29. Exine precursors in prophase I sporocytes of hepatics. **28,** *Haplomitrium hookeri,* exine precursors (EP) that cap projections of the cytoplasm form within the sporocyte wall (SW), x35,300; **29,** *Apotreubia nana* (Hatt. & H. Inoue) Hatt. & Mizut, exine precursors (EP) as in Fig. 28, x87,500.

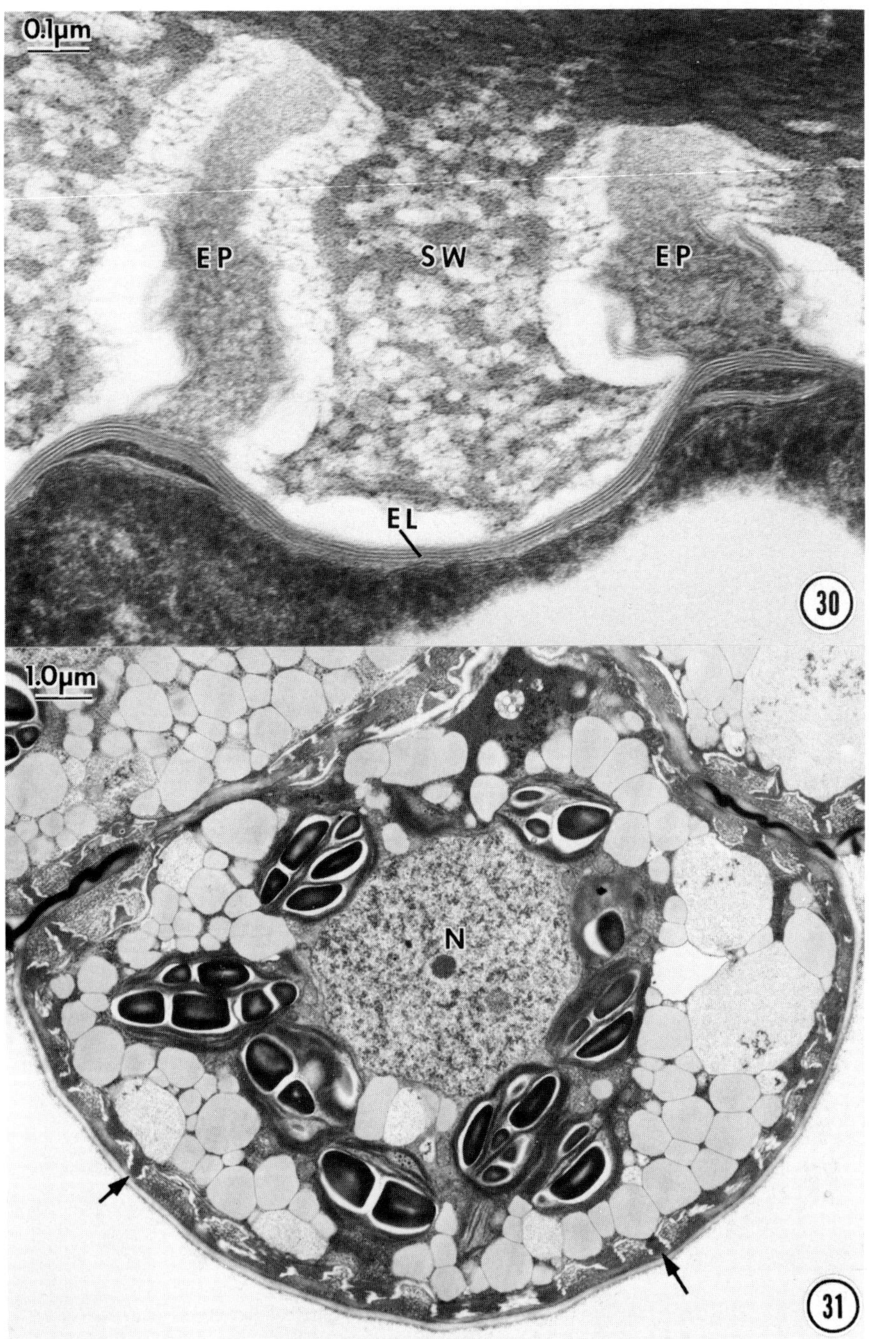

Figures 30, 31. Exine precursors in the hepatic *Pallavicinia lyellii* (Hook.) Carruth. **30,** Exine precursors (EP) within the sporocyte wall (SW) are underlain by exine lamellae (EL), x83,500; **31,** Young spore of a tetrad showing exine precursors (some indicated by arrows) on all surfaces, the central nucleus (N) is surrounded by starch filled plastids, x7,700.

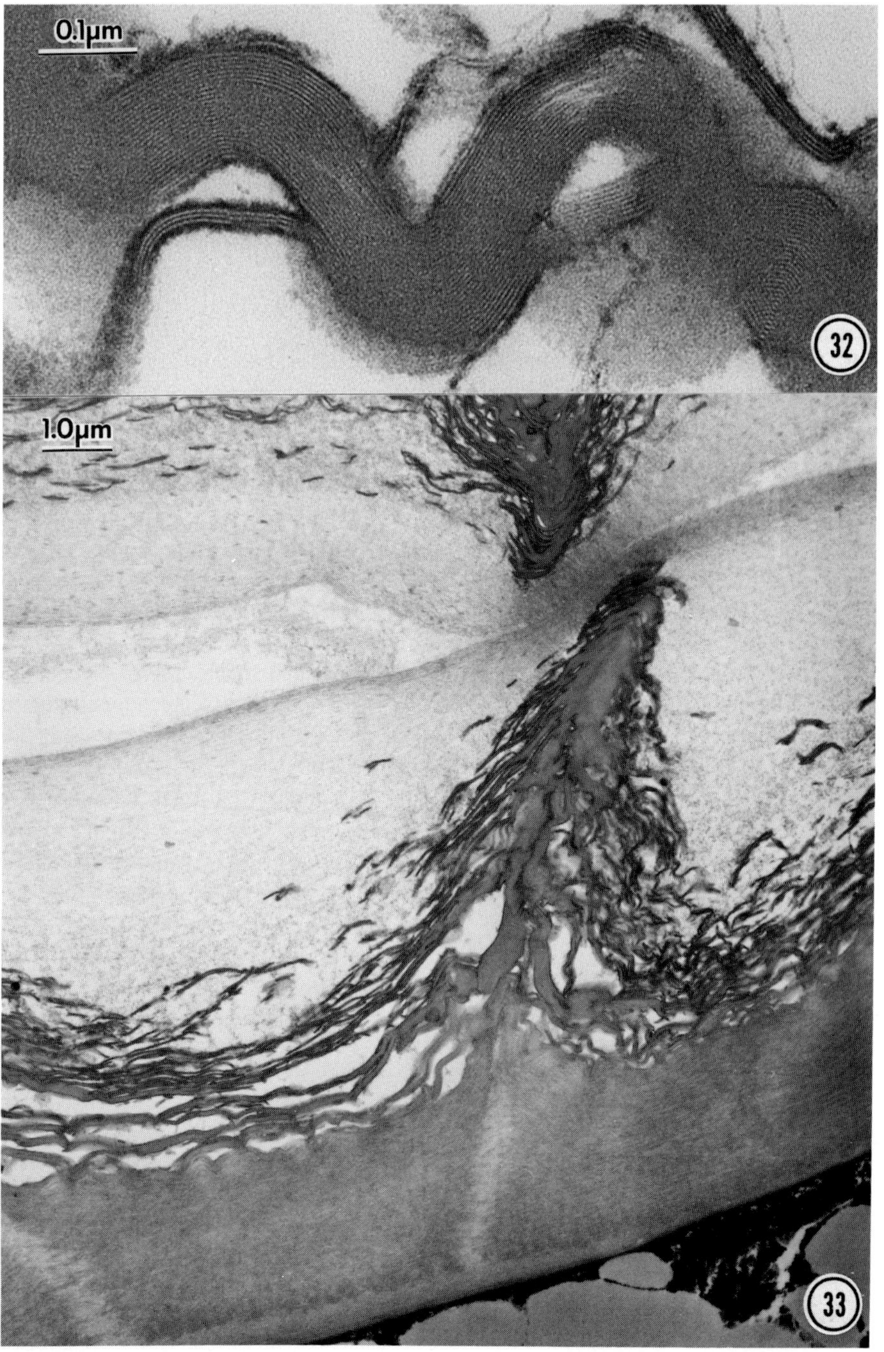

Figures 32, 33. Nearly mature spore wall in the hepatic *Oxymitra paleacea* Bisch. **32,** Multilamellate layer of exine, x133,000; **33,** The sculptured spore exine is compounded of many multilamellate layers of the type illustrated in Fig. 32, x9,500.

lamellae of near unit-membrane dimensions. The appearance of a white line, termed the foundation layer (Brown & Lemmon 1984c, 1985b), at the base of the exine in bryopsid mosses and the well-developed "A" layer in *Sphagnum* are evidence that spore wall deposition is not always strictly centripetal as appears to be the case in hepatics. Exine is deposited outside the foundation layer in centrifugal fashion. Several authors have suggested that this outer homogenous portion of the exine is at least in part of extrasporal contribution. There is some evidence that the tapetum is secretory at the time that the non-lamellate exine is deposited (Seabury 1975, Brown & Lemmon 1985c, Brown et al. 1982a).

The moss spore wall is further characterized by the deposition of perine, which forms an outermost layer that contributes the final spore wall ornamentation. In some moss spores with a thin unsculptured exine, perine may contribute the only pattern to the spore surface. The term perine has been imprecisely used in the palynological literature (Gastony 1974), and little is known about the synthesis of perine. The perine is generally thought to be of extrasporal origin (Mueller 1974, Mogensen 1983, Brown & Lemmon 1984c, 1985b). Structurally, perine appears similar to lipoidal materials and to sporopollenin, which is a complex lipid polymer of carotenoids and carotenoid esters (Shaw 1971, Southworth 1974). The control of species-specific patterns of perine deposition is not understood.

Comparative studies of wall development have revealed an unexpected diversity of developmental patterns and mature form of the moss spore wall.Three major patterns emerge that correlate well with three phyletic lines in mosses, the Andreaeopsida (Fig. 34–37), Sphagnopsida (Fig. 38–43), and Bryopsida (Fig. 44–66).

Sphagnum (Fig. 38–43) has an extremely complex spore wall consisting of five layers: perine, a unique translucent layer of unconsolidated exine lamellae in a matrix of unknown composition, homogeneous outer exine (the "B" layer), inner lamellate exine (the "A" layer), and intine (Brown et al. 1982a, 1982b). The innermost and earliest formed exine layer consists of 20–30 alternating light and dark layers and appears to be of sporic origin, while the thick, later-formed homogeneous layer of exine could be of exogenous origin (Brown et al. 1982b). The outer translucent layer resembles waxy deposits. Although wax is not usually considered a component of spore walls of bryophytes, it has been reported to occur in *Funaria* (Schulz 1972).

Andreaea (Fig. 34, 35) has a luminate or spongy exine that develops in the absence of the trilamellate elements characteristic of most plant groups (Brown & Lemmon 1984a, Filina & Filin 1984). The exine begins with an accumulation of homogeneous globules of medium electron opacity outside the plasma membrane. This accumulation thickens into an irregular layer with many interstitial spaces. Included

Figures 34, 35. Spore wall development in the moss *Andreaea rothii* Web. & Mohr. **34,** Globular exine (E) is deposited upon the plasma membrane (PM) in absence of tripartite lamellae, x74,000; **35,** Nearly mature wall comprises an outermost perine (Pr), a spongy exine (E) and an innermost intine (I), x35,000.

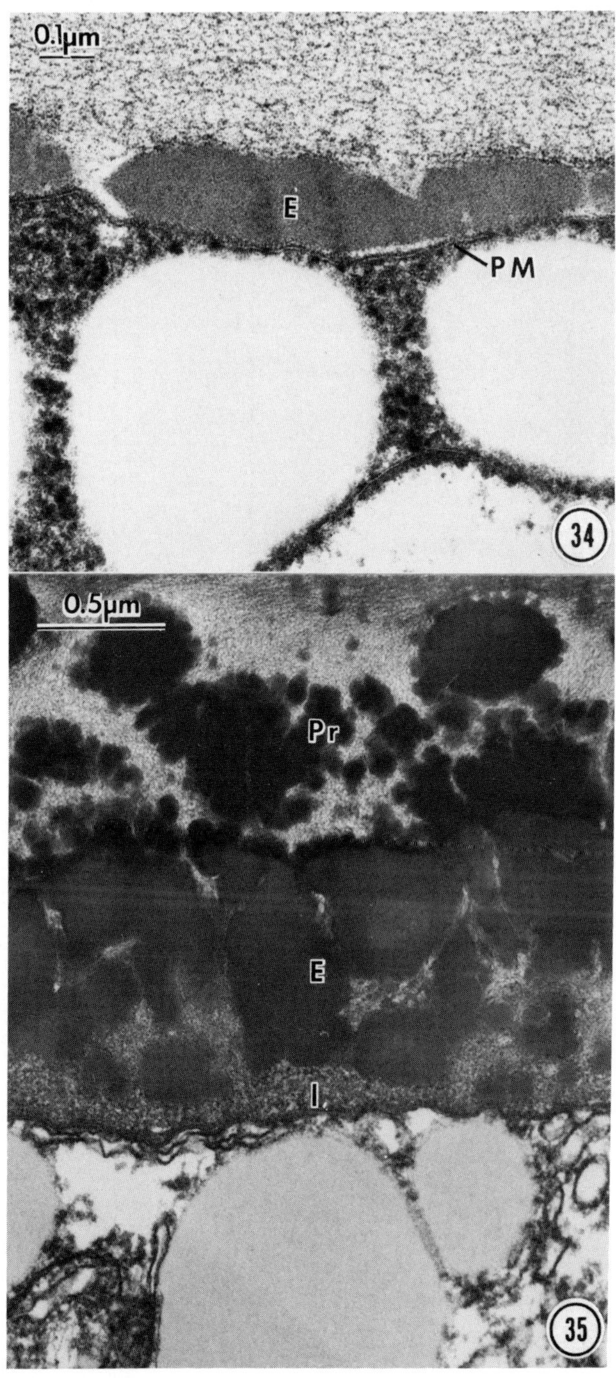
0.1µm
E
PM
34
0.5µm
Pr
E
I
35

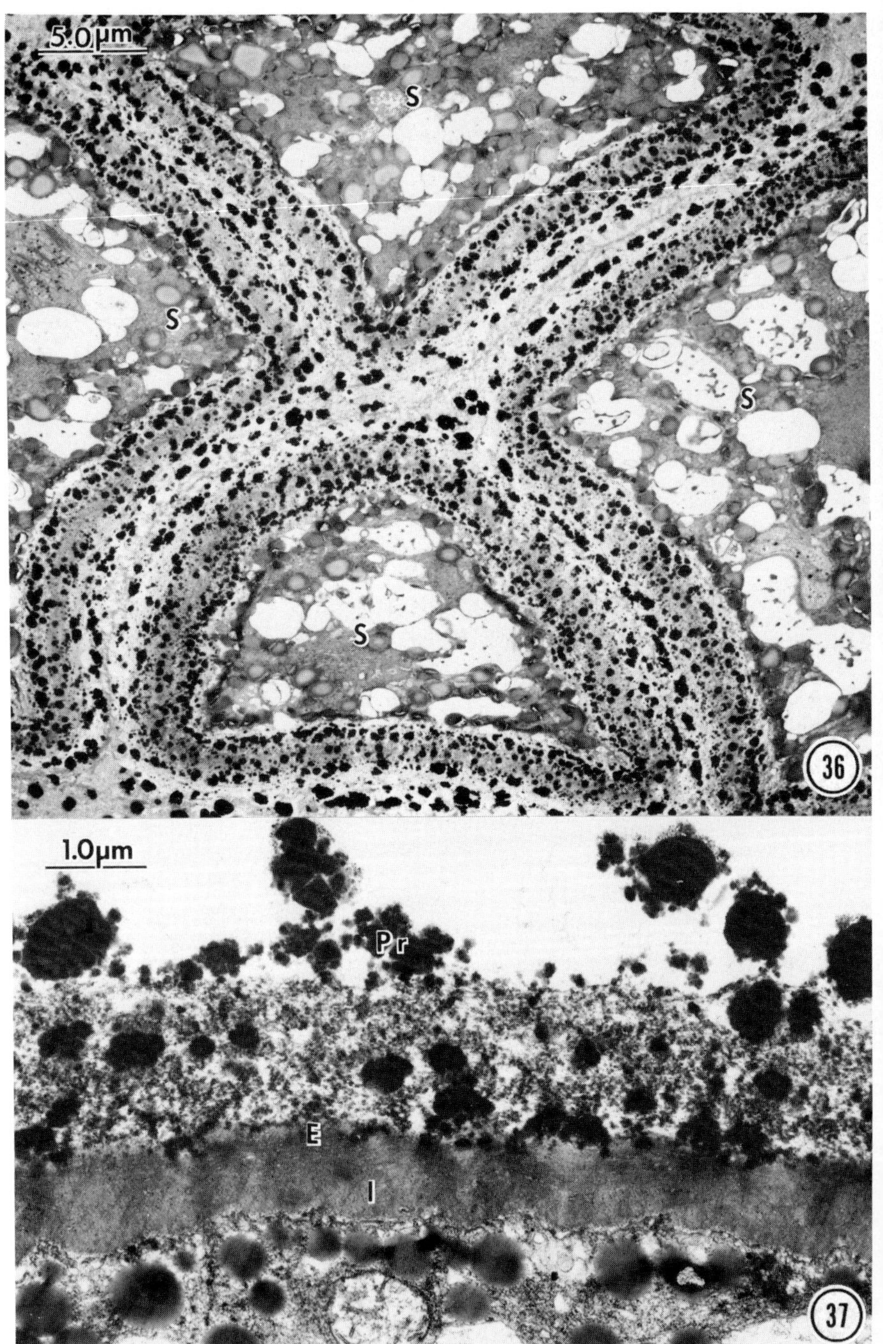

Figures 36, 37. Spore wall development in the moss *Andreaeobryum macrosporum* Steere & B. Murr. **36**, Tetrad of spores (S) showing abundant dark globules of perine, x3,000; **37**, Mature wall consists of outermost globular perine (Pr), scanty exine (E) and innermost intine (I), x13,700.

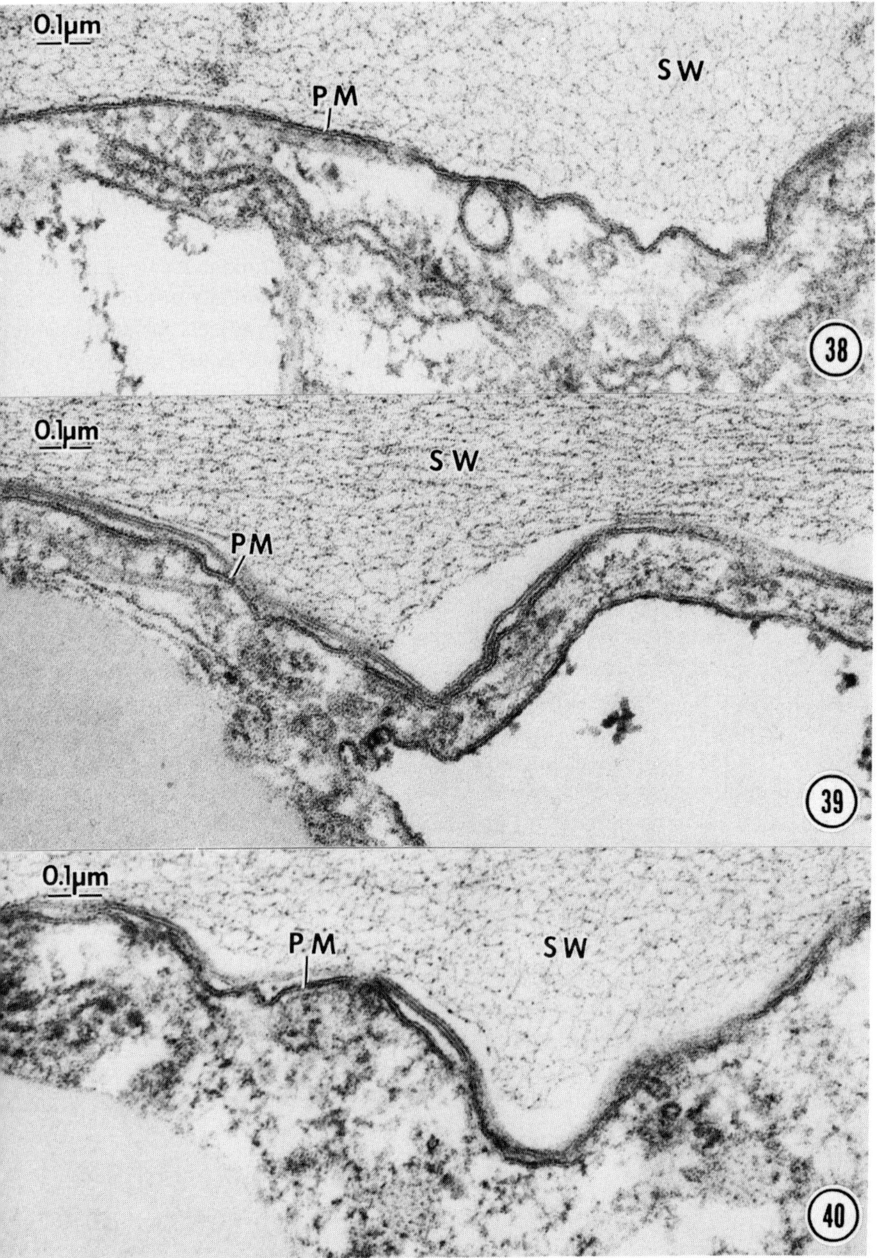

Figures 38–40. Spore wall development in *Sphagnum lescurii* Sull. **38**, Coated plasma membrane (PM) of early spore of a tetrad encased by sporocyte wall (SW), x75,000; **39**, Tripartite strip isolated outside plasma membrane (PM), SW sporocyte wall, x75,000; **40**, Accumulation of tripartite lamellar strips outside the plasma membrane (PM), SW sporocyte wall, x77,000.

in the Andreaeopsida is the genus *Andreaeobryum* (Fig. 36, 37), a taxon of considerable evolutionary significance. Sporophytes of *Andreaeobryum* are considered to be morphologically intermediate between those of the Andreaeopsida and those of the Bryopsida. Our investigation revealed that the mature spore wall of *Andreaeobryum* (Fig. 37), which has a luminate exine and lacks evidence of TPL, is more similar to that of *Andreaea* than that of bryopsid mosses.

The bryopsid-type of exine is typically homogeneous throughout except for an inconspicuous and easily overlooked foundation layer immediately outside the intine (e.g., Fig. 45, 53). The characteristic foundation layer that reflects the participation of tripartite elements in the ontogeny of the bryopsid-type wall has been reported from developmental studies in spores of *Amblystegium* (Brown & Lemmon 1984c) and *Archidium* (Brown & Lemmon 1985b). Additionally, we have found that wall development in *Tetraphis* (Fig. 47–49) of Tetraphididae, *Atrichum* and *Pogonatum* (Fig. 50–54) of Polytrichidae, and *Diphyscium* (Fig. 55–57) of Buxbaumiidae is fundamentally like that in the bryopsid mosses studied to date. Thus, data from this survey support the classification (Crum & Anderson 1981) of Tetraphididae, Polytrichidae, and Buxbaumiidae as subclasses of the Bryopsida. Variation in the mature walls is mainly in thickness and sculpturing of exine, pattern of perine ornamentation, and presence or absence of an aperture. A hilate aperture is characteristic of spores of *Tetraphis, Diphyscium*, and all Polytrichales studied to date. Very similar apertures are found in some bryopsid mosses and appear to be common in spring ephemerals, such as *Archicium, Bruchia, Ditrichum, Trematodon, Weissia* (Seabury 1975, Brown & Lemmon 1980, 1981, 1985b, Rushing 1985).

Microtubules are associated with various aspects of wall development in moss spores, such as exine initiation (Fig. 41, 58), intine ontogeny (Brown & Lemmon 1985b), and aperture development (Fig. 62–66). Microtubules underlie the plasma membrane during both exine (Fig. 41) and intine deposition in *Sphagnum* (Brown et al. 1982a). Microtubules are also associated with the deposition of a unique multilayered intine in the moss *Archidium* (Brown & Lemmon 1985b). The distinct multilayered appearance of the *Archidium* intine is due to "curving" of microfibrils in each layer and straightening at the interfaces of layers. Microtubules in the peripheral cytoplasm are more or less parallel to the microfibrils of the innermost intine. The association of microtubules with wall deposition is well documented in various plant groups, and it seems certain that microtubules are intimately involved with alignment of microfibrils(Robinson & Quader 1982).

Figures 41–43. Spore wall development in *Sphagnum lescurii*. **41,** Early lamellate layer of exine (Ea), the plasma membrane (PM) is underlain by microtubules, SW sporocyte wall, x84,000; **42,** The "B" layer of exine (EB) is deposited outside the earlier formed exine (Ea), intine (I) is deposited adjacent to the plasma membrane, x90,000; **43,** The mature wall consists of perine (Pr), a translucent area (T), "B" layer of exine (Eb), "A" layer of exine (Ea), and an innermost layer of intine (I), scale-like projection of the translucent layer is indicated by an arrow, x40,000.

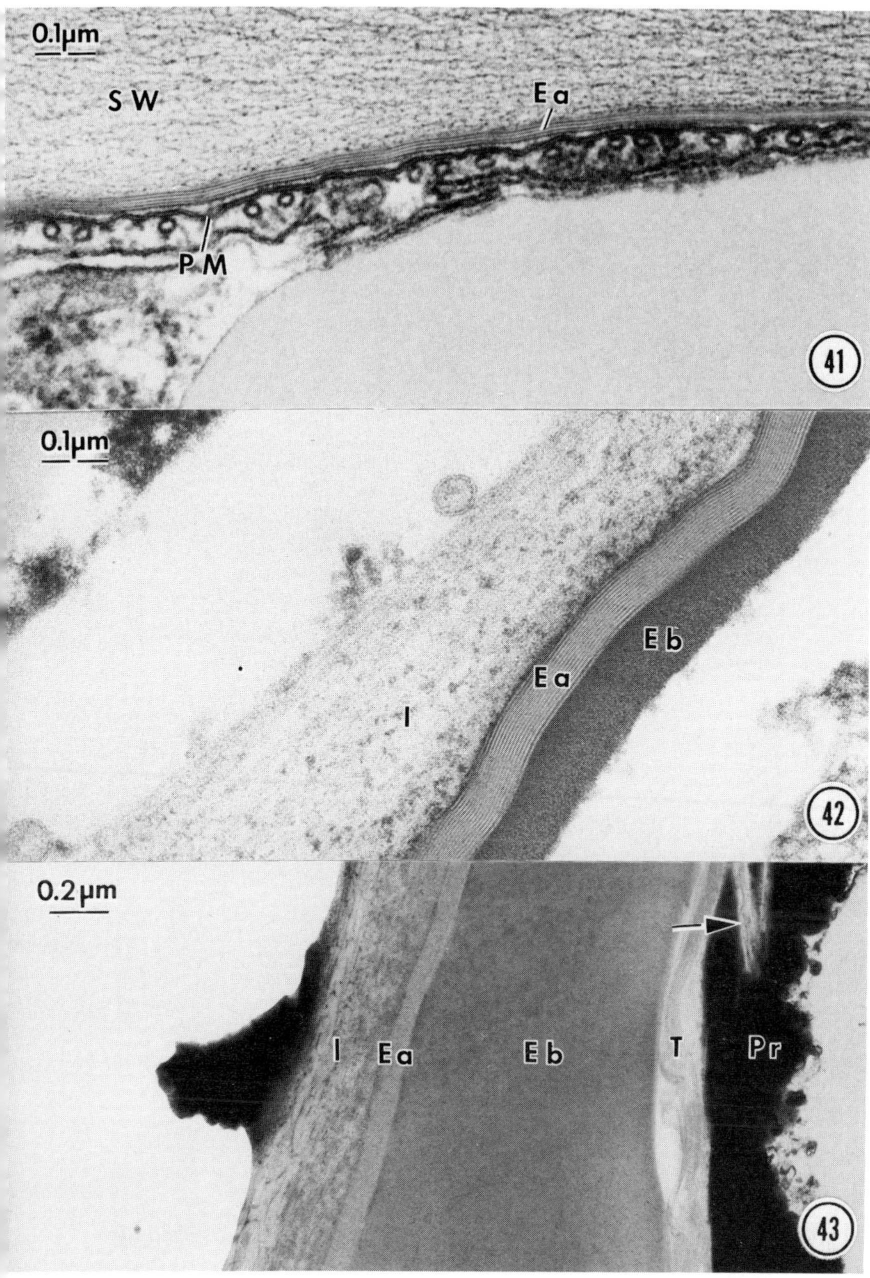
0.1µm
SW
Ea
PM
41
0.1µm
Eb
Ea
I
42
0.2µm
I Ea
Eb
T Pr
43

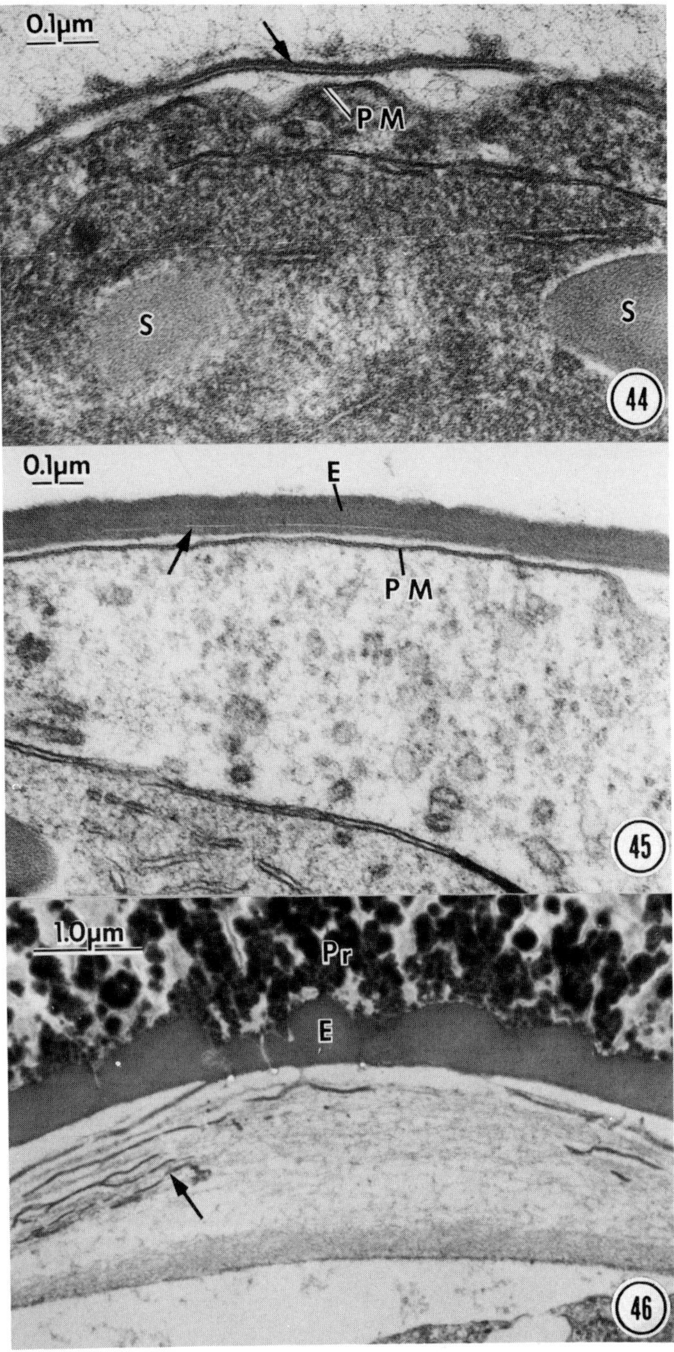

Figures 44–46. Spore wall development in bryopsid mosses. **44,** *Amblystegium ri-parium,* exine initiation is associated with tripartite lamellae (arrow) outside the plasma membrane (PM), S starch, x96,200; **45,** *A. riparium,* exine (E) in young spore released from the tetrad exhibits a single white line, the foundation layer (arrow), PM plasma membrane, x68,800; **46,** *Archidium tenerrimum,* aperture region of nearly mature spore showing splayed exine lamellae (arrow) beneath homogeneous exine (E) coated by perine (Pr) x15,000.

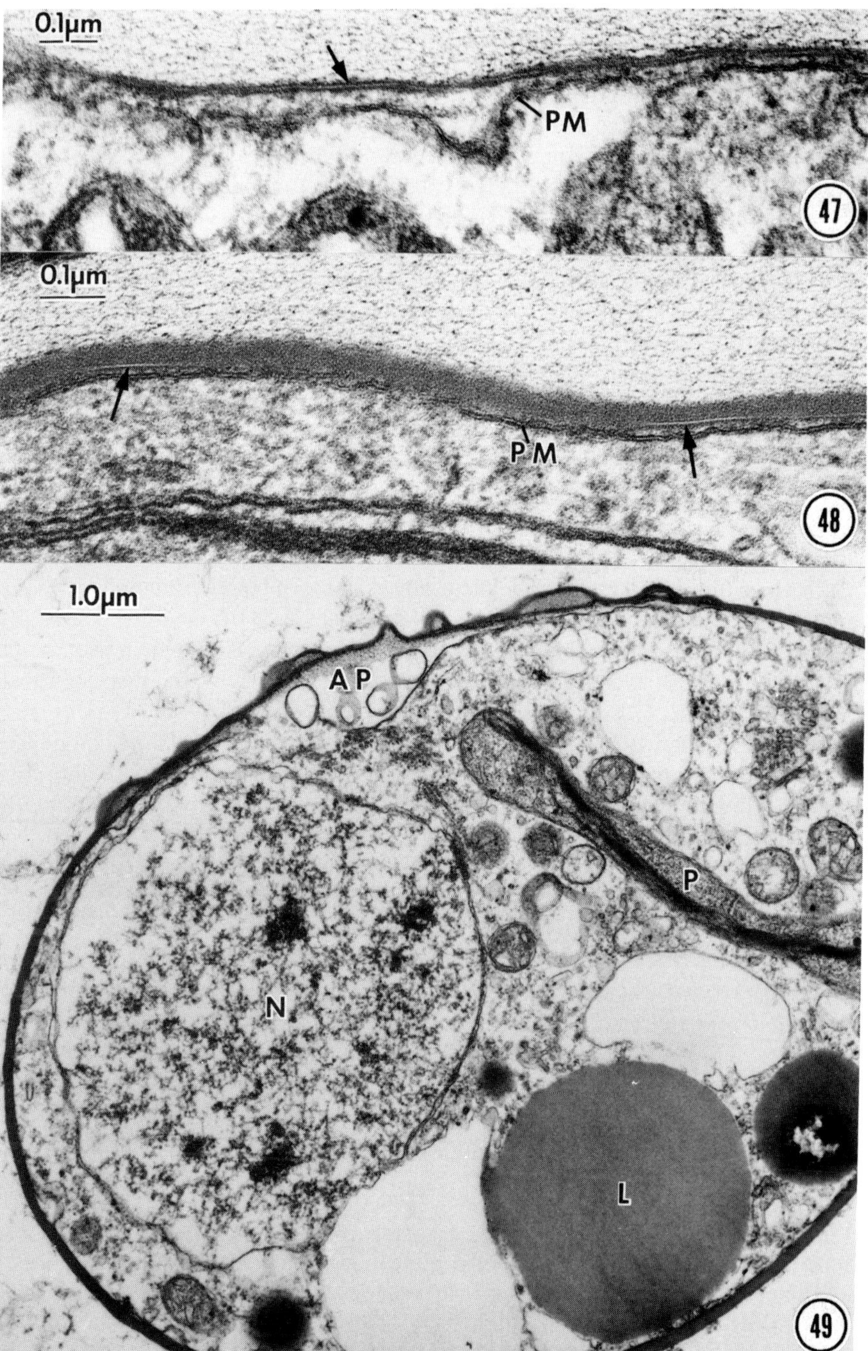

Figures 47–49. Spore wall development in *Tetraphis pellucida*. **47,** The spore wall is in-itiated by tripartite lamellae (arrow) outside the plasma membrane (PM), x81,000; **48,** Early formed homogeneous exine exhibits a foundation layer (arrows), PM plasma membrane, x91,000; **49,** Spore newly released from the sporocyte wall, with proximal aperture (AP), N nucleus, L lipid, P plastid, x17,000.

A radiating system of microtubules in the cortical cytoplasm just below the distal plasma membrane is associated with exine initiation (Fig. 58–61). Immunofluorescence images show that microtubles radiate from an unstained hub like the spokes of a wheel (Fig. 5 *l*, 6i, 60, 61). This radial system of microtubules is associated with a discrete MTOC in the mosses *Tetraphis* (Brown & Lemmon 1983) and *Rhynchostegium* (Fig. 59). Immunofluorescent images reveal unstained regions corresponding to MTOCs (Fig. 6i, 60, 61). This suggests that the MTOC itself does not contain tubulin. A very similar staining pattern has been reported in the blepharoplast of developing sperm where microtubules radiate from unstained MTOCs (Doonan et al. 1986). The concentration of microtubules at the distal pole diminishes only slightly as exine deposition spreads to the spore margins, building up on the distal surface and gradually extending to the proximal surface.

In spores of those mosses in which an aperture develops at the proximal pole, a second highly developed system of microtubules develops after exine deposition advances along the margins of the spore. The sequential development of exine on the distal spore surface followed by development of a complex aperture involving modifications in all three wall layers is an interesting example of polarized morphogenesis in a single cell. The newly organized system of microtubules at the proximal pole signals aperture initiation. In *Tetraphis* the proximal system consists of numerous microtubules that radiate from an electron dense, granular MTOC having the shape of a shallow disc (Fig. 62, 63). In *Ditrichum* (Brown & Lemmon 1980), *Trematodon* (Brown & Lemmon 1981), and *Archidium* (Brown & Lemmon 1985b), microtubules converge at the proximal pole but do not appear to be associated with a discrete MTOC. Microtubules that emanate from the MTOC in *Tetraphis* are organized into more or less discrete bundles.

Numerous spherical vesicles that accumulate in the cytoplasm contribute to expansion of the aperture. At the periphery exine lamellae separate, and the plasma membrane is displaced to form an extensive bowl-shaped region filled with loose fibrillar material. Activity leads to a well-defined aperture consisting of a low annulus surrounding a bowl-like pore underlain by a radiating system of microtubules (Fig. 64–66). It appears that these microtubules may function in directing precursor wall materials to the site of deposition, in a manner similar to microtubules of the phragmoplast.

The mature aperture is a large lens-shaped plug surrounded by an annulus. The exine lamellae splay out at the periphery and separate the plug from the surrounding

Figures 50–54. Spore wall development in the moss *Pogonatum brachyphyllum* (Mx.) P.-Beauv. **50,** Microtubules underlie the plasma membrane during spore wall initiation, x95,200; **51,** Early exine exhibits a foundation layer (arrow), PM plasma membrane, x93,000. **52,** Stacks of lamellae (arrows) occur sporadically at base of homogeneous exine, x93,000. **53,** Considerable amounts of homogeneous exine (E) occur on both sides of the conspicuous foundation layer (arrow), PM plasma membrane, x96,900; **54,** Nearly mature spore wall with perine processes (Pr), homogeneous exine (E) and inner intine (I), x60,000.

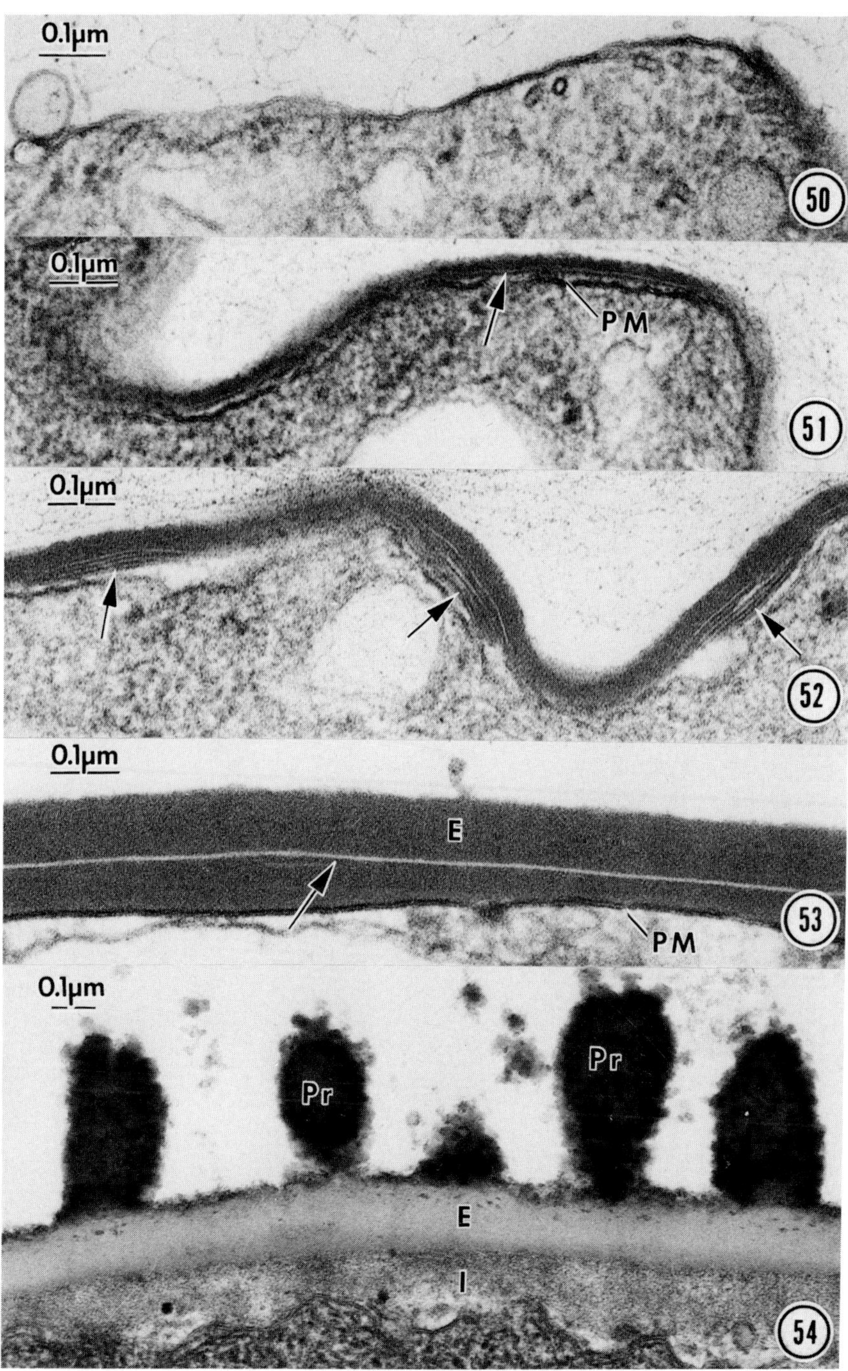
0.1µm
50
0.1µm
PM
51
0.1µm
52
0.1µm
E
PM
53
0.1µm
Pr
Pr
E
I
54

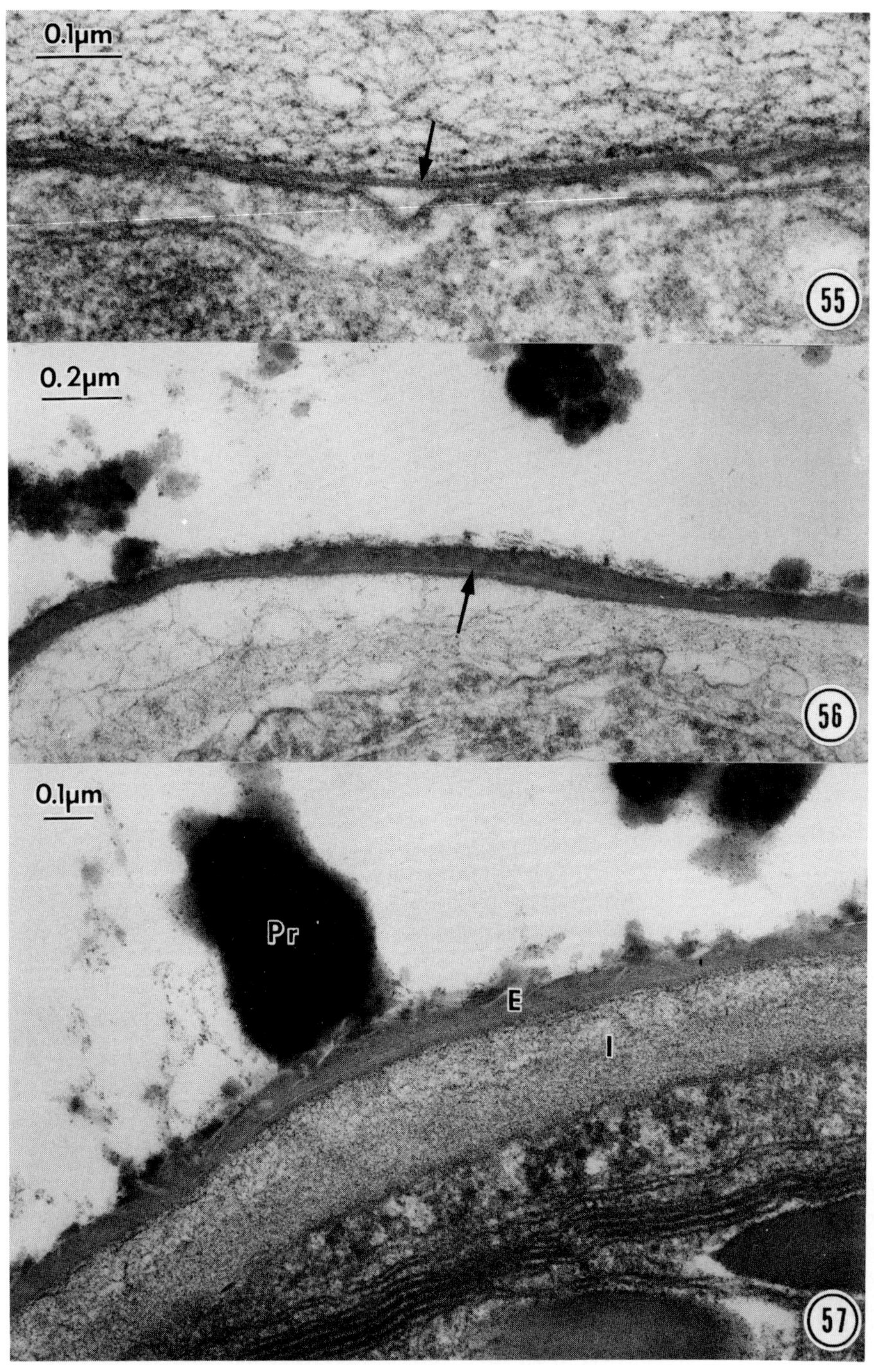

Figures 55–57. Spore wall development in *Diphyscium foliosum* (Hedw.) Mohr. **55,** Exine is initiated by a single lamella (arrow) outside the plasma membrane, x117,000; **56,** The foundation layer (arrow) may appear to consist of a single white line in some places and of several white lines in others, x53,000; **57,** Nearly mature spore wall consisting of perine processes (Pr) outside a thin exine (E), and an innermost intine (I), x70,000.

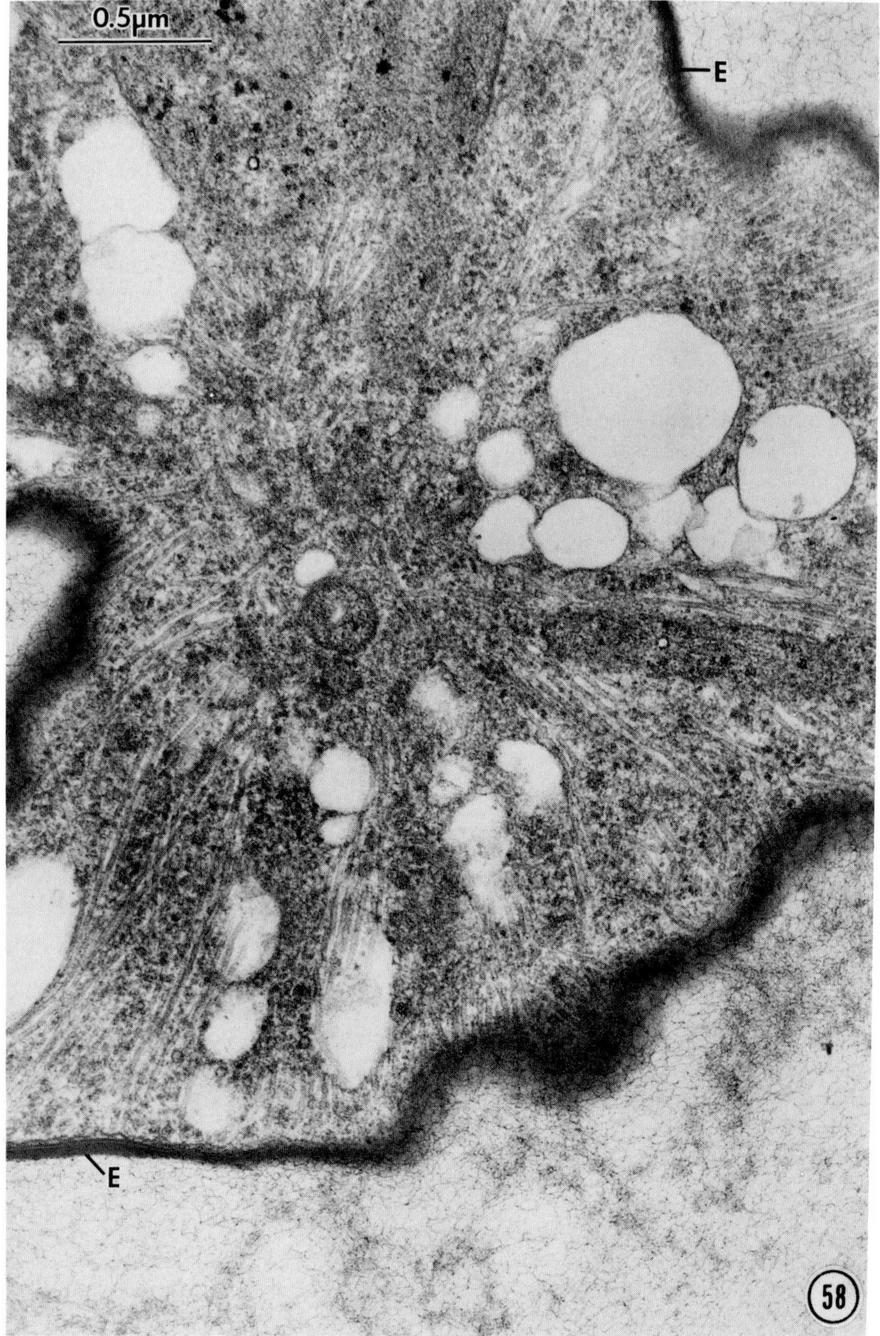

Figure 58. Distal system asociated with exine (E) deposition in young spores of *Trematodon longicollis*. The regular array of microtubules radiates from the region of the plastid, x42,000.

annulus. The aperture is covered by a thin layer of exine. Structural variations have been noticed in the apertures of different mosses. In *Trematodon* (Brown & Lemmon 1981), *Ditrichum* (Brown & Lemmon 1980), *Archidium* (Brown & Lemmon 1985b), and *Tetraphis* (Brown & Lemmon 1983) the aperture is located on the proximal face and is flanked by undulating ridges, which during development were underlain by microtubules radiating from the central pore area. In mature spores of *Trematodon* (Brown & Lemmon 1981), and particularly so in the closely allied genus *Bruchia* (Rushing 1985), SEM images of the proximal surface provided clear evidence of the underlying aperture. In *Atrichum*, *Polytrichum*, and *Pogonatum* a protruding plug is surrounded by a conspicuous annulus. Studies by Reighard (1967) and Olesen and Mogensen (1978) of the mature spores of *Polytrichum* revealed an internal knob-like plug of intine not observed in spores of mosses of the Bryidae. Observations such as these suggest characters of the aperture may prove useful in phylogenetic studies.

The spores of many bryopsid mosses have a simpler wall modification, the leptoma, that serves as the site of germ tube emergence. Ultrastructurally the leptoma consists of an area with a marked thickening of intine, a thin exine, and few or no perine processes (Olesen & Mogensen 1978, Castaldo-Cobianchi et al. 1982). The leptoma is thought to represent an area of spore wettability, allowing imbibition and swelling of the intine to occur. This creates an area of least resistance for germ tube emergence.

Spore Wall Development in Hornworts

The exine of hornworts develops in the absence of tripartite lamellae (Fig. 67–70) that usually characterize the exine of spore walls of hepatics and mosses. This unusual character of exine development in hornworts, first reported by Ridgway (1965), is known in only three other bryophytes, the hepatic *Corsinia* (Wiermann & Weinert 1969) and the mosses *Andreaea* (Brown & Lemmon 1984a) and *Andreaeobryum* (this paper).

Hornworts are regarded as a primitive group of land plants perhaps most closely resembling the advanced charophycean green algae (Crandall-Stotler 1980). Although sporogenesis in hornworts has been studied by Davis (1899), Ajiri and Ueda (1976), and Ridgway (1965), many questions important to the comparative study of

Figures 59–61. Distal system of microtubules associated with exine initiation in bryopsid mosses. **59,** *Rhynchostegium serrulatum*, young spores of the tetrad, a radial system of microtubules (arrows) radiates from a discrete microtubule organizing center (asterisk), N nucleus, P plastid, x28,500; **60,** Indirect immunofluorescent micrograph of distal microtubule system in a young spore of *Bryoandersonia illecebra* (Hedw.) Robins., x2,500; **61,** Same in spores of *Entodon seductrix*, the microtubules radiate from an unstained hub (presumably the MTOC), X2,000.

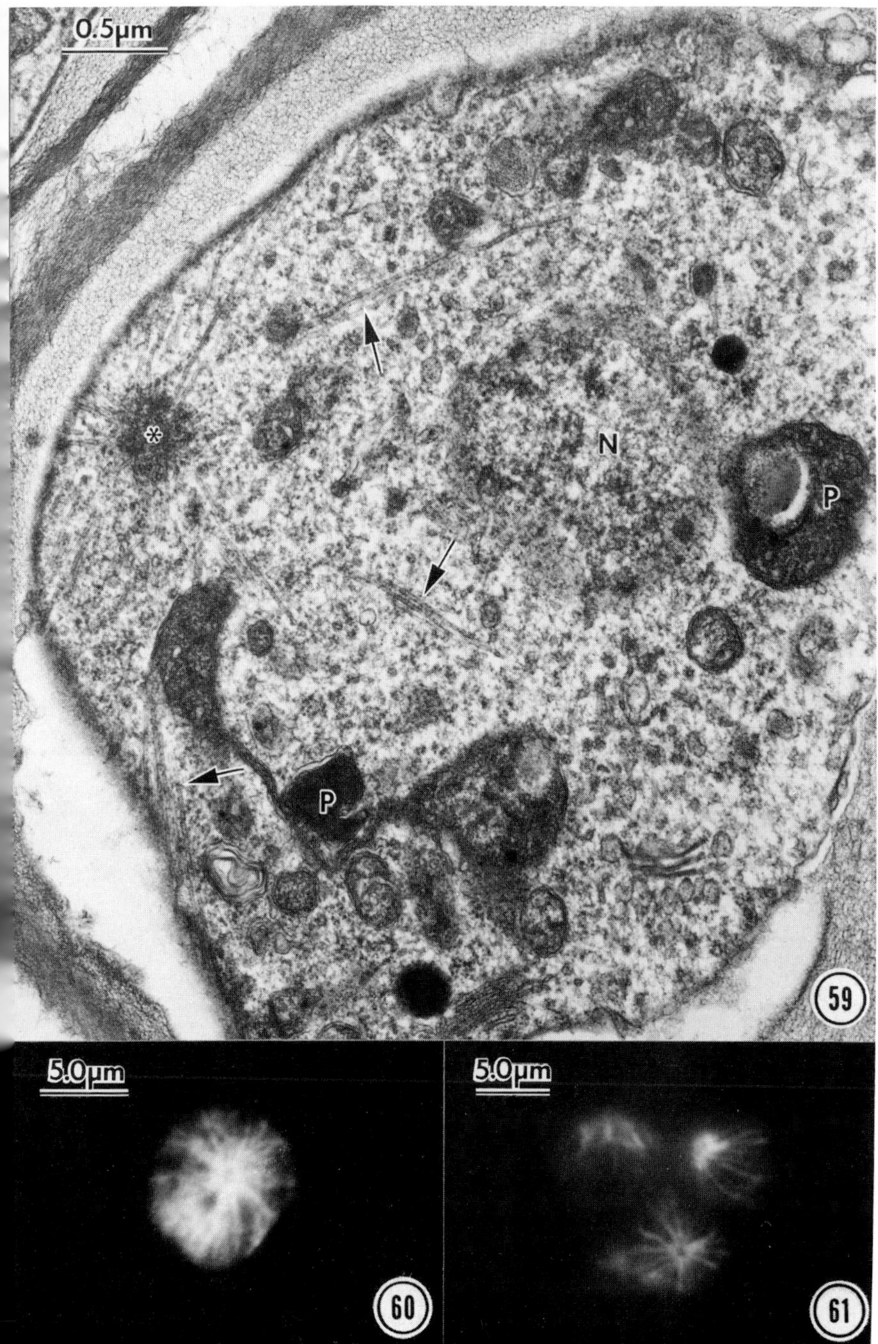
0.5µm
N
P
P
*
59
5.0µm
60
5.0µm
61

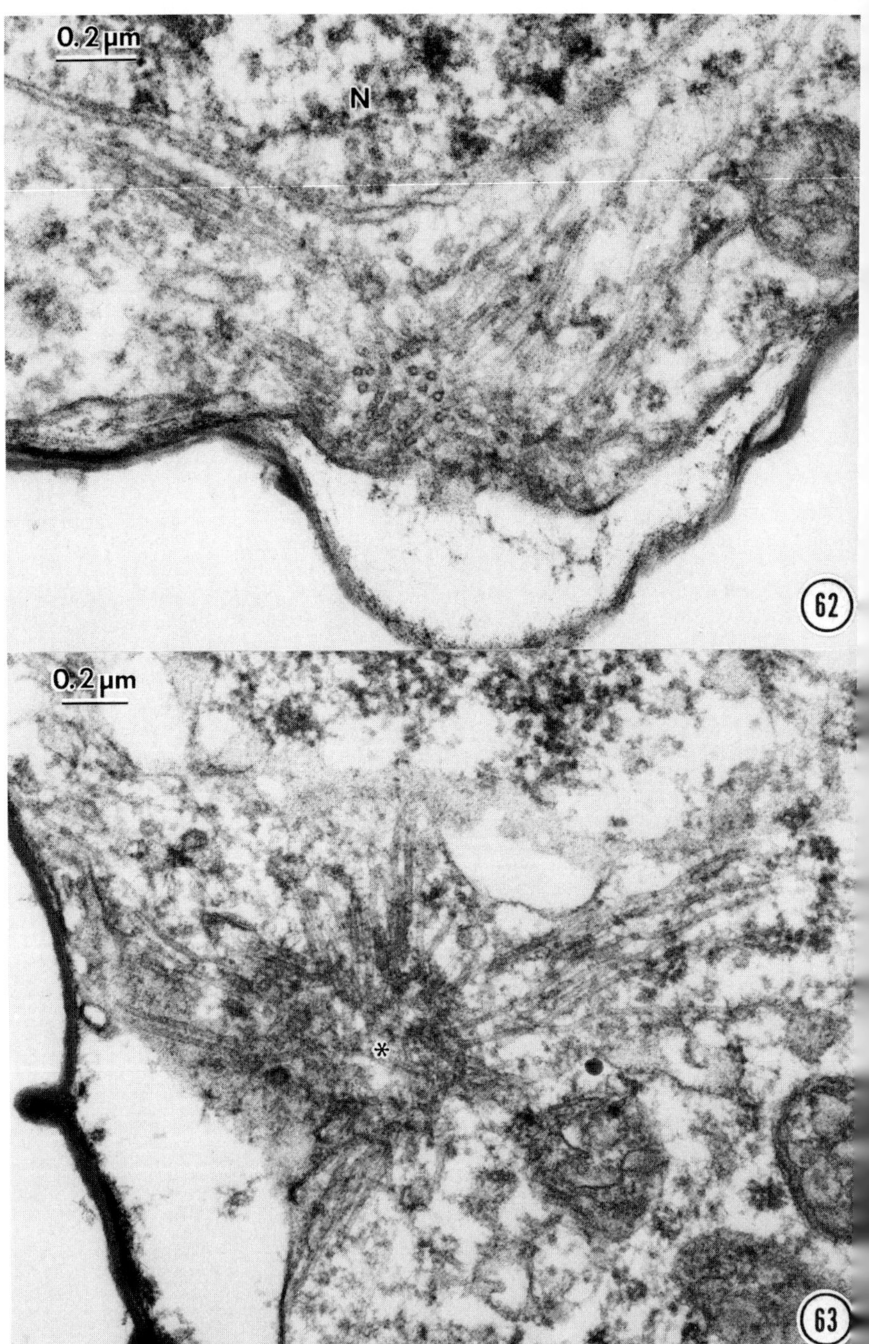

Figures 62, 63. Proximal microtubule system associated with aperture development in spores of *Tetraphis pellucida*. **62,** Microtubules radiate from the proximal pole to ensheath the nucleus (N), x55,000; **63,** Microtubules beneath the developing aperture pore radiate from a discrete MTOC (asterisk), x46,500.

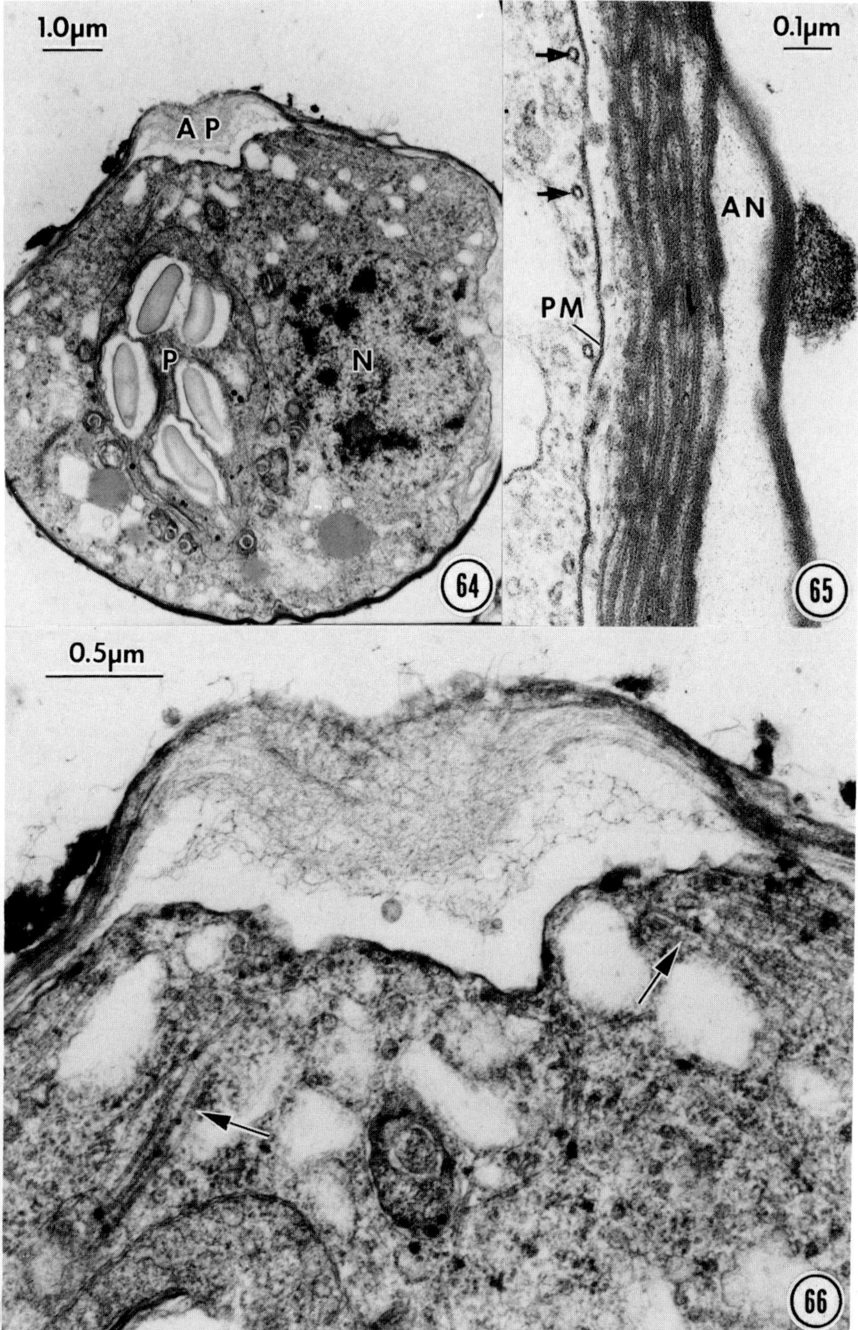

Figures 64–66. Aperture development in spores of *Pogonatum brachyphyllum*. **64,** Immature spore with a proximal aperture (AP), N nucleus, P plastid, x8,000; **65,** Exine lamellae separate annulus (AN) from the aperture pore, microtubules (arrows) underlie the plasma membrane (PM), x65,000; **66,** Enlargement of the aperture pore in Fig. 64 showing the system of microtubules (arrows) associated with development of the pore, x32,000.

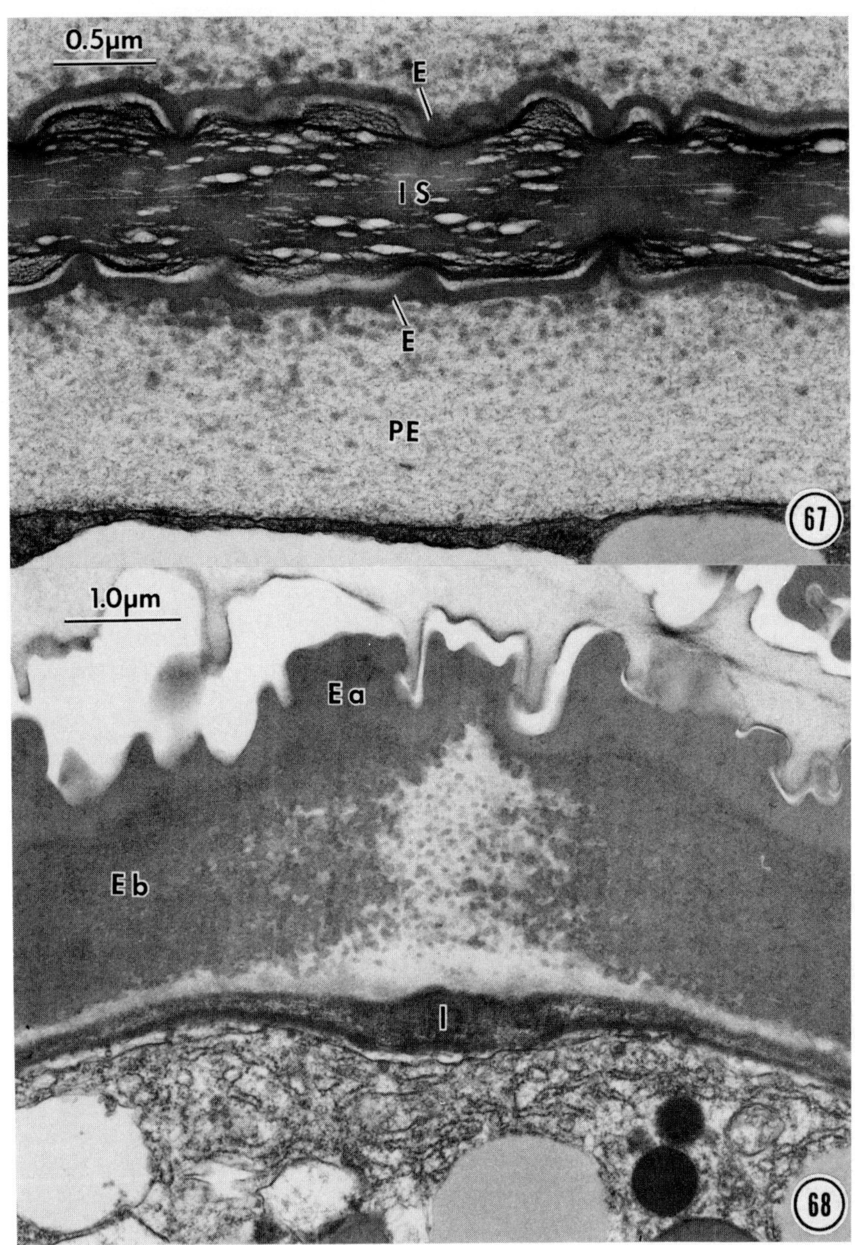

Figures 67, 68. Spore wall development in the hornwort *Notothylas orbicularis*. 67, Portion of two forming spore walls separated by intersporal septum (IS) showing exine (E) consolidated at periphery of the primexine (PE), x28,000; **68,** Later stage in wall development of spores still maintained as a tetrad, exine development is centripetal, the first-formed exine (EA) is the outermost sculptured layer, the later-formed exine (EB) is less consolidated. I intine, x16,000.

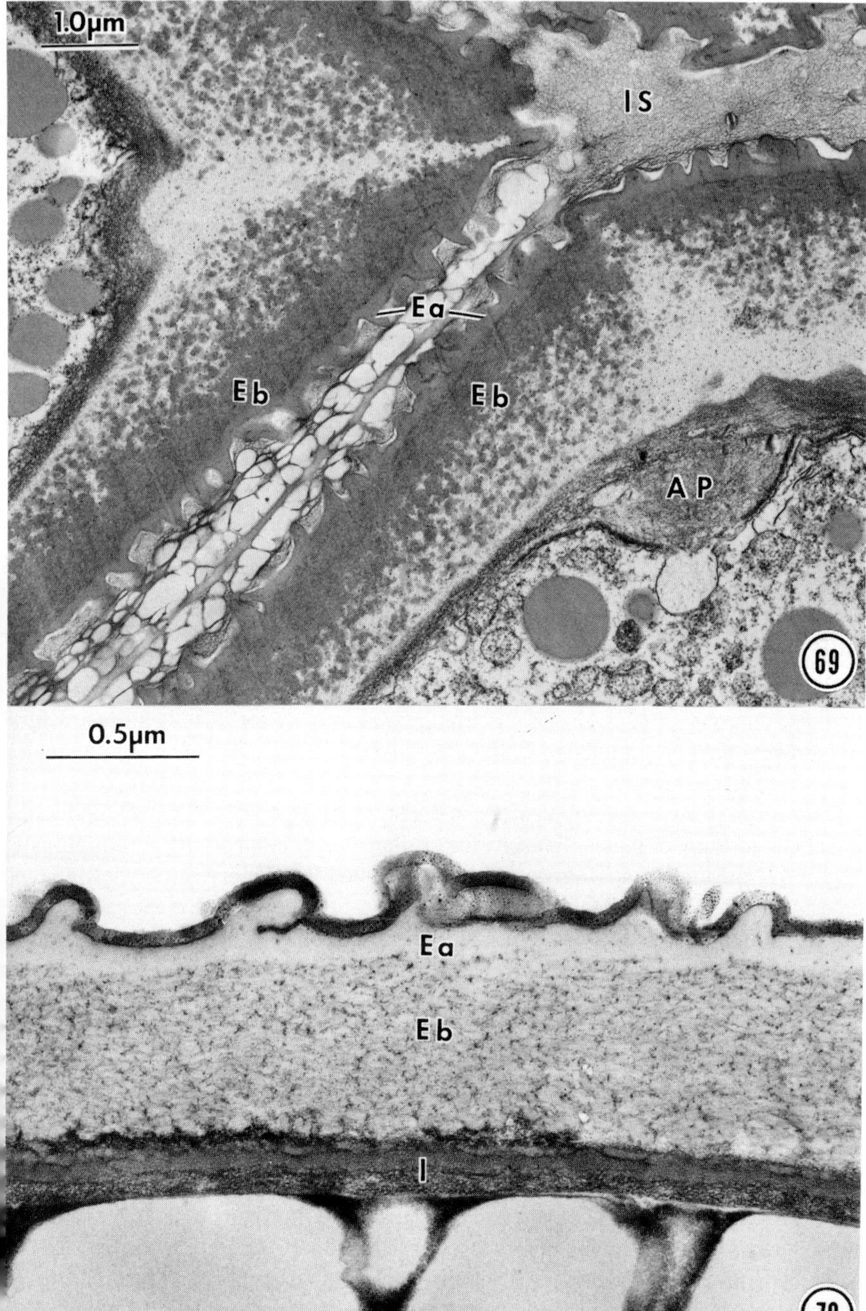

Figures 69, 70. Spore wall development in *Phaeoceros laevis*. **69,** Section through the center of a spore tetrad showing development of an aperture (AP), the exine above the laesurate aperture is poorly consolidated, IS intersporal septum, Ea first-formed exine, Eb later-formed exine, x13,000; **70,** Mature wall consisting of outermost exine (Ea), an inner layer of exine (Eb) that is distinctly textured, and an innermost intine (I), the outer exine (Ea) is evenly coated with an electron dense material resembling perine, x42,000.

sporogenesis remain unanswered. As pointed out by Neidhart (1979), the hornworts seem an ideal system in which to study the entire developmental process of sporogenesis because of the continuous production of sporocytes by a basal meristem. However, problems encountered in fixation and infiltration with resins seriously detract from the value of having all stages of meiosis in a single capsule. Although our observations of hornwort sporogenesis are yet to be completed to our satisfaction, we can report here on at least one character of phylogenetic importance. A spore special wall, which is similar to that of hepatics, is formed after meiotic cytokinesis and serves as a primexine (Fig. 67) in which the exine is deposited. However, exine deposition, which proceeds in the absence of TPL, is not typical of hepatics.

Conclusions

We have attempted to summarize recent progress in understanding the process of sporogenesis in bryophytes. Our approach has been to stress the developmental aspects involved. The resulting developmental information has not yet been integrated into the framework of character states and applied to evolutionary studies. Given the critical evolutionary position of bryophytes as "transitional land plants" between the advanced charophycean algae and vascular cryptogams, we feel that the ultrastructural study of bryophyte reproductive characters will be important in understanding major events in plant evolution.

Only recently have developmental characters of sporogenesis been included in the phylogenetic study of plants. Reasons for this are lack of comparative information on the complex events of sporogenesis, the specialized vocabulary of palynology, and the reluctance of workers to identify homologous features among plant groups. Our developmental approach to the study of sporogenesis has resulted in a better understanding of the comprehensive nature of sporogenesis. The realization that the polarity established early in meiotic prophase is operative throughout sporogenesis has provided a framework of reference for data on subsequent events, e.g., migration of organelles, chromosome movement, and patterning of wall deposition.

One of the most exciting aspects of comparative investigations of bryophyte sporogenesis has been the study of monoplastidy in these simple land plants. The behavior of the single plastid in the early sporocyte and lobing of the cytoplasm in conjunction with plastid migration clearly predicts the planes along which simultaneous cleavage will occur after the two nuclear divisions of meiosis. The complex polarity that gives rise to a tetrahedral arrangement of spores is also operative in the initiation of the exine and aperture. Description of a pre-division microtubule system that predicts the future planes of division has added to understanding of the fundamental nature of the division site in plant cell division. Since monoplastidy is common in algae, all of which lack a pre-prophase band (PPB) in cell division, and is unknown in seed plants, which typically have a PPB, we assume that monoplastidy is the more primitive system. In view of the current interest in the structure and function of the various systems involved in plant cell division, we see great potential for new information from comparative studies of the reproductive cytology of bryophytes.

Spindle ontogeny in monoplastidic sporocytes of mosses (and presumably of hornworts) reflects the unusual quadripolar microtubule system of prophase. Comparative developmental information is needed for hepatics. The sporocytes of Jungermanniidae manifest strong polarity as they are deeply lobed in meiotic prophase. Early reports of a quadripolar origin of the meiotic spindle in liverworts of the Jungermanniidae have not been confirmed by modern methods. Such information is needed so that spindle origin in polyplastidic hepatics can be related to the plastid-based microtubule system that is the precursor of the spindle in mosses. Hepatics of the Marchantiidae are unusual in that they have sporocytes that seem to lack strong polarity. Study of spindle origin in *Conocephalum* showed no evidence of quadripolarity, indicating a basic difference between liverworts of Marchantiidae and the other bryophytes.

Hornworts remain in an isolated position relative to other bryophytes. They share features with both mosses and hepatics and have additional characters, such as a basal meristem, that are peculiarly their own. Monoplastidy, a feature shared by mosses in sporogenesis, occurs throughout the life cycle in hornworts, whereas monoplastidic mitosis has not yet been discovered in mosses. Lobing of the cytoplasm in meiotic prophase is a character shared with mosses and with many hepatics, while the spore special wall is shared with only hepatics.

An unusual system of microtubules centered at the isthmus of the dividing plastid in preprophase of mitosis and the implication of these microtubules in asymmetrical spindle development emphasize the uniqueness of hornworts (Brown & Lemmon 1985a). The division site in hornworts is marked by an axial system of microtubules associated with the dividing plastid, which migrates in preprophase to a position intersecting the future cytokinetic plane. The axially aligned microtubules appear to contribute directly to the formation of the spindle. In monoplastidic mitosis in the vascular cryptogams *Isoëtes* and *Selaginella*, the single plastid similarly migrates to a position intersected by the division site (Brown & Lemmon 1984b). However, the spindle is organized in the perinuclear region in both of these plants and a typical girdling PPB (microtubules in opposite orientation to the majority of those in preprophase of hornworts) marks the division site. The same type of plastid alignment relative to the division site has been reported in preprophase of the monoplastidic green alga *Coleochaete* (Marchant & Pickett-Heaps 1973). Clearly, there is much to be gained by ultrastructural studies of the basic process of mitotic and meiotic cell division in hornworts that will be of fundamental importance in understanding and plant evolution.

A desirable outcome of studies of sporogenesis would be the characterization of the archetypal spore in the bryophytes. Unfortunately this is not yet possible because of the paucity of developmental information on the enormous diversity of bryophyte spores. This is especially apparent from our recent and still incomplete work on the Metzgeriales, Treubiales, and Calobryales. Our studies have shown that the spores of *Haplomitrium* and *Apotreubia* have a multilayered exine patterned by a complex glycocalyx occurring in meiotic prophase. These spores have sporoderms that in many ways approach the complexity of angiosperm pollen grains. We have identified callose in exine precursors of these three members of the Jungermanniidae, a group that was previously thought to lack callose in sporogenesis.

We speculate that the archetypal spore would exhibit the following complex of characters:

1. Lobed sporocytes.
2. Spore special wall functioning as a primexine.
3. Exine initiation associated with production of tripartite lamellae (TPL).
4. Simple sporoderm with a papillate sculpture pattern based on loosely radiating TPL
5. An inner multilaminate layer of one to few TPL.
6. Proximal germination without a specialized aperture.

In Musci three distinct patterns of wall development correlate well with the three major phyletic lines: Sphagnopsida, Andreaeopsida, and Bryopsida. *Andreaea* has a luminate or spongy exine that develops in the absence of TPL. *Sphagnum* has an extremely complex spore wall with two well developed exine layers. The first formed innermost layer is multilaminate, while the thick outer exine is homogeneous. The bryopsid type of exine is typically homogeneous throughout but with a foundation layer that reflects the participation of TPL in exine initiation. Studies of the Tetraphidae, Polytrichidae, and Buxbaumiidae clearly show walls of the bryopsid type.

The greatest diversity in terms of development pattern and mature morphology is found in the Metzgeriales. Complex multilayered spores occur, as well as simple papillate spores similar to those of the Jungermanniales. This group might be expected to hold additional clues to the archetypal bryophyte spore. In many respects the spores of *Riccardia* seem to fit the archetype, although *Riccardia* is not regarded as an especially primitive liverwort. Studies of ontogeny of the large ornate spores of *Fossombronia* and *Petalophyllum* should prove interesting. Recent studies of *Fossombronia* reveal a unique multilayered exine with an inner exine compounded of multilaminate layers and an outer exine that is very similar to the homogeneous exine of mosses and hornworts.

In many respects knowledge of the complex process of sporogenesis in bryophytes has advanced over that of other cryptogams. The recent identification of callose in the walls of meiospores of the green alga *Coleochaete* emphasizes the need for thorough studies of sporogenesis in the charophycean algae to permit comparison with bryophytes. Studies of polarity and the cytoskeleton during development are needed in both algae and vascular cryptogams. Comparative information of this type is essential if we are to use sporogenesis and associated cellular phenomena to further the understanding of land plant evolution.

Acknowlegments

This study was supported in part by NSF grants DEB 8102761 and BSR 8318980 to RCB and BEL and supplemented by an NSF Small College Faculty Opportunity Award to Karen S. Renzaglia for studies of aspects of spore wall development in hepatics. We wish to thank the many bryologists who have generously helped us collect the living bryophytes necessary for developmental studies.

Literature Cited

Ajiri, T. & Ueda, R. 1976. Electron microscope observations on the sporogenesis in the hornwort, *Anthoceros punctatus*. Journal of the Hattori Botanical Laboratory 40: 1–26.

Allen, C. E. 1916. Four-lobed spore mother cells in *Catharinea*. American Journal of Botany 8: 456–460.

Apostolakos, P. & Galatis, B. 1985. Studies on the development of the air pores and air chambers of *Marchantia paleacea* III. Microtubule organization in preprophase-prophase initial aperture cells-formation of incomplete preprophase microtubule bands. Protoplasma 128: 120–135.

Bednara, J., Gielwanowska, I. & Rodkiewicz, B. 1986. Regular arrangements of mitochondria and plastids during sporogenesis of *Equisetum*. Protoplasma 130: 145–152.

Bednara, J. & Rodkiewicz, B. 1985. Distribution of plastids and mitochondria during sporogenesis in *Equisetum hyemale*, pp. 17–19. In: Willemse, M. T. M. & van Went, J. L. (Compilers), Sexual Reproduction in Seed Plants, Ferns and Mosses. Wageningen. Pudoc.

Beer, R. 1906. On the development of the spores of *Riccia glauca*. Annals of Botany (London) 20: 275–291.

Blackmore, S. & Barnes, S. H. 1987. Embryophyte spore walls: Origin, development, and homologies. Cladistics. 3: 185–195.

Blair, M. C. 1926. Sporogenesis in *Reboulia hemisphaerica*. Botanical Gazette (Crawfordsville) 81: 377–400 + 2 plates.

Brown, R. C. & Lemmon, B. E. 1980. Ultrastructure of sporogenesis in a moss, *Ditrichum pallidum*. III. Spore wall formation. American Journal of Botany 67: 918–934.

—— 1981. Aperture development in spores of the moss, *Trematodon longicollis* Mx. Protoplasma 106: 273–287.

—— 1982a. Ultrastructure of meiosis in the moss *Rhynchostegium serrulatum* I. Prophasic microtubules and spindle dynamics. Protoplasma 110: 23–33.

—— 1982b. Ultrastructure of sporogenesis in the moss, *Amblystegium riparium* I. Meiosis and cytokinesis. American Journal of Botany 69: 1096–1107.

—— 1982c. Ultrastructural aspects of moss meiosis: cytokinesis and organelle apportionment in *Rhynchostegium serrulatum*. Journal of the Hattori Botanical Laboratory 53: 41–50.

—— 1982d. Ultrastructural aspects of moss meiosis: Review of nuclear and cytoplasmic events during prophase. Journal of the Hattori Botanical Laboratory 53: 29–39.

—— 1983. Microtubule organization and morphogenesis in young spores of the moss *Tetraphis pellucida* Hedw. Protoplasma 116: 115–124.

—— 1984a. Spore wall development in *Andreaea* (Musci: Andreaeopsida). American Journal of Botany 71: 412–420.

—— 1984b. Plastid apportionment and preprophase microtubule bands in monoplastidic root meristem cells of *Isoëtes* and *Selaginella*. Protoplasma 123: 95–103.

—— 1984c. Ultrastructure of sporogenesis in the moss, *Amblystegium riparium* II. Spore wall development. Journal of the Hattori Botanical Laboratory 57: 139–152.

—— 1985a. Preprophasic establishment of division polarity in monoplastidic mitosis of hornworts. Protoplasma 124: 175–183.

—— 1985b. Phylogenetic aspects of sporogenesis in *Archidium*. Monographs in Systematic Botany from the Missouri Botanical Garden 11: 25–39.

—— 1985c. Development of stomata in *Selaginella*: Division polarity and plastid movements. American Journal of Botany 72: 1914–1925.

—— 1986. Spore wall development in the liverwort, *Haplomitrium hookeri*. Canadian Journal of Botany 64: 1174–1182.

—— 1987a. Division polarity, development and configuration of microtubule arrays in bryophyte meiosis I. Meiotic prophase to metaphase I. Protoplasma 137: 84–99.

—— 1987b. Division polarity, development and configuration of microtubule arrays in bryophyte meiosis II. Anaphase I to the tetrad. Protoplasma 138: 1–10.

Brown, R. C., Lemmon, B. E. & Carothers, Z. B. 1982a. Spore wall development in *Sphagnum lescurii*. Canadian Journal of Botany 60: 2394–2409.

——— 1982b. Spore wall ultrastructure of *Sphagnum lescurii* Sull. Review of Palaeobotany and Palynology 38: 99–107.

Brown, R. C., Lemmon, B. E. & Renzaglia, K. S. 1986. Sporocytic control of spore wall pattern in liverworts. American Journal of Botany 73: 593–596.

Buchen, B. & Sievers, A. 1981. Sporogenesis and pollen grain formation, pp. 349–376. In: Kiermayer, O. (Ed.), Cytomorphogenesis in Plants. Wien and New York. Springer-Verlag.

Castaldo-Cobianchi, R., Giodano, S. & Cafiero, G. 1982. Studies on *Timmiella barbuloides* (Brid.) Moenk, IV. SEM and TEM characterization of spore wall and first germination stages. Journal of Bryology 12: 273–278.

Crandall-Stotler, B. 1980. Morphogenetic designs and a theory of bryophyte origins and divergence. Bioscience 30: 580–585.

Crum, H. A. & Anderson, L. E. 1981. Mosses of Eastern North America. 2 vols. New York. Columbia University Press.

Currier, H. B. 1957. Callose substance in plant cells. American Journal of Botany 44: 478–488.

Davis, B. M. 1899. The spore-mother-cell of *Anthoceros*. Botanical Gazette (Crawfordsville) 27: 89–109 + 2 plates.

— 1901. Nuclear studies on *Pellia*. Annals of Botany (London) 40: 148–180 + 2 plates.

Denizot, J. 1974. Genèse des parois sporocytaires et sporales chez *Targionia hypophylla* (Marchantiales). Justitication de la terminologie utilisée. Pollen et Spores 16: 303–371.

— 1976. Remarques sur l'édification des différentes couches de la paroi sporale à exine lamellaire de quelque Marchantiales et Sphaerocarpales, pp. 185–210. In: Ferguson, I. K.

& Muller, J. (Eds.), The Evolutionary Significance of the Exine. New York and London. Academic Press.

Dickinson, H. G. 1982. The development of pollen. Revue de cytologie et de biologie végétales 5: 5–19.

Dill, F. J. 1964. Dictyotene stage in meiosis of mosses. Science 144: 541–543.

Doonan, J. H., Lloyd, C. W. & Duckett, J. G. 1986. Anti-tubulin antibodies locate the blepharoplast during spermatogenesis in the fern *Platyzoma microphylla* R. Br.: a correlated immunofluorescence and electron-microscopic study. Journal of Cell Science 81: 243–265.

Doyle, W. T. 1962. The morphology and affinities of the liverwort *Geothallus*. University of California Publications in Botany 33: 185–268.

Duckett, J. G. 1986. Ultrastructure in bryophyte systematics and evolution: an evaluation. Journal of Bryology 14: 25–42.

Eleftheriou, E. P. 1985. Microtubules and root protophloem ontogeny in wheat. Journal of Cell Science 75: 165–179.

Farmer, J. B. 1894. Studies in Hepaticae: on *Pallavicinia decipiens* Mitten. Annals of Botany (London) 8: 36–52 + 39 figures.

— 1895. On spore-formation and nuclear division in the Hepaticae. Annals of Botany (London) 9: 469–529 + 74 figures.

— 1904. On the interpretation of the quadripolar spindle in the Hepaticae. Botanical Gazette (Crawfordsville) 37: 63–65.

Farmer, J. B. & Moore, J. E. S. 1905. On the maiotic [sic] phase (reduction divisions) in animals and plants. Quarterly Journal of Microscopical Science 48: 489–557.

Filina, N. I. & Filin, V. R. 1984. The structure and development of the sporoderm in *Andreaea rupestris* Hedw. (Andreaeaceae, Musci). Bjulleten Moskovskogo obscestva ispytatelej prirody. Otdel biologiceskij 89: 86–100. [In Russian, English summary.]

Fowke, L. C. & Pickett-Heaps, J. D. 1978. Electron microscope study of vegetative cell division in two species of *Marchantia*. Canadian Journal of Botany 56: 467–475.

Galatis, B., Apostolakos, P. & Katsaros, C. 1983. Microtubules and their organizing centres in differentiating guard cells of *Adiantum capillus-veneris*. Protoplasma 115: 176–192.

Gastony, G. J. 1974. Spore morphology in the Cyatheaceae. I. The perine and sporangial capacity: General considerations. American Journal of Botany 61: 672–680.

Genevès, L. 1974. Processus comparés d'élaboration des cloisions, au cours de la mitose et de la méiose, chez *Platyhypnidium ripariodes* (Hedw.) Dix. (Hypnacées). Bulletin de la Société Botanique de France 1974: 73–80.

Graham, L. E. & Taylor, C. 1986. Occurrence and phylogenetic significance of "special walls" at meiosporogenesis in *Coleochaete*. American Journal of Botany 73: 597–601.

Gunning, B. E. S. 1982. The cytokinetic apparatus: Its development and spatial regulation, pp. 229–292. In: Lloyd, C. W. (Ed.), The Cytoskeleton in Plant Growth and Development. New York and London. Academic Press.

Gunning, B. E. S., Hardham, A. R. & Hughes, J. E. 1978. Pre-prophase bands in all categories of formative and proliferative cell division in *Azolla* roots. Planta 143: 145–160.

Heckman, C. A. 1970. Spore wall structure in the Jungermanniales. Grana 10: 109–119.

— 1972. Exine structure and formation in the Jungermanniae, with reference to spore and elater development. Ph. D. Dissertation. University of Massachusetts. Amherst.

Hepler, P. K. 1976. The blepharoplast of *Marsilea*: its *de novo* formation and spindle association. Journal of Cell Science 21: 361–390.

Heslop-Harrison, J. 1963. Ultrastructural aspects of differentiation in sporogenous tissue. Symposia of the Society of Experimental Biology 17: 315–340.

— 1968. Wall development within the microspore tetrad of *Lilium longiflorum*. Canadian Journal of Botany 46: 1185–1192.

Hogan, C. J. 1985. Absence of pre-prophase band microtubules in meiotic higher plant cells. Journal of Cell Biology 101 (5,2): 149a.

Horner, H. T., Jr., Lersten, N. R. & Bowen, C. C. 1966. Spore development in the liverwort *Riccardia pinguis*. American Journal of Botany 53: 1048–1064.

Inoue, H. & Hibino, R. 1984. Studies on spore morphology of hepatics (1). Metzgeriales. Bulletin of the National Science Museum Tokyo, Series B, 10: 177–189.

Klasterska, I. 1976. A new look on the role of the diffuse stage in problems of plant and animal meiosis. Hereditas 82: 193–204.

Klasterska, I. & Ramel, C. 1979. Prophase of plant meiosis: Sequences and interpretation of stages. Genetica 51: 15–20.

Kreitner, G. L. & Carothers, Z. B. 1976. Studies of spermatogenesis in the Hepaticae V. Blepharoplast development in *Marchantia polymorpha*. American Journal of Botany 63: 545–557.

Lal, M. & Chauhan, E. 1982. Cytochemical studies on sporogenesis in *Physcomitrium cyathicarpum* Mitt. Nature of the spore mother cell wall. Cryptogamie, Bryologie et Lichénologie 3: 51–57.

Lambert, A.-M. 1974. Ultrastructure de l'appareil fusorial en méiose chez la mousse *Mnium hornum* L. Bulletin de la Société Botanique de France 121: 93–96.

— 1977. La méiose chez les Bryophytes: ultrastructure et dynamique du fuseau chez une mousse, *Mnium hornum* Hedw. Bryophytorum Bibliotheca 13: 113–145.

— 1980. The role of chromosomes in anaphase trigger and nuclear envelope activity in spindle formation. Chromosoma 76: 295–308.

Ledbetter, M. C. & Porter, K. R. 1963. A "microtubule" in plant cell fine structure. Journal of Cell Biology 19: 239–250.

Lee, K. W. 1982. Ultrastructural study of sporogenesis in *Psilotum*. Botanical Society of America Miscellaneous Publications 162: 17–18.

Luomajoki, A. 1985. Duration of meiosis in *Ceratodon purpureus* (Ditrichaceae, Musci). Annales Botanici Fennici 22: 361–365.

Marchant, H. J. & Pickett-Heaps, J. D. 1973. Mitosis and cytokinesis in *Coleochaete scutata*. Journal of Phycology 9: 461–471.

Marengo, N. P. 1977. Ultrastructural features of the dividing meiocyte of *Onoclea sensibilis*. American Journal of Botany 64: 600–601.

Meyer, K. I. 1929. Die Entwicklung des Sporogons bei *Fagetella conica*. (Untersuchungen über den Sporophyt der Lebermoose IV). Planta 8: 36–54.

Mogensen, G. S. 1983. The spore, pp. 325–342. In: Schuster, R. M. (Ed.), New Manual of Bryology. Vol. I. Nichinan, Japan. Hattori Botanical Laboratory.

Mohl, H. von. 1839. Ueber die Entwicklung der Sporen von *Anthoceros laevis*. Linnaea 13: 273–290 + 1 plate. (Reprinted 1845. In: Vermischte Schriften botanischen Inhalts. Tübingen. L. F. Fues).

Moore, A. C. 1905. Sporogenesis in *Pallavicinia*. Botanical Gazette (Crawfordsville) 40: 81–96.

Mueller, D. M. J. 1974. Spore wall formation and chloroplast development during sporo-
    genesis in the moss *Fissidens limbatus*. American Journal of Botany 61: 525–543.
Nageli, C. 1844. Zellkerne, Zellenbildung, und Zellenwachstum bei den Pflanzen. Zeit-
    schrift für Wissenschaftliche Botanik 1: 34–118.
Neidhart, H. V. 1979. Comparative studies of sporogenesis in bryophytes, pp. 251–280.
    In: Clark, G. C. S. & Duckett, J. G. (Eds.), Bryophyte Systematics. New York and Lon-
    don. Academic Press.
Newton, M. 1983. Cytology of the Hepaticae and Anthocerotae, pp. 117–148. In: Schust-
    er, R. M. (Ed.), New Manual of Bryology. Vol. I. Nichinan, Japan. Hattori Botanical La-
    boratory.
Olesen, P. & Mogensen, G. S. 1978. Ultrastructure, histochemistry and notes on ger-
    mination stages of spores in selected mosses. The Bryologist 81: 493–516.
Oltmann, O. 1974. Licht- und Elektronenmikroskopische Untersuchungen zur Sporogenese
    von *Lophocolea heterophylla* (Schrad.) Dum. I. Die Entwicklung von der Sporenmutter-
    zelle bis zum Tetradenstadium. Pollen et Spores 16: 5–25.
Parihar, N. S. 1965. An Introduction to Embryophyta. Vol. I. Bryophyta. 377 pp. Alla-
    habad, India. Central Book Depot.
Pickett-Heaps, J. D. & Northcote, D. H. 1966. Organization of microtubules and en-
    doplasmic reticulum during mitosis and cytokinesis in wheat meristems. Journal of Cell
    Science 1: 109–120.
Ramsay, H. 1983. Cytology of mosses, pp. 149–221. In: Schuster, R. M. (Ed.), New
    Manual of Bryology. Vol. I. Nichinan, Japan. Hattori Botanical Laboratory.
Reighard, J. A. 1967. Light and electron microscopic studies on spore germination and
    bud apical meristems in *Polytrichum juniperinum* Hedw. and *P. ohioense* Ren. & Card.
    Ph. D. Dissertation. University of Illinois. Urbana.
Ridgway, J. E. 1965. Some aspects of morphogenesis, biotic coaction, and ultrastructure
    in the genus *Anthoceros*. Ph. D. Dissertation. University of Texas. Austin.
Robbins, R. R. 1984. Origin and behavior of bicentriolar centrosomes in the bryophyte
    Riella americana. Protoplasma 121: 114–119.
Robinson, D. G. & Quader, H. 1982. The microtubule-microfibril syndrome, pp.
    109–126. In: Lloyd, C. W. (Ed.), The Cytoskeleton in Plant Growth and Development.
    London and New York. Academic Press.
Rodkiewicz, B., Kudlicka, K. & Stobiecka, H. 1984. Patterns of amyloplast distri-
    bution during microsporogenesis in *Tradescantia, Impatiens* and *Larix*. Acta Societatis
    Botanicorum Poloniae 53: 437–441.
Roland, J.-C., Lembi, C. A. & Morré, D. J. 1972. Phosphotungstic-chromic acid as
    a selective electron-dense strain for plasma membranes of plant cells. Stain Technology
    47: 195–200.
Rowley, J. R., Dahl, A. O., Sengupta, S. & Rowley, J. S. 1981. A model of ex-
    ine substructure based on dissection of pollen and spore exines. Palynology 5: 107–152.
Rowley, J. R. & Dunbar, A. 1967. Sources of membranes for exine formation. Svensk
    Botanisk Tidskrift 61: 49–64 + 6 plates.
Rowley, J. R. & Southworth, D. 1967. Deposition of sporopollenin on lamellae of
    unit membrane dimensions. Nature 213: 703–704.
Rushing, A. E. 1985. Spore morphology in the genus *Bruchia* Schwaegr. (Musci). Amer-
    ican Journal of Botany 72: 75–85.

Schnepf, E. 1974. Microtubules and cell wall formation. Portugaliae Acta Biologica, Serie A, 14: 451–462.

Schulz, D. 1972. Darstellung der submikroskopischen Strukturen lufttrockener Moossporen. Berichte der Deutschen Botanischen Gesellschaft 85: 193–202.

Schuster, R. M. 1966. The Hepaticae and Anthocerotae of North America. Vol. 1. 802 pp. New York and London. Columbia University Press.

— 1983. New Manual of Bryology. Vol. I. 626 pp. Nichinan, Japan. Hattori Botanical Laboratory.

— 1984. Evolution, Phylogeny and Classification of the Hepaticae, pp. 892–1070. In: Schuster, R. M. (Ed.), New Manual of Bryology. Vol. II. Nichinan, Japan. Hattori Botanical Laboratory.

Seabury, F., Jr. 1975. Sporogenesis in selected genera of the Musci. Ph. D. Dissertation. Texas A&M University. College Station.

Shaw, G. 1971. The chemistry of sporopollenin, pp. 305–348. In: Brooks, J., Grant, P. R., Muir, M., van Gijzel, P. & Shaw, G. (Eds.), Sporopollenin. New York. Academic Press.

Sheffield, E. & Bell, P. R. 1979. Ultrastructural aspects of sporogenesis in a fern, *Pteridium aquilinum* (L.) Kuhn. Annals of Botany (London) 44: 393–405.

Southworth, D. 1974. Solubility of pollen exines. American Journal of Botany 61: 36–44.

Steer, M. W. 1984. Mitosis in bryophytes. Advances in Bryology 2: 1–23 + 27 figures.

Steinkamp, M. P. 1973. Spore wall ultrastructure in the Hepaticae with special reference to the Sphaerocarpales. Ph. D. Dissertation. University of California. Santa Cruz.

Steinkamp, M. P. & Doyle, W. T. 1979. Spore wall ultrastructure in four species of the liverwort *Riccia*. American Journal of Botany 66: 546–556.

— — 1981. Spore wall ultrastructure in the liverwort *Athalamia hyalina*. American Journal of Botany 68: 395–401.

— — 1984. Spore wall ultrastructure in the liverwort *Fossombronia longiseta*. Canadian Journal of Botany 62: 1871–1879.

Stotler, R. & Crandall-Stotler, B. 1977. A checklist of the liverworts and hornworts of North America. The Bryologist 80: 405–428.

Strasburger, E. 1880. Zellbildung und Zellteilung. pp. 364–365. Jena. Gustav Fischer.

Suire, C. 1970. Recherches cytologiques sur deux hépatiques: *Pellia epiphylla* (L.) Corda (Metzgériale) et *Radula complanata* (L.) Dum. (Jungermanniale). Ergastome, sporogénèse et spermatogénèse. Le Botaniste 53: 125–392.

Thaithong, O. 1982. Fine structure of spore wall in fourteen species of *Riccia*. Journal of the Hattori Botanical Laboratory 53: 133–146.

Thiéry, J. P. 1967. Mise en évidence de polysaccharides sur coupes fines en microscopie électronique. Journal de Microscopie 6: 987–1018.

Vasil, I. K. & Aldrich, H. C. 1970. A histochemical and ultrastructural study of the ontogeny and differentiation of pollen in *Podocarpus macrophyllus* D. Don. Protoplasma 71: 1–37.

Wiermann, R. & Weinert, H. 1969. Untersuchungen zur Sporodermentwicklung bei *Corsinia coriandrina* (Sprengl) Lindb. Berichte der Deutschen Botanischen Gesellschaft 82: 175–182.

Wilson, E. B. 1937. The Cell in Development and Heredity. 3rd Ed. with corrections. New York. Macmillan.

Wolniak, S. M. 1976. Organelle distribution and apportionment during meiosis in the microsporocyte of *Ginkgo biloba* L. American Journal of Botany 63: 251–258.
Wyatt, R. 1977. Spatial pattern and gamete dispersal distances in *Atrichum angustatum*, a dioicous moss. The Bryologist 80: 284–291.

Advances in Bryology 3: 225 – 274 (1988).

The Sporophyte-Gametophyte Junction in Bryophytes

Roberto Ligrone and Raffaele Gambardella

Dipartimento di Biologia Vegetale, Universitá di Napoli
Via Foria, 223, I-80139 Napoli, Italy.

With 42 Figures

Abstract: The histological region that connects the two generations in bryophytes is comprised of the sporophyte foot and the surrounding gametophyte vaginula. In acrocarpous mosses the foot usually penetrates deeply into the leafy stem tissue early in development and forms a tip of collapsed cells. Penetration of the gametophyte tissue in leafy hepatics rarely occurs, and a collapsed foot tip has not been observed. The region of contact between the foot and vaginula is called the placenta. A wall labyrinth is generally found in both sporophyte and gametophyte placental cells. Cells with wall labyrinths are called transfer cells. Gametophyte transfer cells are absent from the placenta in polytrichaceous mosses, while in all hepatics examined so far, gametophyte transfer cells form several layers around the foot. In the anthocerote *Phaeoceros laevis* the placenta consists of sporophyte haustorial cells lacking wall ingrowths, gametophyte transfer cells, and an intervening system of spaces and lacunae containing protein crystals. Placental cells in bryophytes are rich in mitochondria and endoplasmic reticulum, and their plastids are variable in structure. Moreover, these cells exhibit high enzyme activities, including adenosine triphosphatase, acid phosphatase, succinate dehydrogenase, and cytochrome oxidase. Photosynthetic and radioactive tracer studies reveal that (a) the sporophyte is photosynthetically active, but (b) it depends on nutrients supplied by the gametophyte for growth, which (c) are actively absorbed through the transfer cells of the foot. Placental transfer cells provide the driving force for nutrient and water transport from the gametophyte to the sporophyte. This activity requires close functional interaction with the symplasmic and apoplasmic compartments in the adjoining tissues. Some observations indicate that hydroids, which are common in the foot and seta of mosses, play a role in directing nutrient transport towards the capsule. With the onset of spore maturation, the wall labyrinth is obliterated by deposition of new wall material, followed by cytoplasmic degeneration of transfer cells.

Zusammenfassung: Die historische Zone, die die beiden Generationen bei Bryophyten verbindet, enthält den Fuß des Sporophyten und die umgebende gametophytische Vaginula, welche durch Wachstum des Archegoniumstiels entsteht. Bei akrokarpen Moosen dringt der Fuß schon am Anfang der Entwicklung gewöhnlich tief in das Gewebe des beblätterten Stämmchens ein und bildet eine Spitze aus kollabierten Zellen. Bei Lebermoosen wurde dieses Eindringen gametophytischen Gewebes selten beobachtet und kollabierte Fußspitzen wurden noch nie festgestellt. Die Kontaktzone zwischen Fuß und Vaginula wird als Placenta bezeichnet. Im allgemeinen besitzen sowohl die sporophytischen wie die gametophytischen Placentazellen, welche dann Transferzellen genannt werden, ein Wandlabyrinth. Gametophytische Transferzellen fehlen in der Placenta von Polytrichaceae, während in allen bisher untersuchten Lebermoosen die gametophytischen Transferzellen mehrere Schichten um den Fuß herum bilden. Im Hornmoos *Phaeoceros laevis* besteht die Placenta aus sporophytischen Haustorienzellen ohne Wandeinwachsungen, gametophytischen Transferzellen und einem dazwischenliegenden System von Lücken und Vertiefungen, die Proteinkristalle enthalten. Bryophyten-Plazentazellen sind reich an Mitochondrien und endoplasmatischem Reticulum und ihre Plastiden zeigen veränderliche Strukturen. Außerdem besitzen diese Zellen eine hohe Enzymaktivität (Adenosin-Triphosphatase, saure Phosphatase, Succinat-Dehydrogenase und Cytochromoxidase). Photosynthese-Untersuchungen und solche mit radioaktiver Markierung ergaben, daß 1) der Sporophyt photosynthetisch aktiv ist, aber 2) für das Wachstum von den vom Gametophyten gelieferten Nährstoffen abhängt, die 3) aktiv von den Transferzellen des Fußes absorbiert werden. Die Placentatransferzellen ermöglichen den Nährstoff- und Wassertransport vom Gametophyten zum Sporophyten. Hierfür ist eine enge funktionelle Wechselwirkung zwischen den symplasmatischen und apoplasmatischen Kompartimenten der umgebenden Gewebe erforderlich. Einige Beobachtungen deuten darauf hin, daß die in Fuß und Seta der Laubmoose häufig anzutreffenden Hydroiden von Bedeutung für die Ausrichtung des Nährstofftransports zur Kapsel hin sind. Zu Beginn der Sporenreife wird das Wandlabyrinth durch Ablagerung von neuem Wandmaterial zerstört, das Cytoplasma der Transferzellen degeneriert und der Wasser- und Nährstofftransport hört auf.

Keywords: bryophytes, mosses, hepatics, anthocerotes, placenta, anatomy, ultrastructure, development, nutrient transport, transfer cells, sporophyte-gametophyte relationships.

Contents

Introduction

"Bryophytes can be best defined as those archegoniate plants that have the determinate sporophyte generation attached to and partially dependent on the gametophyte generation for the duration of the life cycle" (Vitt 1981). This close relationship is reflected by a high degree of cellular and subcellular specialization existing in the region of contact between the sporophyte and gametophyte.

The lower portion of the sporophyte of bryophytes is inserted deeply in the gametophyte tissue and is usually referred to as the sporophyte foot. The gametophyte forms a multilayered sheath (vaginula) around the foot. The occurrence of a distinctive specialization in this region has been known for a long time. Lorch (1925a, 1925b) first noticed polysaccharide deposits lining the outer tangential walls in the foot epidermal cells in mosses. He also described fine protoplasmic strands penetrating these deposits. Similar findings were reported by Blaikley (1933) who produced accurate drawings of the sporophyte foot in several bryophytes and emphasized the considerable variability in the overall structure of this organ. More recently, Roth (1969) published a fundamental study of the anatomy and development of the sporophyte-gametophyte junction in mosses.

Introduction of the concept of transfer cells by Gunning and co-workers (Gunning et al. 1968, Gunning & Pate 1969) coincided with the first ultrastructural description of cells with a wall labyrinth at the interface between the sporophyte and gametophyte in two moss species (Maier 1967, Eymé & Suire 1967) and a liverwort (Kelley 1969). With additional ultrastructural research it has become evident that transfer cells are almost invariably present in this region, although a great deal of variation has been noticed in their structure, development, and distribution.

Greater knowledge of the structure of the sporophyte-gametophyte junction in bryophytes stimulated study of the nutritional relationships and the pattern of nutrient flow between the two generations. In recent years physiological and autoradiographical studies have demonstrated that nutrients are actively translocated from the gametophyte to the sporophyte and that transfer cells play a central role in this process. Transfer cells are generally restricted to the placenta, a specialized region that includes the epidermal cells of the foot, the adjoining gametophyte cells, and the intervening space between them. Transfer cells are closely integrated structurally and functionally with their neighbors, i.e., the sporophyte and gametophyte tissues on both sides of the placenta. Thus, an attempt to describe the anatomical bases of the nutritional relationships between gametophyte and sporophyte in bryophytes should consider the whole histological complex connecting the two generations. In addition to the placenta this complex comprises the inner tissue(s) of the sporophyte foot and the peripheral cells of the vaginula. In those species of mosses in which the foot penetrates the upper part of the leafy stem, this, too, is to be considered as a part of the sporophyte-gametophyte junction.

The sporophyte-gametophyte junction is a matter of great interest because of its functional, taxonomic, and phyletic implications. However, electron microscopic studies of this histological region are few, especially in hepatics and anthocerotes, Paradoxically, for some bryophytes, physiological information about sporophyte-gametophyte relationships is sometimes greater, though far from exhaustive, than is morphological information. Physiological data are essential for a comprehensive approach to the topic and will be discussed in a separate section of this paper.

Brief summaries of the topic have been published by Hébant (1977), Bopp (1981, 1983), and Frey (1981).

With regard to the names of the taxa considered, we have followed Vitt (1984) and Corley et al. (1981) for mosses, Grolle (1983) for hepatics, and Renzaglia (1978) for anthocerotes.

Mosses

The Mature Sporophyte Foot

Among bryophytes, mosses exhibit the most advanced tissue differentiation in both the gametophyte and sporophyte, notably in conducting tissues (Hébant 1977, 1979). It is not surprising, therefore, that they also possess the most complexly organized sporophyte foot. The foot is generally elongate in shape with a tapering end in the Bryidae and in the Andreaeidae (Blaikley 1933, Roth 1969), whereas a bulbous foot is found in the Sphagnidae (Vaizey 1888, Blaikley 1933, Roth 1969, Crandall-Stotler 1984).

There is a great deal of variation in the inner anatomy of the foot among the species examined. However, the available information is by far too scarce, especially at the ultrastructural level, to give a comprehensive view of the topic. Electron microscopic studies published so far pertain to only eight genera in five families of the Bryidae. Additional information about other taxa is presented in this paper. There is currently no information on the ultrastructure of the foot of the Andreaeidae and the Sphagnidae. Light microscope observations indicate that a wall labyrinth is present in the epidermal cells of the foot in *Andreaea* (Blaikley 1933, Roth 1969), while apparently transfer cells are lacking in the bulbous foot of *Sphagnum* (Blaikley 1933, Roth 1969, Gunning & Pate 1974).

Typically, the foot in the Bryidae is 1–3 mm long and consists of an epidermal layer of tightly appressed cells, a middle parenchyma, and a central strand of conducting tissue (Fig. 1–4). In the lower half of the foot, the epidermal cells possess a prominent wall labyrinth on their outer tangential walls and are referred to as transfer cells. Generally transfer cells are lacking or little specialized in the upper part of the foot, which is similar in structure to the seta because of the presence of a robust peripheral sterome. Major features of foot constitution include (a) the presence of a specialized foot tip, (b) the ultrastructure of transfer cells, (c) the reduction or disappearance of conducting tissues, and (d) the presence of wall ingrowths in inner cortical cells. Another feature to be considered is the level of penetration of the foot into the leafy stem tissue (Roth 1969).

The most complex foot is found in polytrichaceous mosses (Maier 1967, Maier & Maier 1972, Hébant 1975, Caussin et al. 1983). It possesses a conspicuous central strand of hydroids surrrounded by a sheath of elongate living cells that have been interpreted as poorly differentiated leptoids (Hébant 1975). The tapering tip of the foot penetrates deeply into the central strand of the leafy stem and terminates in an appendage of cells that degenerates and collapses precociously. Therefore, no symplasmic barrier is interposed between the water-conducting systems ot the two generations at this point. A prominent wall labyrinth is found in the epidermal cells in the middle and basal region of the foot. Distinctive features of the foot in the polytrichaceous mosses are (a) the presence of abundant intercellular spaces in the inner cortical tissue (Fig. 1), and (b) the presence of a wall labyrinth in the two to three layers of cortical cells beneath the epidermis (Fig. 5). Isolated wall ingrowths have been seen in the inner cortical cells too, generally along the cell walls toward the intercellular spaces (Maier & Maier 1972, Hébant 1975). In the basal region of the foot the hydroids are shorter and wider than in the upper region. Hébant (1975) suggested that these hydroids correspond to the "water-absorbing organ" described by Vaizey (1888). The cells adjoining these hydroids have wall ingrowths on their inner tangential walls and do not show characteristics of leptoids, as expected from their position, but rather of transfer cells because of dense cytoplasm and abundance of rough endoplasmic reticulum and mitochondria (Fig. 6).

In *Buxbaumia piperi* Best the foot parenchyma is rich in intercellular spaces and lacunae, but wall ingrowths are found in the epidermis only. Leptoids are absent and the central strand of hydroids is ensheathed by elongate cells of conducting parenchyma (Ligrone et al. 1982b). In the closely allied *Diphyscium foliosum* (Hedw.) Mohr, the foot does not differentiate a central strand and consists of elongate parenchyma cells with scanty cytoplasm and a peripheral sheath of tubular outgrowths, probably of epidermal origin, divided by thin cross-walls and containing a wall labyrinth

(Blaikley 1933, Roth 1969). In both *Buxbaumia* and *Diphyscium* the foot lacks a tip of necrotic cells and is entirely surrounded by the parenchyma tissue of the vaginula, which is supported by an extremely short gametophyte stem.

Most information about the ultrastructure and development of the sporophyte-gametophyte junction in mosses comes from studies of *Funaria hygrometrica* Hedw. (Wiencke & Schulz 1975, 1978, Browning & Gunning 1979a). In the basal part of the foot parenchyma cells are lacking, and a central strand of hydroids enlarges and comes into contact with the epidermal cells. Cells recognizable as leptoids are present in the upper part of the foot only, where they form a sheath around the hydroids. Hébant (1979) considered these cells as residual leptoids. A wall labyrinth is found in the epidermal cells along most of the length of the foot. Wall ingrowths have not been reported in the inner foot parenchyma. The lower part of the foot penetrates into the gametophyte central strand. Early degeneration of the cells at the extreme tip of the foot has been described (Browning & Gunning 1979a), resulting in a situation similar to that found in polytrichaceous mosses.

In *Bryum capillare* Hedw. the lower end of the hydroid central strand is somewhat enlarged at the base of the foot, and a tip of necrotic cells is formed (Fig. 2), producing apoplastic continuity between the sporophyte and gametophyte water-conducting systems. The epidermal cells above the tip exhibit a prominent wall labyrinth, and in the upper half of the foot these elongate radially, penetrating the adjoining vaginula tissue (Fig. 2, 7). A similar structure was reported in *B. caespiticium* Hedw. and *Rhodobryum roseum* (Hedw.) Limpr. (Lorch 1925b).

In *Dicranum undulatum* Ehrh. the gametophyte apex bears several sporophytes. The upper extremity of the gametophyte central strand enlarges and divides into smaller strands, each of which surrounds a sporophyte foot (Roth 1969).

Some mosses are known in which a well developed central strand occurs in both generations, but the foot does not penetrate the gametophyte stem appreciably and a tip of necrotic cells is not observed (Fig. 3; Roth 1969). Conversely, penetration of the leafy stem has been reported in species lacking a conducting central strand (Roth 1969). In such mosses a tip of necrotic cells seems to be lacking.

In *Physcomitrium cyathicarpum* Mitt. the foot has a persistent tip of four cells that are broader than the other epidermal cells, possess a wall labyrinth, and possibly have some specialized function (Chauhan & Lal 1984).

Figures 1–4. Light micrographs of the sporophyte-gametophyte junction in mosses. **1,** *Pogonatum aloides*, transverse section at the level of the middle region of the foot, note the abundant intercellular spaces in the cortical tissue, gametophyte vaginula (GV), bar = 100 μm; **2,** *Bryum capillare*, longitudinal section at the level of the basal region of the foot, a tip of collapsed cells is clearly visible (arrow), the epidermal cells elongate radially in the surrounding vaginula tissue, hydroids (H), bar = 100 μm; **3,** *Plagiomnium undulatum* (Hedw.) T. Kop., longitudinal section of the foot tip, which consists of living cells, vaginula (GV), hydroids (H), bar = 100 μm; **4,** *Brachythecium velutinum*, longitudinal section of the foot tip, made of living cells, hydroids (H), bar = 100 μm.

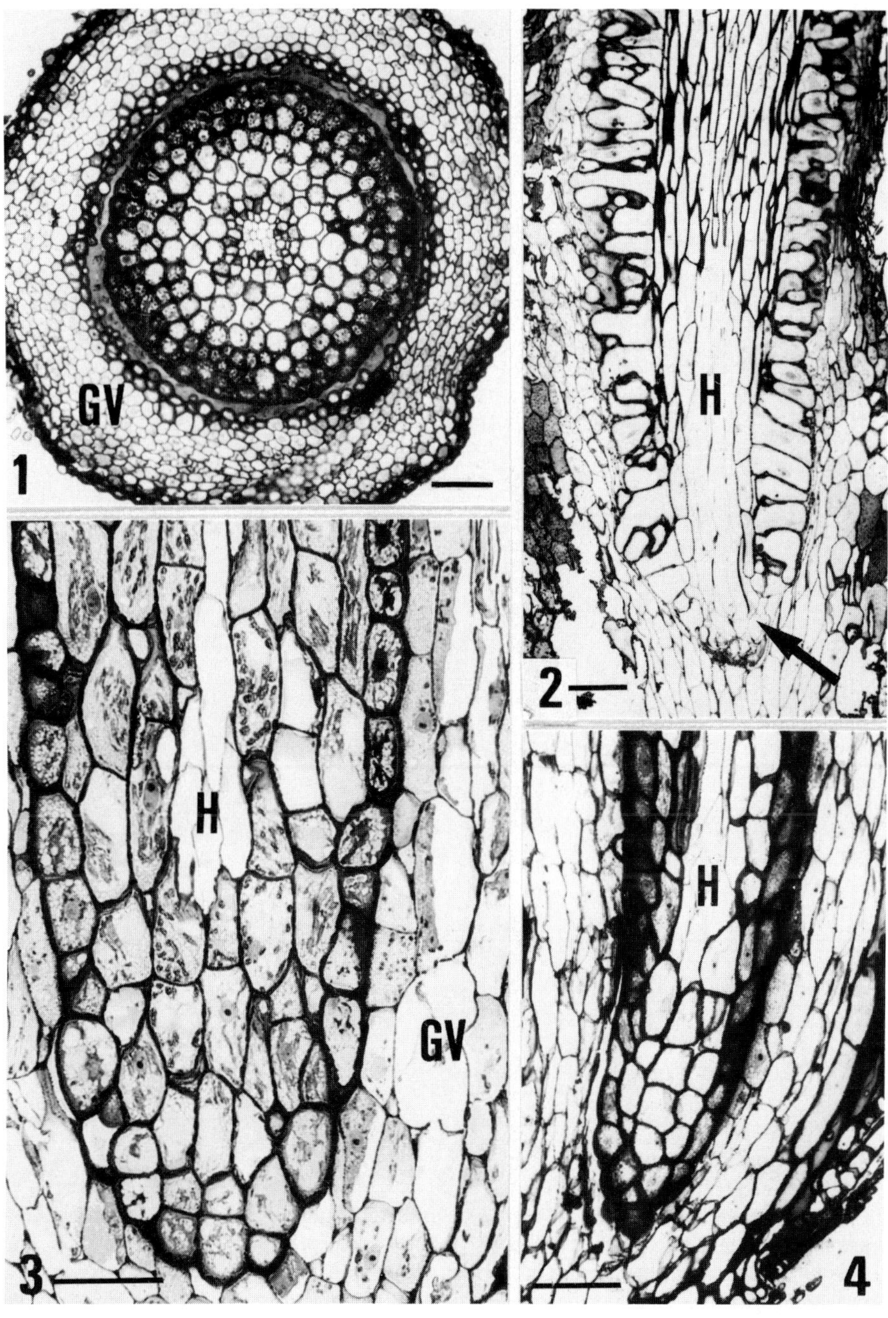
1
GV
2
H
3
H
GV
4
H

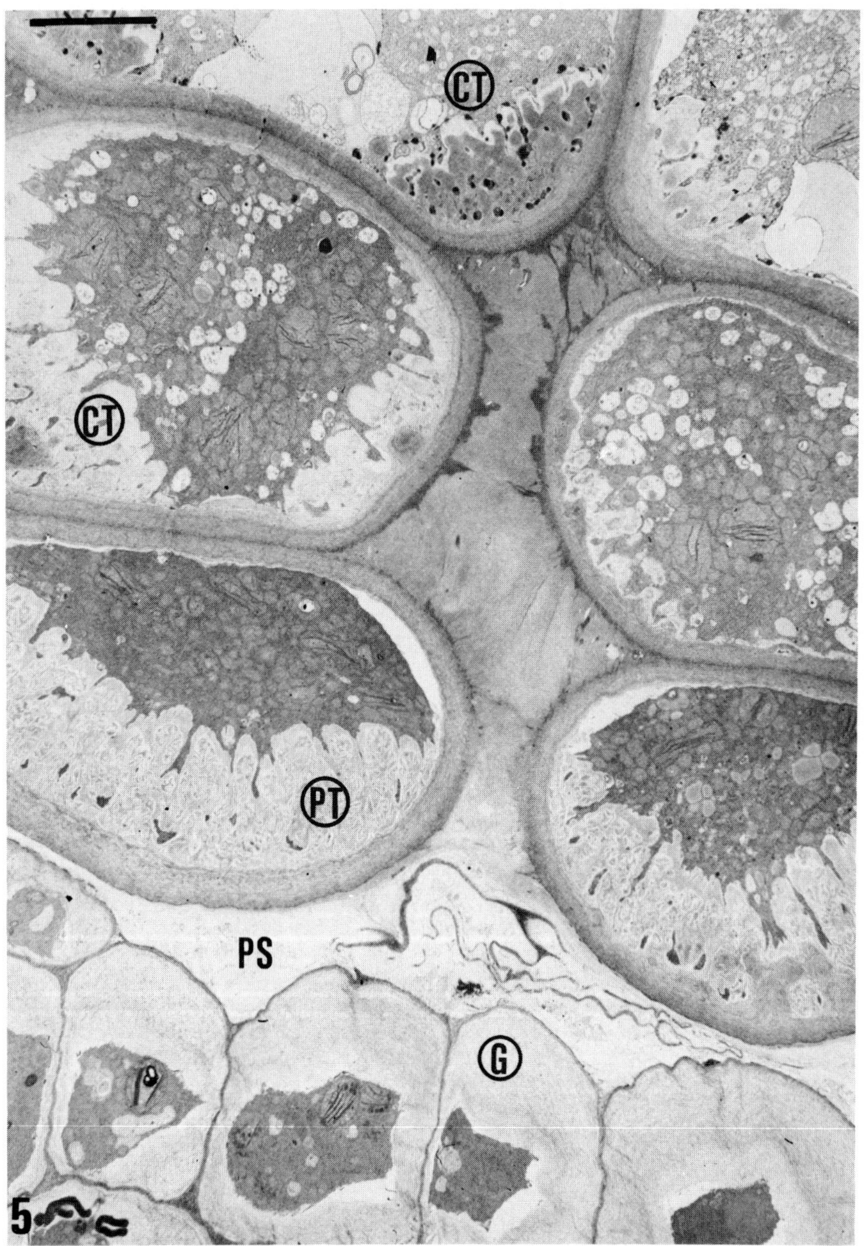

Figure 5. *Pogonatum aloides*. Several layers of cortical transfer cells (CT) are present beneath the placental transfer cells (PT) in the foot. The placental space (PS) appears to be continuous with an intercellular space of the foot. Gametophyte placental cells (G) lack wall ingrowths. Signs of cytoplasmic degeneration are visible in both sporophyte and gametophyte placental cells. The figure is of the middle region of the foot in a sporophyte with a fully expanded, green capsule. Bar = 5 μm.

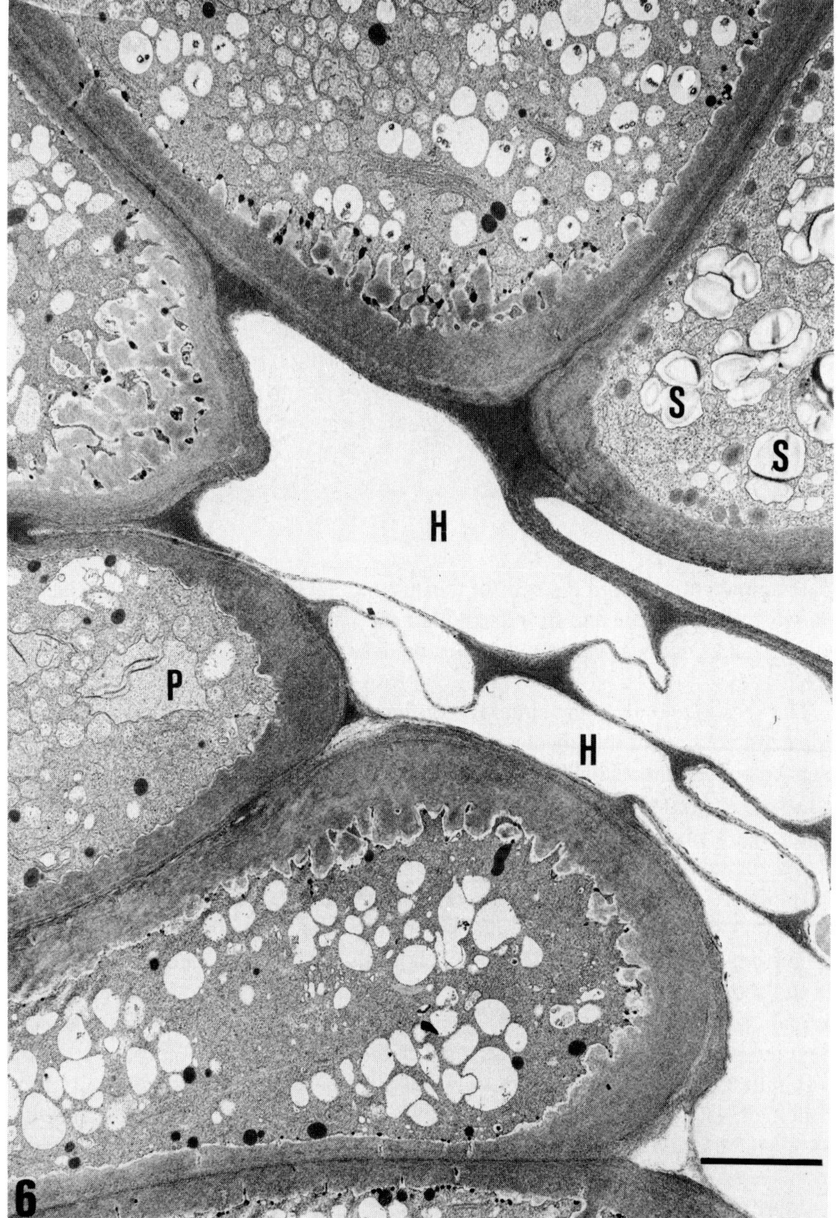

Figure 6. *Pogonatum aloides.* Transverse section at the level of the lower region of the foot. The central strand of hydroids (H) is surrounded by cells with wall ingrowths. Plastids (P) in these cells lack starch. Note that plastids rich in starch (S) occur in the cell on the right top corner, which lacks wall ingrowths. Same stage as Fig. 5. Bar = 5 μm.

In *Brachythecium velutinum* (Hedw.) B.S.G. (Fig. 4) and *Isopterygium pulchellum* (Hedw.) Jaeg., the only pleurocarpous species investigated so far, a persistent tip is found. In both species, as is typical of pleurocarpous mosses (Roth 1969), sporophytes develop at the apex of short lateral ramifications emerging from the main gametophyte stem. The foot is closely ensheathed by the vaginula and does not penetrate the leafy stem tissue appreciably.

The above observations seem to indicate that the presence of a tip of degenerated cells is associated with the penetration of the foot downwards into the central strand of the gametophyte stem. An apparent exception is *Timmiella barbuloides* (Brid.) Moenk. in which the lower part of the foot penetrates a strand of hydroids in the gametophyte stem but has a persistent tip of healthy cells (Ligrone et al. 1982a). The foot of *Timmiella* contains a conspicuous central strand of hydroids. In the lower tapering end of the foot the central strand enlarges due to increases in the number and width of hydroids, and hydroids and epidermal transfer cells are in close contact (Ligrone et al. 1982a).

The Vaginula

The vaginula arises from the cells of the archegonial stalk, which are stimulated to divide after fertilization, and from the perichaetial parenchyma below (Roth 1969). The vaginula does not bear vegetative leaves but is often ensheathed by perichaetial leaves (Roth 1969).

The sporophyte foot may either be confined to the region of the vaginula (Fig. 3, 4) or extend downward into the leafy stem tissue (Fig. 2). The first situation is typical of pleurocarpous mosses but is also common in acrocarps. The second situation is typical of acrocarps with a well developed central strand of conducting tissue in the gametophyte stem.

The vaginula consists of isodiametric or slightly elongate cells. The most peripheral cells have thickened walls and probably play a mechanical role. Inward from these are found thin-walled cells with prominent vacuoles and large plastids with a well developed lamellar system and abundant starch. Cells of vaginula nearer the foot are smaller and show denser cytoplasm than those to the outside. Stacks of rough endoplasmic reticulum are rather common (Fig. 10; Ligrone et al. 1982a). Intercellular spaces are reported in the inner vaginula tissue in *Timmiella barbuloides* (Ligrone et al. 1982a). Vaginula cells in *Pogonatum aloides* (Hedw.) P. Beauv. are rich in lipid globules, which are scattered throughout the cytoplasm, while vacuoles are nearly absent (Fig. 9). Plastids with a rudimentary inner membrane system and lacking starch were found in the vaginula cells of *Buxbaumia piperi* (Ligrone et al. 1982b).

Starting from the early stages of sporophyte development, empty (dead) cells are found intermingled with living cells in the vaginula of *Isopterygium pulchellum* (Fig. 10).

A wall labyrinth is generally formed in the innermost layer of vaginula cells only. In *Funaria hygrometrica* isolated wall protuberances are found also along the inner

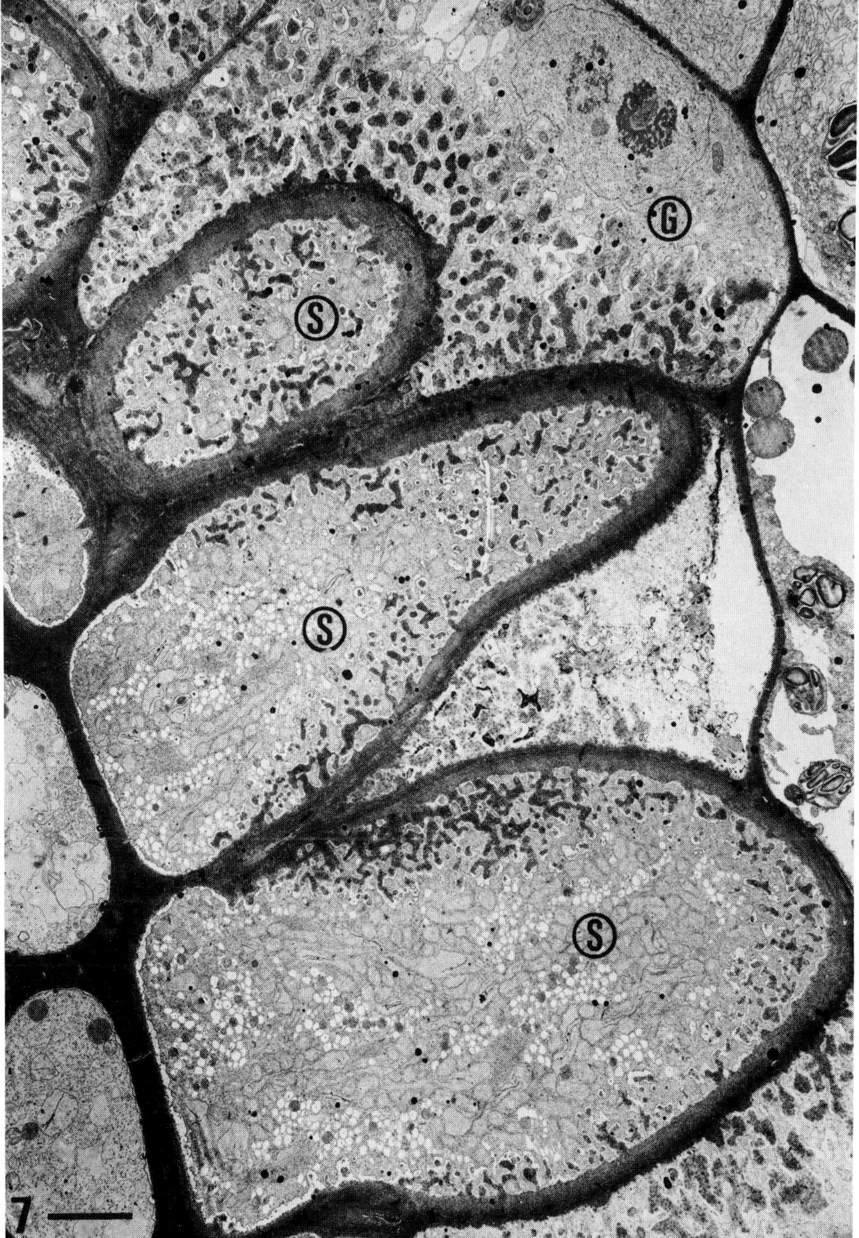

Figure 7. *Bryum capillare.* Transverse section of the placenta (middle part of the foot) of a plant with fully expanded, green capsule. A prominent wall labyrinth is present in both sporophyte (S) and gametophyte (G) placental cells. The elongate shape of sporophyte transfer cells increases the area of surface contact between the two generations. Bar = 5 μm.

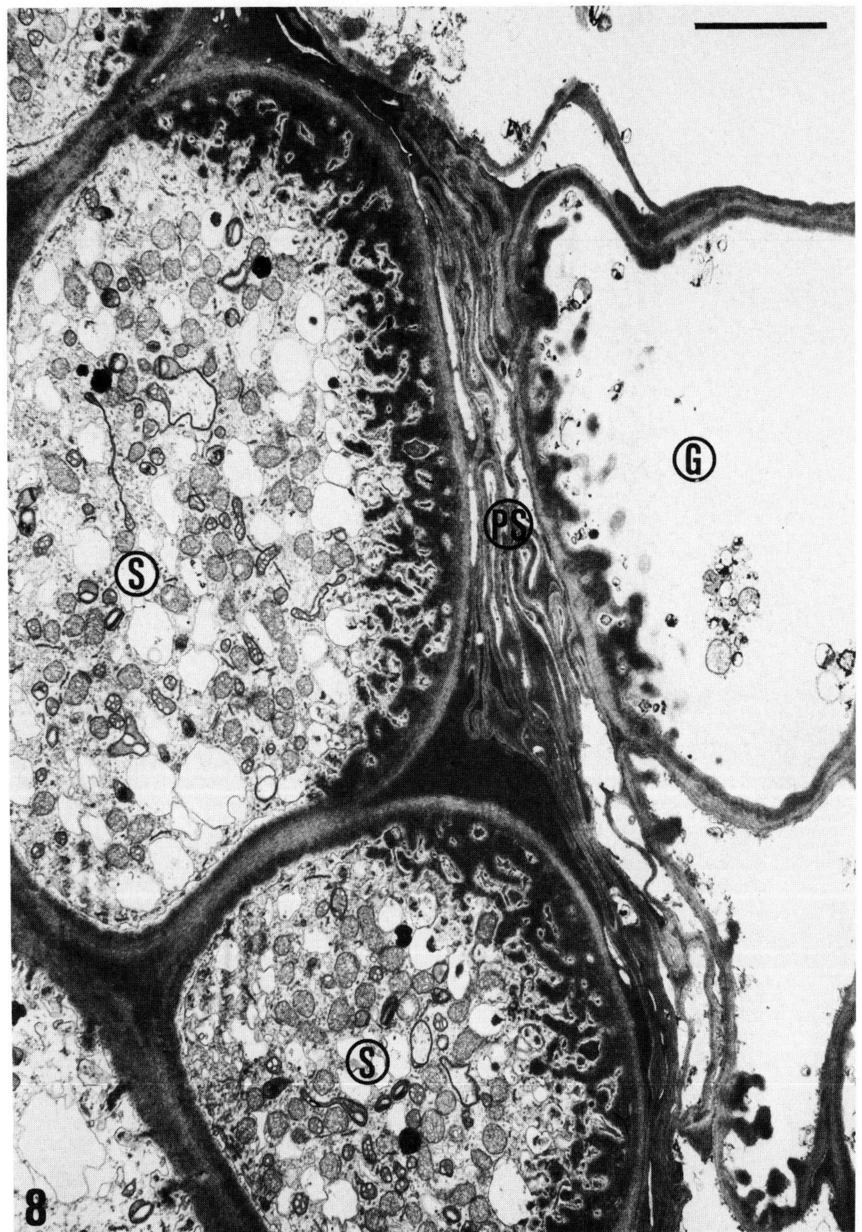

Figure 8. *Brachythecium velutinum*. Transverse section (lower part of the foot) of a sporophyte with fully expanded, green capsule. The sporophyte transfer cells (S) exhibit dense cytoplasm and prominent wall labyrinth. Gametophyte placental cells (G) show a less complex wall labyrinth and are completely degenerated. Placental space (PS). Bar = 2 μm.

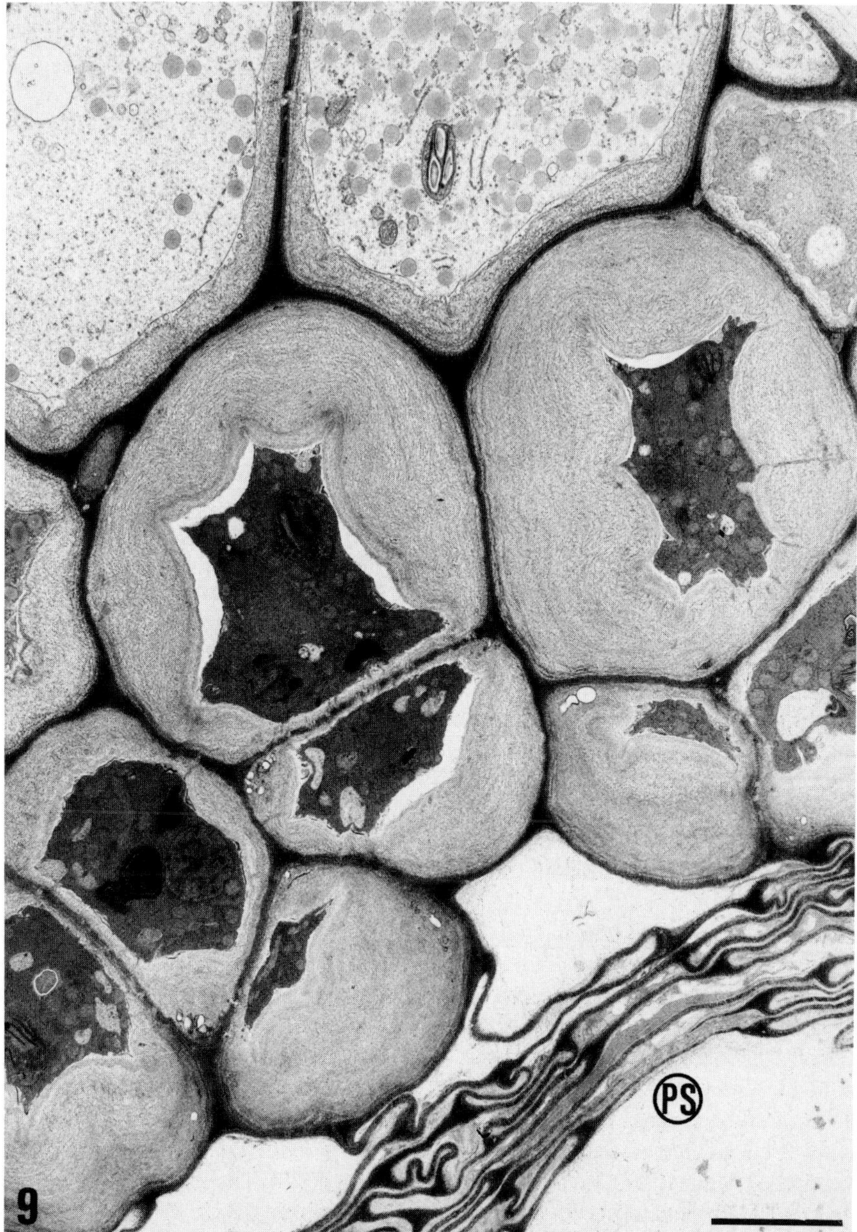

Figure 9. *Pogonatum aloides*. Detail of the gametophyte vaginula at the level of the basal part of the foot. The cellular lumen in the inner vaginula cells is nearly completely obliterated by the thickening of the cell walls. Abundant lipids are present in the more external cells (top). Placental space (PS). Same stage as Fig. 5. Bar = 4 μm.

tangential wall of neighboring cells, their frequency and size decreasing with distance from the foot (Browning & Gunning 1979a). In *Buxbaumia piperi* a well developed wall labyrinth was found in several layers of vaginula cells adjacent to the foot (Ligrone et al. 1982b). Conversely, in polytrichaceous mosses transfer cells are absent from the gametophytic side of the placenta (Roth 1969, Maier & Maier 1972, Hébant 1975). Roth (1969) reports that transfer cells are lacking in the vaginula of several other mosses including *Sphagnum palustre* L., *Andreaea rupestris* Hedw., and *Tetraphis pellucida* Hedw.

In *Pogonatum aloides* the three to four layers of vaginula cells adjoining the foot show exceedingly thick walls with a loose fibrillar texture (Fig. 9), recalling the "nacreous" walls in leptoids (Hébant 1977, Stevenson 1977). Fluorescence microscopy of specimens treated with aniline blue at high pH to reveal callose (which is present in leptoid walls) yielded negative results. On the basis of the micrographs published, thick-walled cells apparently similar to those of *Pogonatum* seem to form the innermost layers of the vaginula in *Dawsonia* (Hébant 1975) but are absent from *Dendroligotrichum* (Hébant 1975) and *Polytrichum* (Maier 1967).

Sporophyte and gametophyte are separated by an intervening gap, the placental space, which contains carbohydrate-rich mucilage and abundant cellular remnants, mainly flattened walls (Fig. 7-9). In those species in which the foot penetrates the leafy stem tissue, the lower end of the foot is in contact with the hydroids of the gametophyte central-strand. This region may be of varying length according to the species considered. It is worth noting that the neighboring parenchyma cells of the gametophyte stem develop wall ingrowths along their inner tangential walls, even though they are separated from the foot by intervening gametophyte hydroids (Wiencke & Schulz 1978, Ligrone et al. 1982a).

Ultrastructure of Transfer Cells

Generally, sporophyte transfer cells have a more elaborate wall labyrinth than gametophyte ones, but the reverse was reported (Lal & Chauhan 1981) for *Physcomitrium cyathicarpum*. In accordance with Gunning and Pate (1974), the wall ingrowths may be considered specialized secondary walls. In specimens fixed and embedded conventionally the ingrowths appear to consist of an inner fibrillar core and a clear interfacial zone of varying thickness outlined by the plasmalemma. This zone, because it was absent from wall ingrowths in freeze-substituted transfer cells of *Funaria hygrometrica*, was considered an artifact caused by the expansion of wall ingrowth during conventional chemical fixation (Browning & Gunning 1977). The surface-to-volume ratio of wall labyrinth in freeze-substituted cells is more than double that in conventional preparations, but the surface amplification brought about by wall ingrowths is only a little greater (ibid.). Conventional fixation probably also causes some alteration in the internal structure of wall ingrowths. The fibrillar material of the wall ingrowths in freeze-substituted cells appears as compact as that in the primary wall, and the

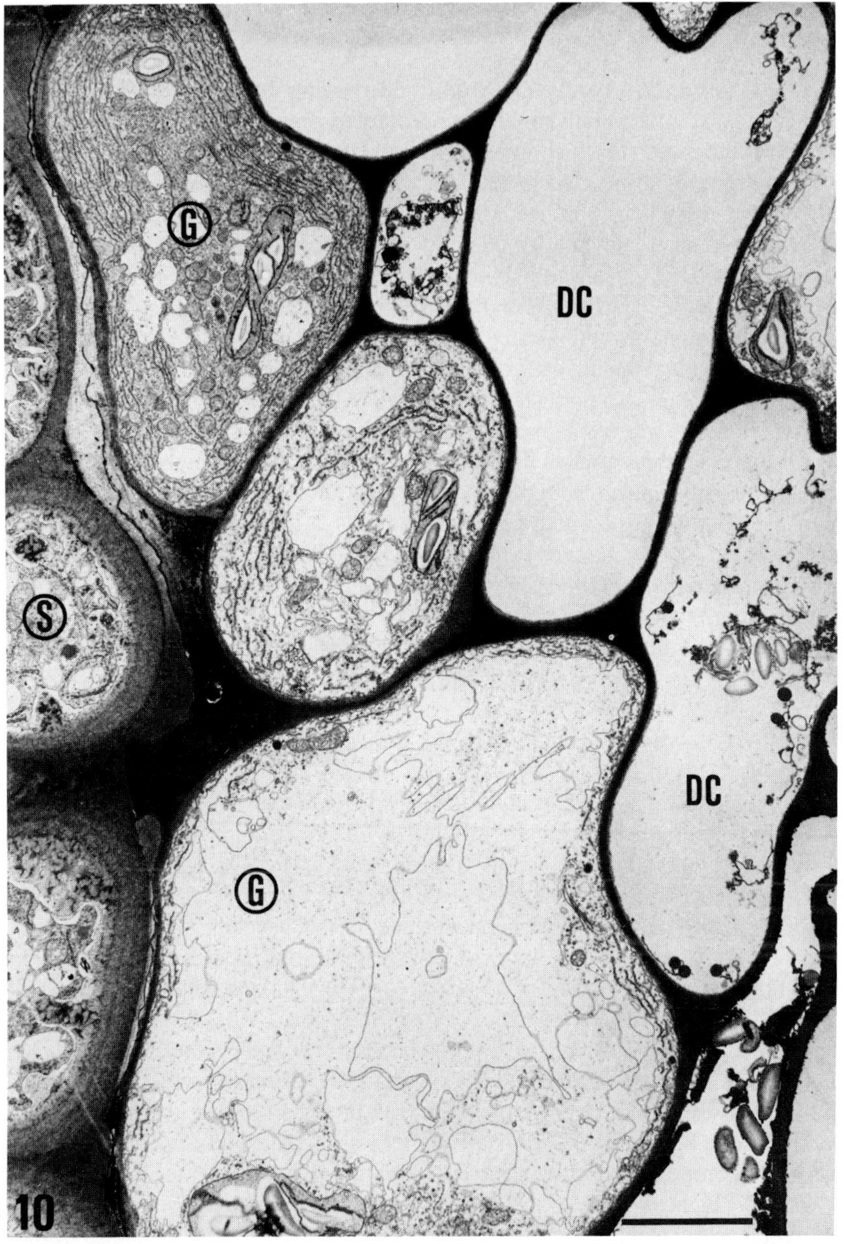

Figure 10. *Isopterygium pulchellum.* Transverse section (middle part of the foot) of a sporophyte with unexpanded capsule. The vaginula tissue contains dead cells (DC) intermingled with living ones. Gametophyte placental cells (G) lack wall labyrinth and may contain stacks of rough endoplasmic reticulum. Sporophyte placental cells (S) contain little complex wall labyrinth. Bar = 4 μm.

individual fibrils are barely discernible (Browning & Gunning 1977). Freeze-substitution also reveals polygonal "decorations" on coated vesicles and on localized portions of the plasmalemma outlining the wall labyrinth (ibid.).

The wall ingrowths in gametophyte transfer cells of mosses are more compact and stain more densely than those in sporophyte transfer cells. The wall labyrinth is mostly restricted to outer and inner tangential walls in sporophyte and gametophyte transfer cells, respectively. Isolated wall ingrowths may be seen along the other cell walls. Transverse walls of sporophyte transfer cells lack wall ingrowths and may contain numerous plasmodesmata (Wiencke & Schulz 1975). Plasmodesmata also connect the transfer cells with adjoining cells of the parental generation (Wiencke & Schulz 1975, Caussin et al. 1983). No plasmodesmata connect sporophyte and gametophyte cells. In *Funaria hygrometrica* those sporophyte transfer cell walls not covered by ingrowths have a loosely packed fibrillar texture that contrasts strongly with the densely staining compact walls in adjoining cortical cells (Browning & Gunning 1979a). An extensive labyrinth covers both radial and outer walls in the sporophyte transfer cells of *Buxbaumia piperi* (Ligrone et al. 1982b). The external walls of the epidermal transfer cells in *Polytrichum formosum* Hedw. consist of two different layers in addition to the wall labyrinth layer (Caussin et al. 1983).

The wall labyrinth exhibits metachromatic reaction with the basic dye toluidine blue, denoting the presence of carbohydrates with carboxyl groups. A thin cuticular layer outlining the outer wall of sporophyte transfer cells was observed in *Funaria hygrometrica* (Browning & Gunning 1979a) and *Timmiella barbuloides* (Ligrone et al. 1982a).

Major features of both sporophyte and gametophyte transfer cells are an abundance of mitochondria (which are always provided with numerous cristae) and rough endoplasmic reticulum (which may form extensive stacks) (Fig. 10, 12, 13; Eymé & Suire 1967, Ligrone et al. 1982a). Moreover, transfer cells contain several vacuoles (generally of small size) and a prominent nucleus with a well developed nucleolus (Fig. 13). Transfer cells of mosses may show, in transverse section, polarized cytoplasmic organization, with mitochondria, plastids, and other organelles mostly concentrated near the wall labyrinth, and vacuoles occupying the opposite side of the cell. The same characteristics were found in transfer cells in a variety of plants other than bryophytes (Gunning & Pate 1969, 1974, Pate & Gunning 1972). Two populations of mitochondria were observed side by side, one type with undilated cristae, the other type with much denser matrix and dilated cristae, in freeze-substituted transfer cells of the relatively old sporophyte foot in *Funaria* (Browning & Gunning 1977).

Some differences in the extent of the wall labyrinth are found in transfer cells at different levels of the foot. In *Funaria hygrometrica* (Wiencke & Schulz 1975, 1978, Browning & Gunning 1979a) and *Timmiella barbuloides* (Ligrone et al. 1982a), the most extensive wall labyrinth is found in transfer cells at the level of the middle part of the foot. A reduced labyrinth occurs in the transfer cells in the basal and upper part of the foot. In *T. barbuloides*, sporophyte transfer cells in the lower part of the foot have larger vacuoles and less dense cytoplasm compared with transfer cells in the

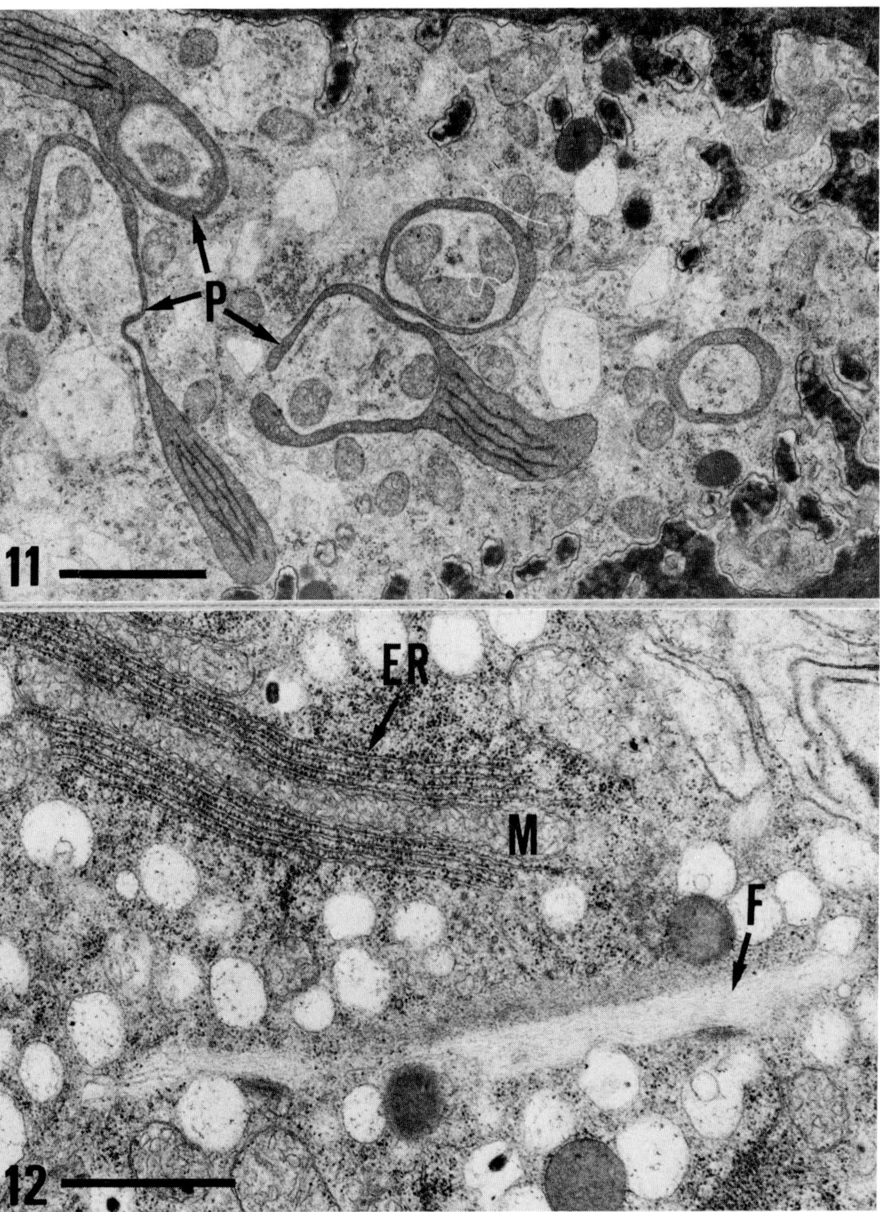

Figures 11–12. Details of sporophyte transfer cells in *Bryum capillare*, same stage as Fig. 7. **11**, Highly pleomorphic plastids (P), bar = 2 μm; **12**, Very close association of endoplasmic reticulum (ER) and mitochondria (M), note a deposit of fibrillar material (F), bar = 2 μm.

middle of the foot and are often in contact at their opposite sides with both gametophyte and sporophyte hydroids (Ligrone et al. 1982a). Wiencke and Schulz (1975, 1978) found in *F. hygrometrica* highly vacuolate cells with short wall-ingrowths (less than 1 mm long) and plastids with a rudimentary thylakoid system in the lower part of the foot, while transfer cells in the middle of the foot showed dense cytoplasm rich in organelles and a prominent wall labyrinth. These authors proposed that while the main function of basal transfer cells is to pump water (because these cells are in contact with the gametophyte hydroids), the middle transfer cells are mainly concerned with assimilate uptake because of their contact with vaginula transfer cells. The latter assumption was confirmed by the results of Browning and Gunning (1979a) who performed a careful study of the sporophyte-gametophyte junction in the same species. However, they noted that the highly vacuolate cells with wall ingrowths in the basal part of the foot (Wiencke & Schulz 1975) might arise from early degeneration of more typical transfer cells. This process would eventually produce the cluster of empty collapsed cells that connects the hydroid central strand of the two generations. Collapsed cells with wall ingrowths also occur in the foot tip of *Pogonatum aloides*. Apparently, the degeneration of tip cells in *Bryum capillare* is not preceded by the formation of wall ingrowths (Fig. 16). In *Brachythecium velutinum* and *Isopterygium pulchellum* the sporophyte foot does not penetrate the leafy stem. Transfer cells with a prominent wall labyrinth occur on both sides of the placenta at the level of the lower part of the foot, whereas the cells above have a reduced labyrinth or none at all. Persistent transfer cells with dense cytoplasm and a prominent wall labyrinth occur at the level of the foot-tip in *Physcomitrium cyathicarpum* (Lal & Chauhan 1981, Chauhan & Lal 1984). The presence of vacuolate transfer cells with short wall ingrowths in the foot tip of *T. barbuloides* (Ligrone et al. 1982a) is rather puzzling because these cells differ from the foot-tip cells of *F. hygrometrica* by remaining alive to advanced stages of sporophyte development.

Plastids in gametophyte transfer cells do not differ markedly from those found in the other cells of the vaginula. They usually have a well developed thylakoid system and contain varying amounts of starch. Conversely, plastids in sporophyte transfer cells are frequently modified compared to the typical chloro-amylopasts that are by far the most common kind of plastid found in mosses. In *Funaria hygrometrica* (Wiencke & Schulz 1975, Browning & Gunning 1979a), *Timmiella barbuloides* (Ligrone et al.

Figures 13–14. *Pogonatum aloides*, details of sporophyte transfer cells of the middle of the foot, sectioned longitudinally, same stage as Fig. 5. **13,** The nucleus (N), endoplasmic reticulum (arrows), plastids (P), and fibrillar deposits (F) appear to be arranged parallel to the vertical axis of the cell, prominent wall labyrinth (WL) is visible on the outer tangential wall, bar = 2 μm; **14,** High magnification of the fibrillar material in Fig. 13, bar = 0.5 μm.

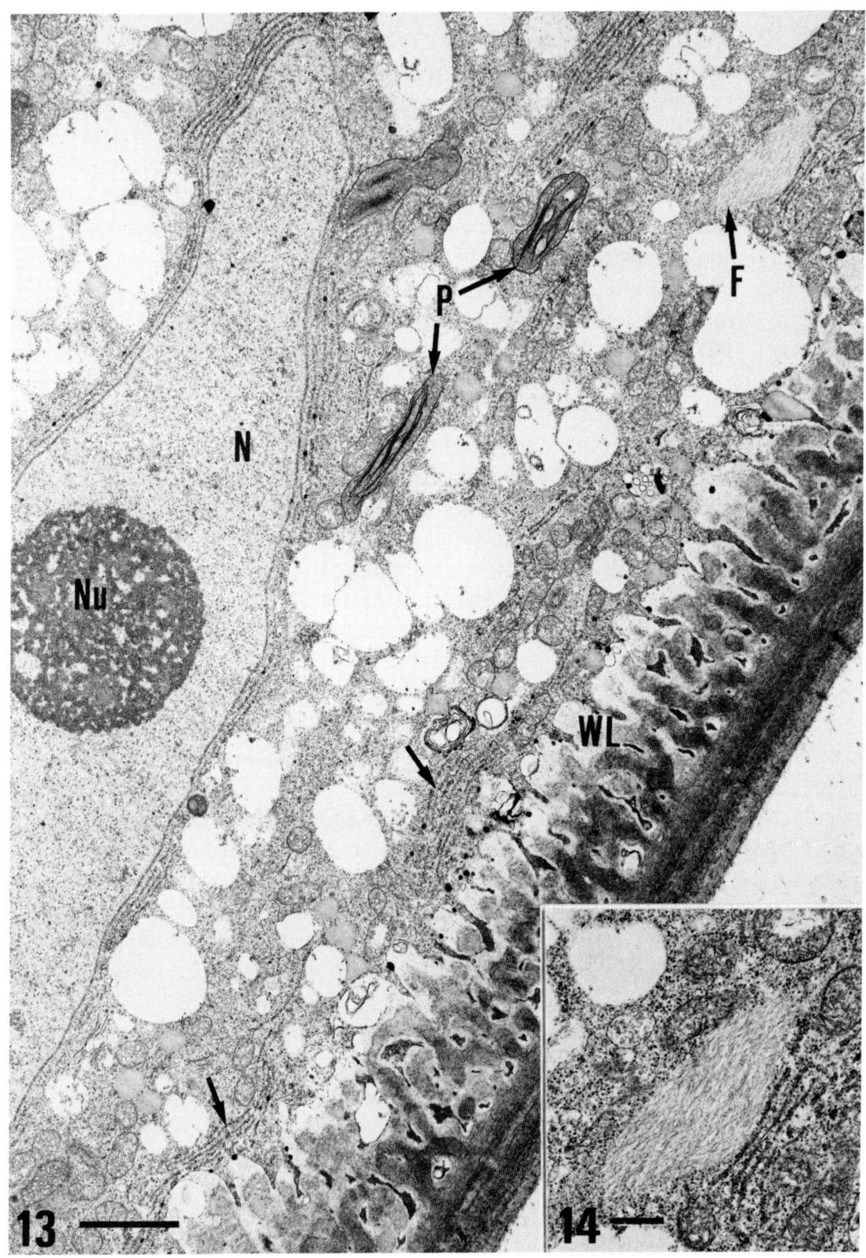
N
Nu
P
F
WL
13
14

1982a), *Plagiomnium cuspidatum* (Hedw.) T. Kop. (Eymé & Suire 1967), and *Pogonatum aloides* (Fig. 13) plastids with a well developed thylakoid system occur in transfer cells, at least those in the middle part of the foot. Plastids with a rather extensive inner membrane system were also reported in the foot tip cells in *Physcomitrium cyathicarpum* (Chauhan & Lal 1984). Except in *P. cuspidatum* these plastids contain very little or no starch. This supports the hypothesis of Browning and Gunning (1979a) that plastids in sporophyte transfer cells are mainly concerned with photophosphorylation because they are located beneath many layers of gametophyte cells where they receive light at low intensity and minimal amounts of CO_2. However, abundant starch deposits can be found in plastids of nearby gametophyte cells (Ligrone et al. 1982a), hence the presence or absence of starch is not likely to be a reliable indication of plastid photosynthetic competence but rather reflects the balance between sugar import and export activity of the whole cell. Extremely pleomorphic plastids with a poor inner thylakoid system are found in the sporophyte transfer cells in *Bryum capillare* (Fig. 11) and *Brachythecium velutinum* (Fig. 15), while in *Buxbaumia piperi* (Ligrone et al. 1982b) and *Isopterigium pulchellum*, the plastids are very small and sometimes barely distinguishable from mitochondria. Plastids with a rudimentary thylakoid apparatus were also reported in placental transfer cells in a number of polytrichaceous mosses (Hébant 1975, Caussin et al 1983). These modifications in plastid structure appear related to transfer cell specialization because typical chloro-amyloplasts are found in epidermal cells higher up the foot and in cortical cells, which lack wall ingrowths.

Lipid deposits are of common occurrence in both sporophyte and gametophyte transfer cells, but usually they are not prominent. Abundant lipid accumulations are found in gametophyte vaginula cells in *Pogonatum aloides*. Lipids are also abundant in gametophyte transfer cells in *Buxbaumia piperi*, in which they assume the form of small droplets associated closely with the wall labyrinth. On the opposite side of the placenta, the sporophyte transfer cells contain numerous microbodies associated with lipid globules and mitochondria in close proximity to the wall labyrinth. This suggests that in *B. piperi*, in which the gametophyte lacks starch and presumably is not active in photosynthesis, lipids or lipid derivatives are primarily involved in gametophyte-sporophyte nutritional relationships (Ligrone et al. 1982b).

Compact aggregates of fibrillar material are found in sporophyte transfer cells in *Bryum capillare* (Fig. 12) and *Pogonatum aloides* (Fig. 13, 14). Apparently they lack a surrounding membrane, and their fibrillar component is preferentially oriented paral-

Figure 15. *Brachythecium velutinum.* Detail of sporophyte transfer cell in the lower part of the foot. Plastids (P) are highly pleomorphic and contain many vesicles but lack thylakoids. Mitochondria (M). Same stage as Fig. 8, bar = 1 μm.

Figure 16. *Bryum capillare.* Degenerating cells of the foot tip (S) without wall ingrowths. This figure is of a very young sporophyte still enclosed in the calyptra, bar = 10 μm.

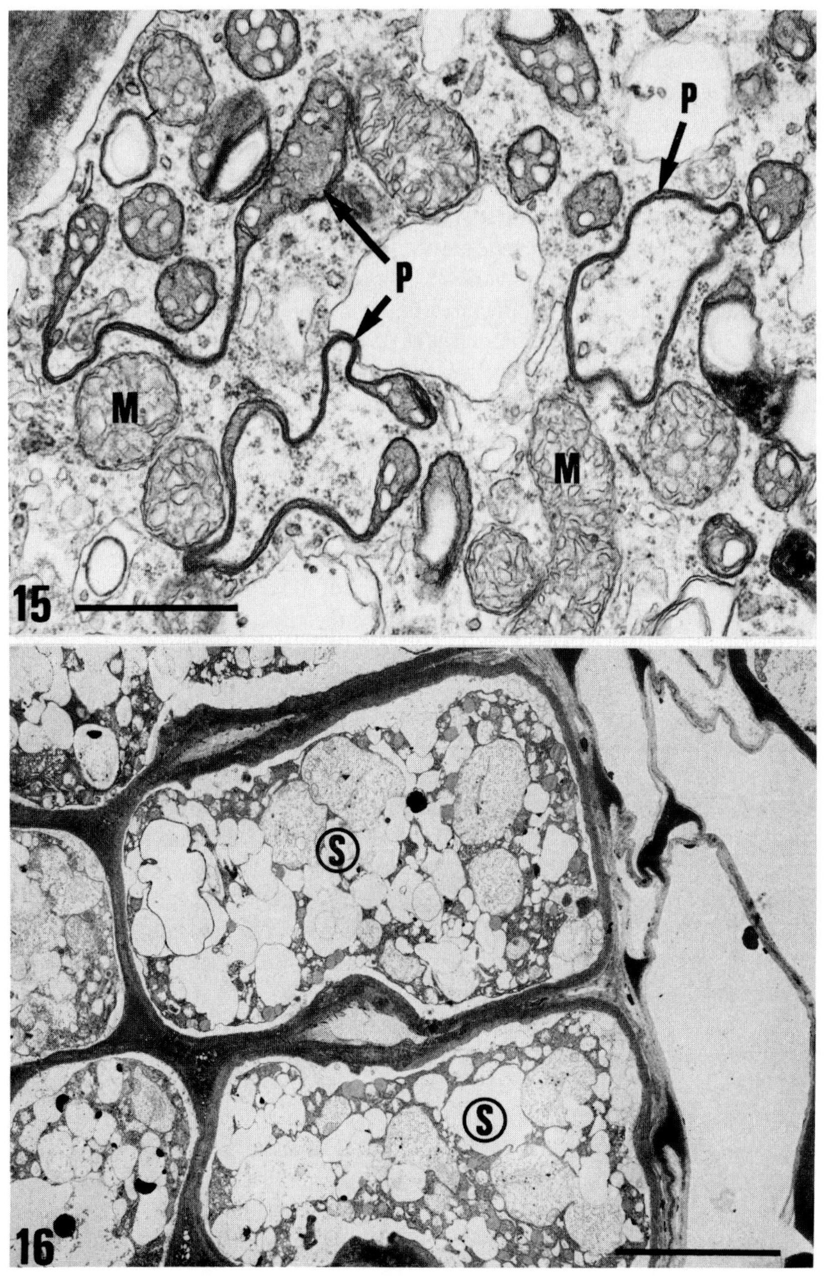

lel to the long axis of the cells. Hence, in *B. capillare* the fibrillar structure can be best seen in transverse sections of the foot (in which transfer cells are elongated in a radial direction), while in *P. aloides* longitudinal sections best demonstrate fibrillar structure. Cytoplasmic deposits of fibrillar or granular material, bordered by clusters of ribosomes, were observed in the sporophyte transfer cells of *Funaria hygrometrica* (Browning & Gunning 1979a). It is possible that the granular deposits correspond to fibrillar ones that had been cut transversely. Prominent fibrillar inclusions were also found in gametophyte transfer cells in the liverwort *Reboulia hemisphaerica* (L.) Raddi (see later). A "fibrous body" similar to these inclusions was observed in generative and vegetative cells of the microgametophyte in higher plants and was related to the assembly of microtubules (Brighigna et al. 1980). Such a function for the fibrillar inclusions of bryophyte placental cells is unlikely because, notwithstanding their apparent polarization, these structures do not seem to be associated with microtubules.

Browning and Gunning (1979a) put forward the attractive hypothesis that the fibrillar deposits in *Funaria hygrometrica* transfer cells are involved in osmotic protection against drought stress. A physiological necessity to protect placental tissue from drought damage may be widespread in bryophytes, most of which experience water deficits daily in their natural habitats. Future work should consider the presence and chemical composition of cytoplasmic inclusions of the type described above and their possible relation to drought tolerance during the reproductive phase.

Acid phosphatase and adenosine triphosphatase (ATPase) activities were localized in the cell organelles and plasmalemma of the transfer cells in *Polytrichum piliferum* Hedw. (Maier & Maier 1972). The highest activity was found in epidermal cells that had the most elaborate wall labyrinth. The precipitation of the reaction product along the plasmalemma was generally more intense in the region of the wall labyrinth than along the other cell sides, denoting a polarity in the distribution of enzymes. Activity was lower in the inner transfer cells. The subepidermal cells showed a sharp increase in enzyme activity as a result of the degeneration of the adjacent epidermal cells, but the shape and extent of their wall labyrinth remained unchanged. The gametophyte cells facing the sporophyte foot gave positive responses to both enzymes. This might depend in part on diffusion of the enzymes or their reaction products from adjacent sporophyte cells (Maier & Maier 1972).

Histochemical investigations in several polytrichaceous mosses, as well as in *Dicranum scoparium* Hedw. and in the liverwort *Scapania undulata* (L.) Dum. (Hébant 1973, Hébant & Suire 1974) revealed high acid phosphatase, peroxidase, and respiratory enzyme activities in foot epidermal cells. These enzyme activities were considerably lower in the adjacent cells of the vaginula. The authors ascribed this to developmental and/or environmental factors, but based on current ultrastructural information, lower enzyme activity in gametophyte placental cells is more likely to be related to their lesser specialization compared to sporophyte placental cells. Hébant and Suire (1974) also found high enzyme activity in the transfer cells (Maier & Meier 1972, Hébant 1975) surrounding the central strand in the basal part of the foot in *Polytrichum commune* Hedw. and *Dawsonia papuana* F. Muell.

The activity of Ca^{2+}-activated ATPase (Chauhan & Lal 1980), acid and alkaline phosphatases and esterases, and succinate dehydrogenase (Chauhan & Lal 1982) has been reported recently in the placental cells of *Physcomitrium cyathicarpum*, confirming that these cells are a center of intense metabolic activity and ATP consumption.

Development of the Sporophyte-Gametophyte Junction

According to Roth (1969), an intercalary meristem that forms in the middle of the young embryo is responsible for the basipetal development of the foot and the lower part of the seta and the acropetal development of the upper part of the seta and the capsule base. Development of the foot is preceded by the formation of a suspensor-like structure derived from randomly oriented divisions of the hypobasal cell. The depth of penetration and dimensions of the foot depend on the rate and number of cell divisions in the intercalary meristem of the embryo. Consequently, mosses like *Sphagnum*, *Archidium*, and *Ephemerum*, in which the intercalary meristem is either not produced or barely active, possess only a hypobasally derived suspensor-like bulbous mass or a very small conical foot (Crandall-Stotler 1984). In most other cases the suspensor-like structure degenerates very early, and the nutrient uptake function is carried out by the foot. Unfortunately, no electron microscope study of the earliest stages of the sporophyte development in mosses or any other bryophyte has been undertaken.

The development of the sporophyte-gametophyte junction, starting with more advanced stages, was investigated in detail in *Funaria hygrometrica* (Wiencke & Schulz 1978, Browning & Gunning 1979a). It is worth noting that cells with a prominent wall labyrinth and dense cytoplasm were first found at the tip of the foot in the very young (1 mm long) sporophyte. At this stage (primary stage of Browning & Gunning 1979a), the epidermal cells higher up in the foot had larger vacuoles, fewer organelles, and a less complex wall labyrinth than the tip cells. Cellular disruption was observed in the gametophyte tissue close to the young tip. The expansion of the periplasmic space between cell wall and plasmalemma, with consequent disruption of the plasmodesmata, was seen to precede the degeneration of the cytoplasmic contents. The emergence of the upper part of the sporophyte coincided with the greening and cessation of growth in the sporophyte foot. Concomitantly, the tip cells collapsed and formed an appendage of dead empty cells (Browning & Gunning 1979a). These authors suggested that the tip cells might produce and secrete factors responsible for the dissolution of the surrounding gametophyte tissue. The dissolution would provide nutrients to meet the early nutritional requirements of the embryo and would also facilitate the penetration of the young foot. The placental space between the foot and vaginula probably arises from gametophyte cells that collapse during the growth of the foot. Interestingly, acid phosphatase activity was found in this space, as well as in the necrotic cells of the appendage in *Polytrichum commune* (Hébant & Suire 1974). An involvement of hydrolytic factors in sporophytic origin was suggested (Browning

& Gunning 1979a) because in gametophyte cells wall degeneration apparently preceded cytoplasmic degeneration. The results of Chauhan and Lal (1980, 1982), which demonstrate the presence of high hydrolase activity in the tip cells of *Physcomitrium cyathicarpum*, sustain this hypothesis. However, wall hydrolyzing enzymes should not pass through sporophyte cell walls without affecting them, unless these cell walls were quite different from gametophyte cell walls, which is unlikely. Cellular degeneration was frequently observed in the leafy stem tissue at some distance from the growing sporophyte foot (Roth 1969). Thus, autolysis of gametophyte cells, mediated by endogenous hydrolases and possibly induced by some factor of sporophytic origin, seems to be a more plausible mechanism. Closer cytochemical study is needed to make this point clearer. Mechanical pressure by the growing sporophyte foot may also be involved in gametophyte cell degeneration. Vaizey (1888) noted that when the young sporophyte lengthens, the calyptra prevents it from pushing upward for some time, perhaps allowing the foot to be forced down into the gametophyte.

A prominent wall labyrinth develops on epidermal cells of the foot above the tip during the phase of seta elongation that precedes the expansion of the capsule (secondary stage of Browning & Gunning 1979a). A wall labyrinth was also found in the gametophyte cell adjoining the foot at this stage. Wall labyrinths in gametophyte transfer cells enlarge progressively by the deposition of an amorphous, lightly staining material. The main organelle involved is the Golgi apparatus. This organelle produces vesicles and multivesicular bodies containing secondary vesicles that fuse with the plasmalemma (Wiencke & Schulz 1978). Multivesicular structures associated with the wall labyrinth are of common occurrence in sporophyte transfer cells (Ligrone et al. 1982a, Caussin et al. 1983). Coated vesicles also seem to be involved in the development of wall ingrowths (Browning & Gunning 1977), whereas there is no indication of the involvement of microtubules.

A similar pattern of the wall labyrinth development was described for the sporophyte epidermal cells in several species of *Polytrichum* (Maier 1967, Maier & Maier 1972). The first wall ingrowths are rather thin and smooth-surfaced. With the addition of new wall material the ingrowths enlarge and become more and more irregular in outline, forming a complex wall-membrane apparatus. The addition of sugars to the culture medium of isolated sporophytes of *Polytrichum piliferum* drastically increased the volume of wall material added to existing wall ingrowths (Maier & Maier 1972). In *Isopterygium pulchellum* the secondary phase of wall labyrinth growth begins at the tip of the foot and then proceeds upwards.

Occlusion of the wall labyrinth was seen to start with the onset of capsule differentiation in *Funaria hygrometrica* (tertiary stage of development of Browning & Gunning 1979a). Wall ingrowths merged with one another with the deposition of new, lightly staining material. The same material was also deposited along the radial and inner walls of transfer cells and in the adjoining cortical cells (Browning & Gunning 1979a). At the end of the tertiary stage, the wall labyrinth of sporophyte transfer cells was completely obliterated, and cytoplasmic remnants were visible in the resulting thickened wall (Wiencke & Schulz 1978). The inner cytoplasm of these modified sporophyte transfer cells persisted intact for a rather long time, whereas the gametophyte transfer cells had degenerated before the tertiary stage was reached (Browning & Gunning 1979a). Early degeneration of gametophyte transfer cells has been observed in other moss species including *Brachythecium velutinum* (Fig. 8), *Pogonatum aloides* (Fig. 9), and *Bryum capillare*.

Degeneration of sporophyte transfer cells in *Funaria hygrometrica* was observed to take place simultaneously with spore maturation. The degeneration involved reduction and disappearance of polysomes, rough endoplasmic reticulum, and dictyosomes. Mitochondria and plastids swelled and lost their inner membrane systems. Accumulation of osmiophilic globules within plastids was also observed (Wiencke & Schulz 1978). In the sporophyte foot of *Polytrichum piliferum* epidermal transfer cells occasionally degenerated during sporophyte growth and were replaced in function by adjoining cortical cells (Maier & Maier 1972).

Liverworts

The sporophyte-gametophyte junction in hepatics, as in mosses, exhibits a great deal of variation in morphology and ontogenetic pattern (Schuster 1966, 1984, Crandall-Stotler 1984). The mature foot may be a few-celled convex disc in *Riccia* and the Lejeuneaceae or a quite massive organ of variable shape in the Marchantiales (Fig. 17, 18, 19, 21, 22) and Metzgeriales (Fig. 20). In the Jungermanniidae, the foot frequently has a haustorial collar, which increases the area of contact with the gametophyte. The distinctive foot of *Jubulopsis* consists of radiating finger-like extensions that are longer, septate, and in part filamentous in the collar region. Ontogenetic patterns of the foot depend on the fate of the epibasal and hypobasal cells. These result from the initial transverse division of the zygote. In the Marchantiidae, the foot derives from the hypobasal cell, while in the Jungermanniidae the epibasal cell may contribute to its formation too. In all cases the foot develops synchronously with the seta and capsule while the sporophyte is still enclosed within gametophyte tissue. In the Jungermanniidae an ephemeral, few-celled filamentous suspensor, which functions as an absorbing organ, is formed before foot development. The foot of the jungermannioid sporophyte penetrates the gametophyte stem to a variable depth, but in most cases foot penetration is not real, being simulated by the proliferation of stem tissue around the foot. Gametophyte tissues ensheathing the foot include the archegonium-derived vaginula and in some instances, as in the Jungermanniidae, the basal part of stem-derived structures such as marsupia, which protect the developing embryo and sporophyte.

The anatomy of the foot and surrounding vaginula is much simpler in hepatics than in mosses, because of the absence of specialized conducting or mechanical tissues. Both histological compartments consist of moderately large, vacuolate parenchymatous cells. The main cellular specialization is transfer cells in the placental region where the sporophyte and gametophyte are separated by a more or less wide space.

Generally, intercellular spaces are lacking in both foot and vaginula of hepatics (Fig. 20, 21, 22). However, rather extensive intercellular spaces are found in the foot of *Conocephalum conicum* (L.) Dum. (Fig. 19). These spaces are probably of lysigenic origin and are sometimes delimited by cells with wall ingrowths (Fig. 27).

The available ultrastructural information pertains to the placental region in members of the Sphaerocarpales (Kelley 1969), Marchantiales (Gambardella & de Lucia Sposito 1981-82, 1983), and Metzgeriales (Fig. 36), which mostly have been investigated during advanced stages of capsule maturation. Transfer cells with variously elaborated wall labyrinths are present on both sides of the placenta. As in mosses, the wall labyrinth is more extensive in the sporophyte than in the gametophyte. Conversely, gametophyte transfer cells typically form several layers around the foot (Fig. 25, 26, 30). This seems to be rare in mosses.

The wall ingrowths of hepatics are similar in structure to those found in mosses, with an inner fibrillar component and a peripheral light area of variable extent. Multivesicular structures are frequently associated with the wall labyrinth (Fig. 31). Cytoplasmic specializations accompanying wall labyrinth development include proliferation of endoplasmic reticulum (which is often arranged in stacks, Fig. 28), dictyosomes, and mitochondria. In all species so far examined, the vacuolar system of sporophyte transfer cells comprises numerous, small, closely spaced vesicles (Fig. 25, 30; Gambardella & de Lucia Sposito 1981-82, 1983). The same feature is found in cells of plants exposed to drought stress, and it is considered to be a morphological device to minimize water-deficit damage (Oliver & Bewley 1984, Hallam & Luff 1980, Bergtrom et al. 1982). Vacuoles are larger and less numerous in gametophyte transfer cells.

A centripetally increasing gradation of cytoplasmic degeneration is visible in gametophyte transfer cells (Fig. 25, 26, 36) and probably reflects their progressive inactivation in transport function. Cell degeneration is probably induced by the foot and appears to be closely related to the deposition of "tertiary" wall material and consequent obliteration of the wall labyrinth and plasmodesmata (Gambardella & de Lucia Sposito 1981-82). This is not the case in *Reboulia hemisphaerica*, in which living gametophyte transfer cells remain to an advanced stage of sporophyte development (Fig. 30).

Figures 17–19. *Conocephalum conicum.* **17,** An isolated, mature sporophyte seen by SEM, the foot, below the capsule (C), is closely invested by the gametophyte vaginula tissue (GV), bar = 0.2 mm; **18,** Sporophyte-gamatophyte junction in longitudinal section, the cylindrical foot (F) is connected with the capsule (C) by a short seta, gametophyte vaginula (GV), bar = 0.2 mm; **19,** Transverse section of the foot, showing intercellular spaces (arrows), bar = 100 μm.

Figure 20. *Fossombronia echinata* Macv. The bulbous foot (F) is connected to the capsule by a seta (Se) that elongates conspicuously once spores mature, pushing the capsule out of the gametophyte involucre. The placental tissue is marked by intense staining. Bar = 100 μm.

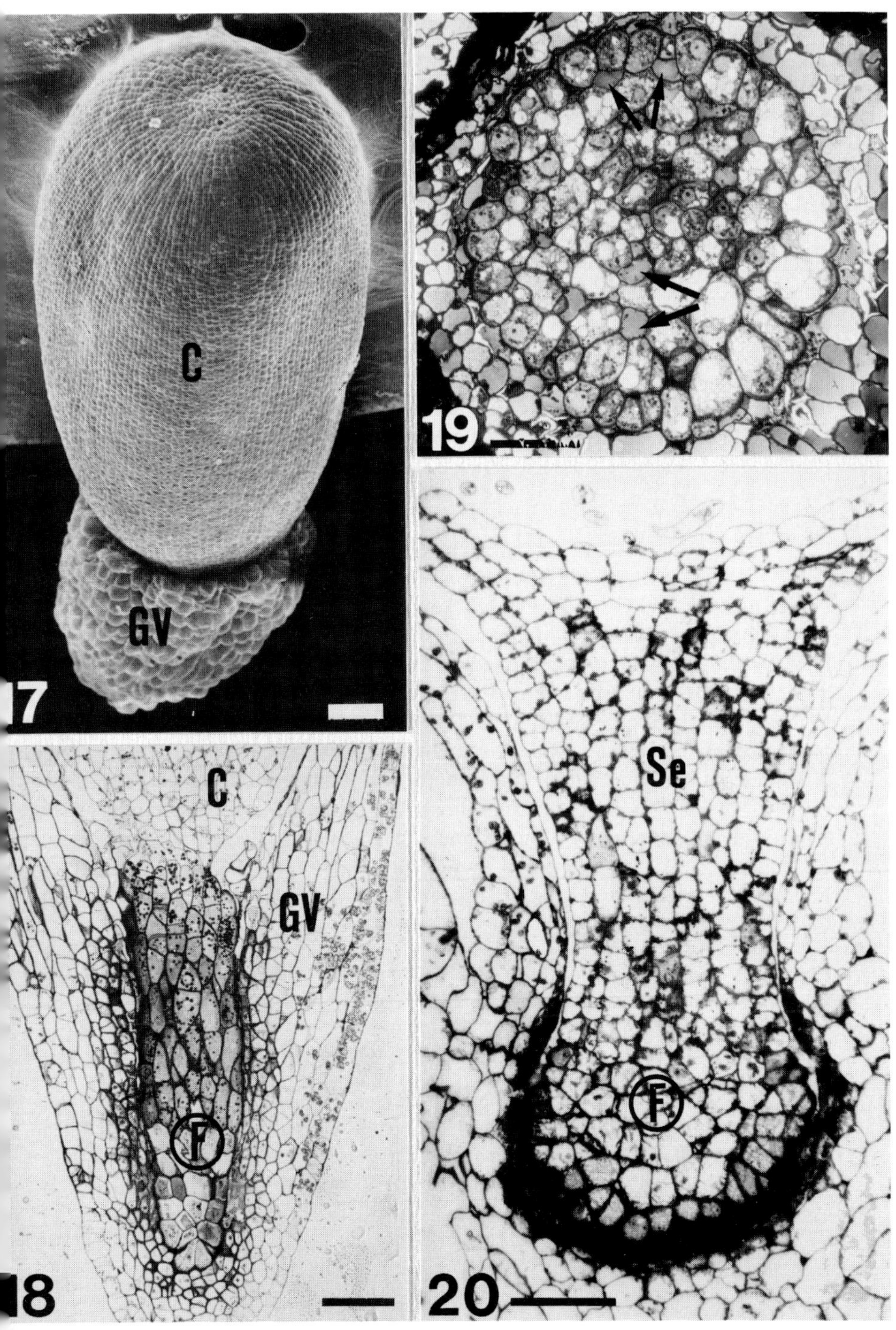
C
GV
17
19
Se
F
C
GV
F
18
20

are (a) giant, ring-shaped mitochondria arranged concentrically and frequently associated with endoplasmic reticulum and dictyosomes (Fig. 32), and (b) deposits of fibrillar material apparently located within dilated elements of rough endoplasmic reticulum (Fig. 34). Mitochondria comparable to those found in *R. hemisphaerica* were observed in the meristematic cells of gemmae in *Marchantia polymorpha* L. (Berrie & Webster 1982). Fibrillar material is also seen in the form of prominent cytoplasmic deposits without a surrounding membrane (Fig. 35).

Plastids of gametophyte transfer cells generally contain a well developed thylakoid system. Plastids with an extensive inner membrane system also occur in sporophyte transfer cells, for example in *Reboulia hemisphaerica* (Fig. 30) and *Sphaerocarpos donnelli* Aust. (Kelley 1969). On the other hand, plastids with a rudimentary thylakoid system occur in the sporophyte transfer cells in other species such as *Conocephalum conicum* (Fig. 29), *Mannia androgyna* (L.) Evans, and *Plagiochasma rupestre* (Forst.) Steph. (Gambardella & de Lucia Sposito 1981-82, 1983). Plastids in the gametophyte transfer cells of *R. hemisphaerica* contain a stromal bundle of very thin parallel fibrils, seemingly associated with the plastid envelope(Fig. 33).

Anthocerotes

Unlike mosses and liverworts, the initial division of the zygote is vertical in anthocerotes. Subsequent divisions produce a nearly cilindrical, three-tiered embryo (Renzaglia 1978). A massive foot is formed by the lower two tiers, and the upper tier produces both a basal meristematic zone and acropetally differentiating sporangial tissues. In *Notothylas* the foot arises from the lower tier only, and the middle and upper tiers produce the basal meristem and the sporangial initials, respectively (Renzaglia 1978). A somewhat different pattern of embryo development is described by Schuster (1984b). In *Phaeoceros laevis* (L.) Prosk. the epidermal cells of the foot produce haustorial projections, with or without cross walls, that penetrate the adjoining

Figure 21. *Reboulia hemisphaerica.* The round foot (F), seta (Se) and a part of the capsule (C) with mature spores and elaters, bar = 100 μm.

Figure 22. *Preissia quadrata* (Scop.) Nees. The placental tissue is restricted to the lower side of the convex disc-shaped foot (F), bar = 100 μm.

Figure 23. *Phaeoceros laevis.* Very young sporophyte (S) still enclosed in the gametophyte involucre, bar = 100 μm.

Figure 24. *Phaeoceros laevis.* Elongating sporophyte with nearly mature spores at the upper extremity. The bulbous parenchymatous foot (F) is surrounded by the placenta (PL) in which sporophyte and gametophyte cells are barely discernible, bar = 100 μm.

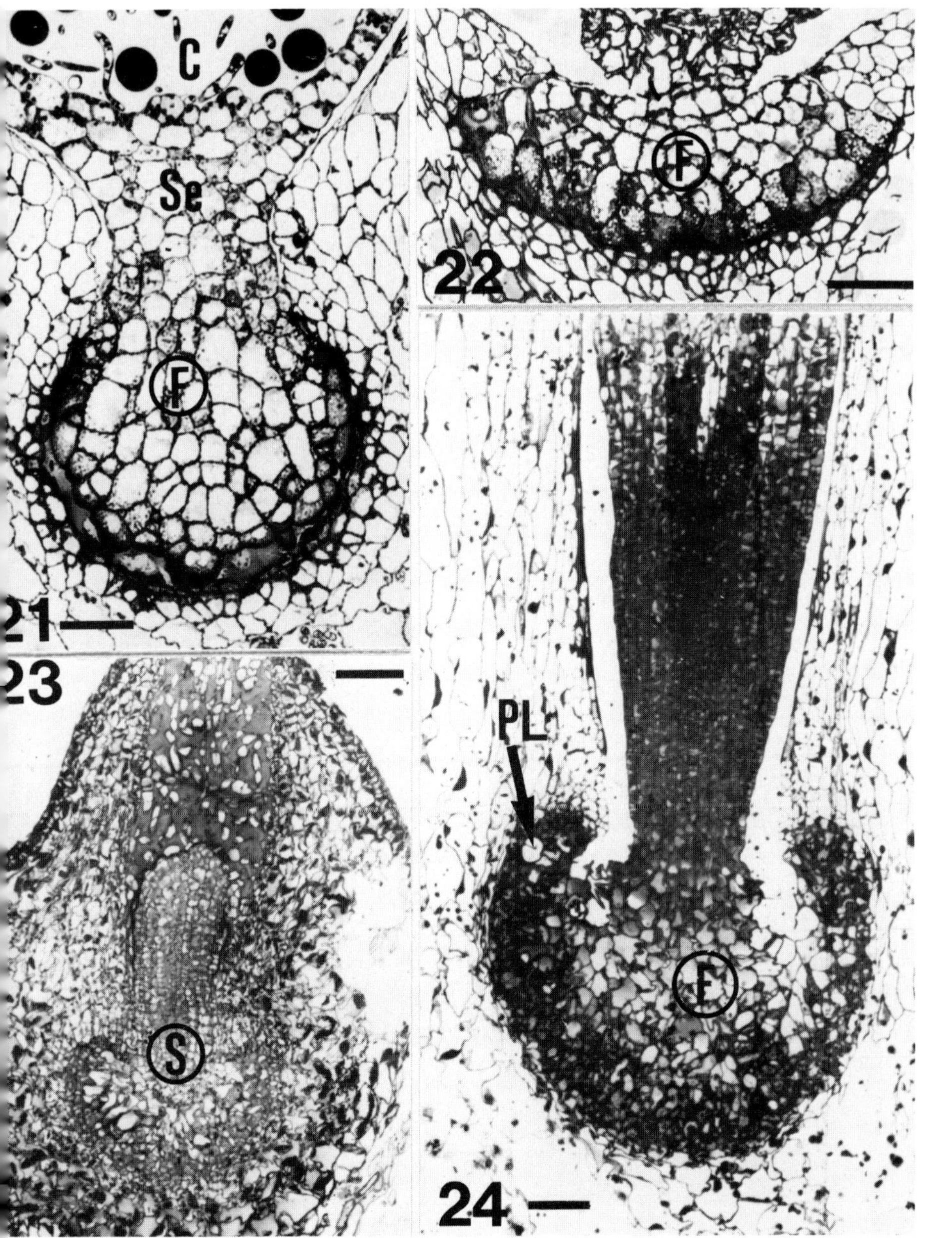
C
Se
F
21
22
F
23
S
PL
F
24

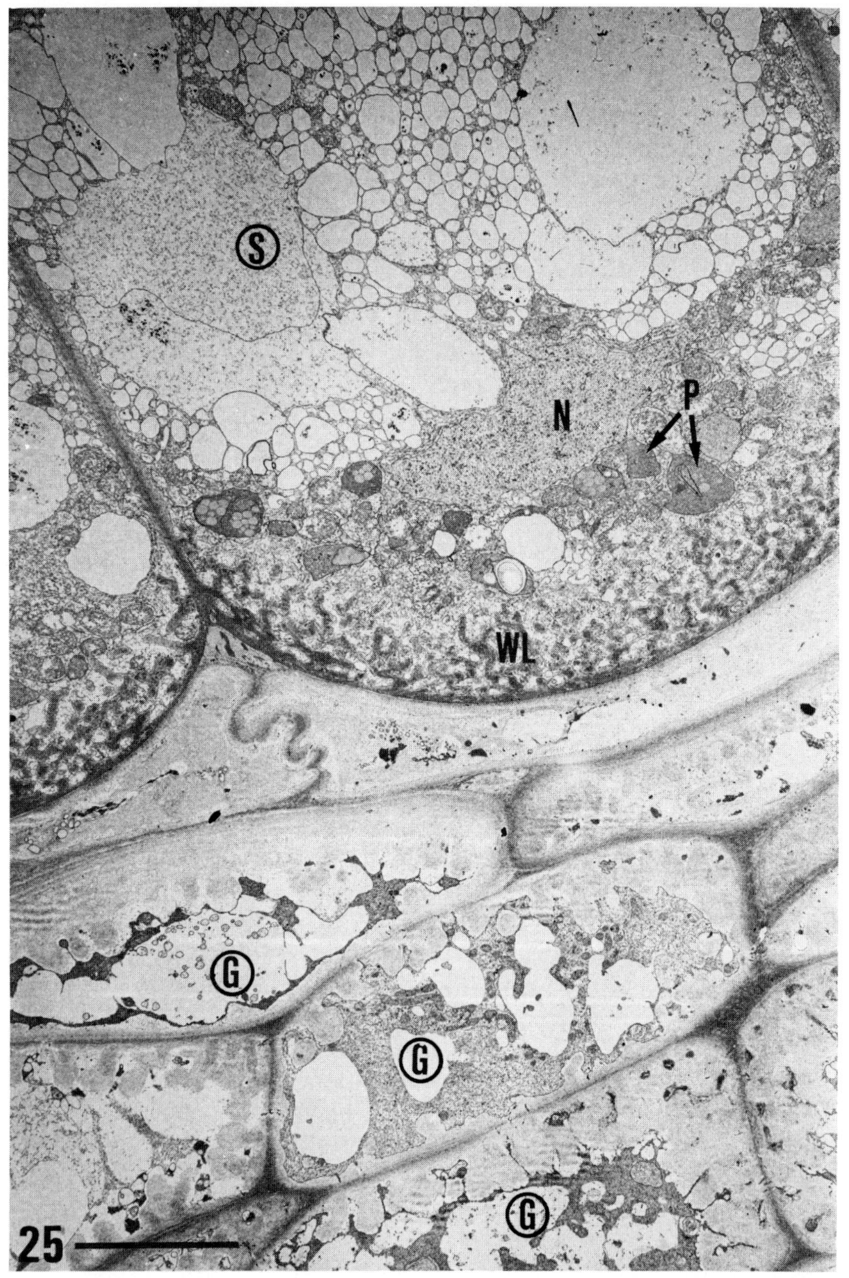

Figure 25. *Conocephalum conicum.* Sporophyte transfer cells (S) have an elaborate wall labyrinth (WL) and contain numerous vacuoles of small size. Several layers of gametophyte transfer cells (G) are visible, the innermost of which have degenerated. The gametophyte transfer cells have short, coarse wall ingrowths. Plastids (P). Nucleus (N). From a sporophyte with nearly mature spores. Bar = 10 μm.

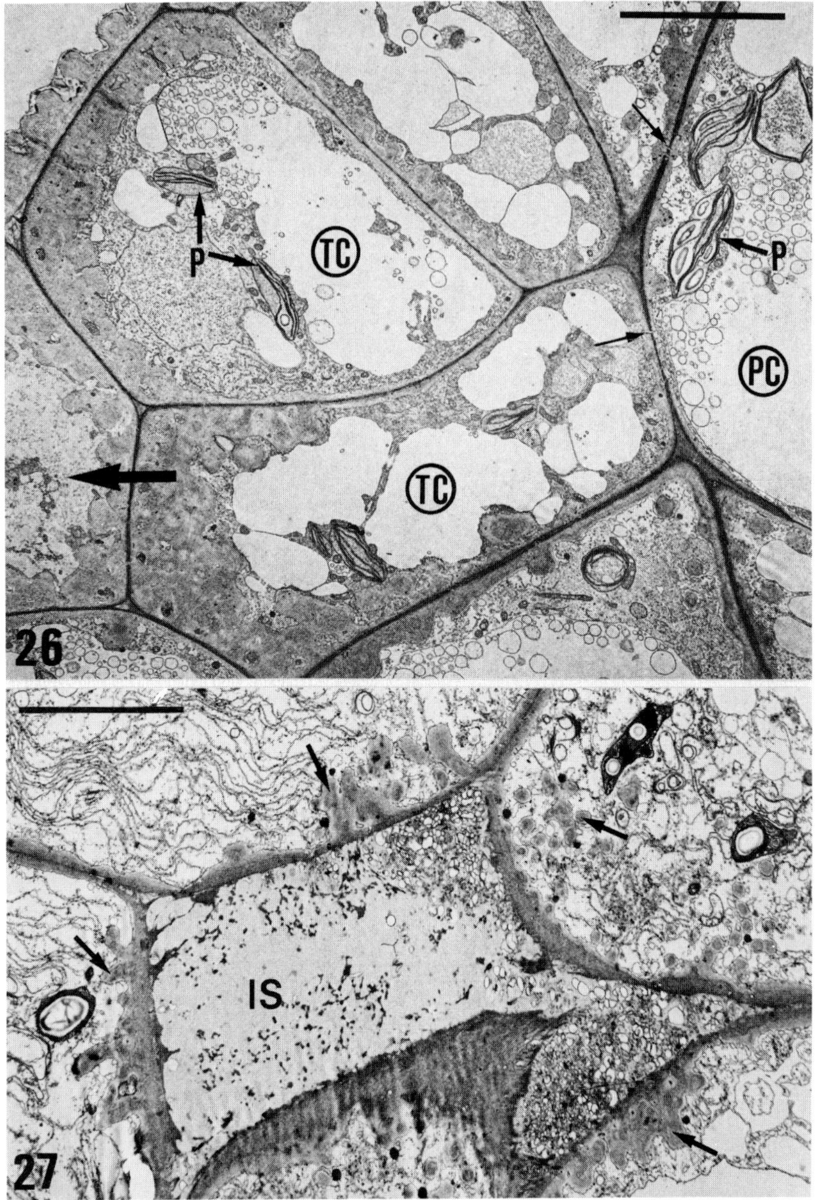

Figures 26–27. *Conocephalum conicum.* **26,** Detail of the gametophyte vaginula showing the most peripheral transfer cells (TC) connected by plasmodesmata (arrows) to an adjacent parenchyma cell (PC), note the marked difference in plastid (P) size and starch content in the two cell types, large arrow indicates the direction of the foot, same stage as Fig. 25, bar = 10 μm; **27,** Wall ingrowths (arrows) in the foot cells facing an intercellular space (IS), bar = 10 μm.

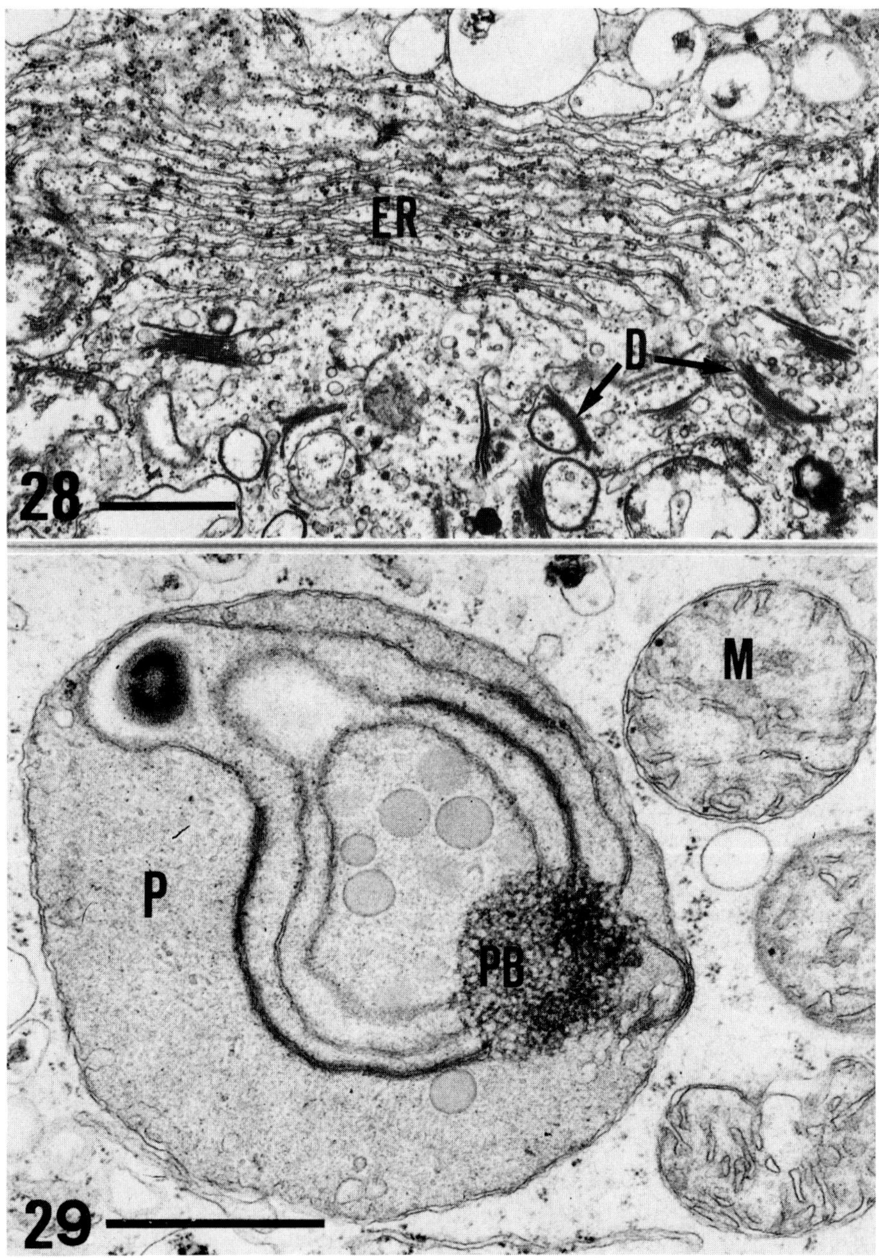

Figures 28–29. Details of sporophyte placental cells in *Conocephalum conicum*. **28**, Rough endoplasmic reticulum (ER) and dictyosomes (D), bar = 1 μm; **29**, Mitochondria (M) and a plastid (P) with a prolamellar body-like structure (PB), bar = 1 μm.

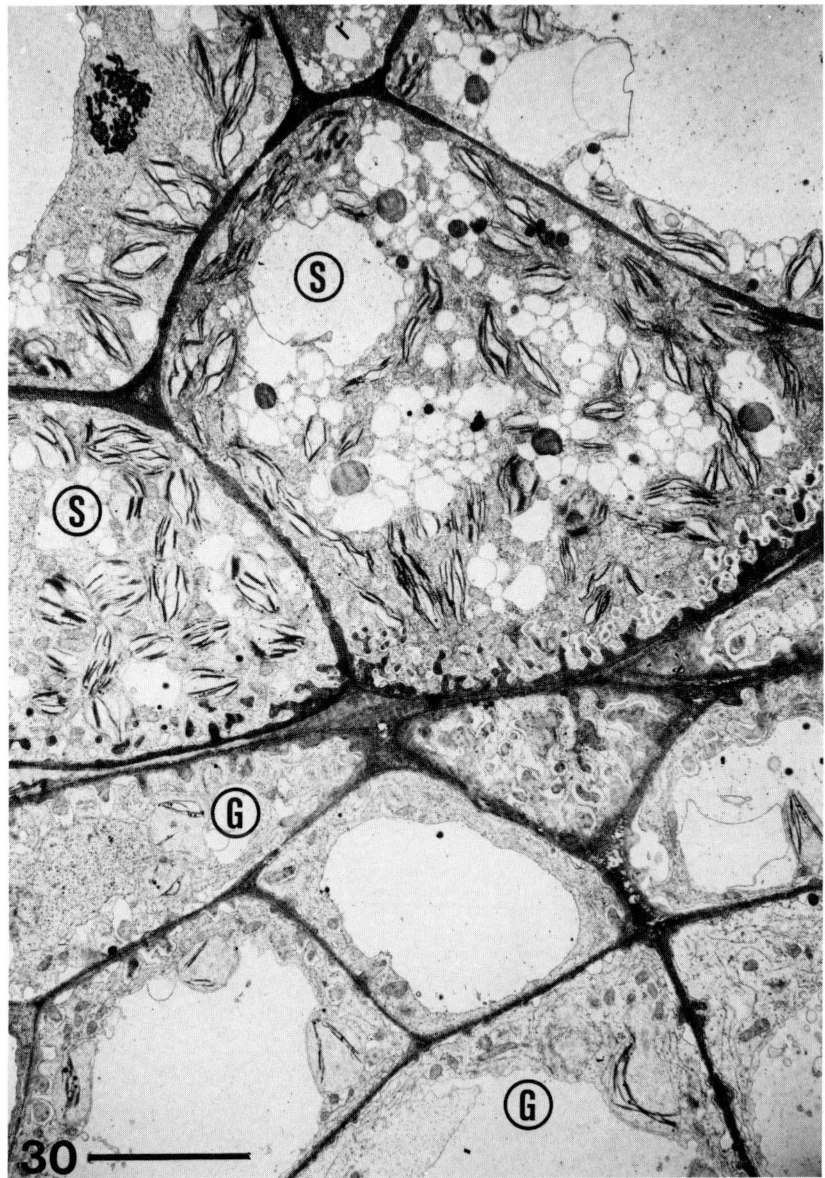

Figure 30. *Reboulia hemisphaerica.* Sporophyte transfer cells (S) are rich in plastids and small vacuoles. The wall labyrinth appears to be less complex than in other species. Gametophyte transfer cells (G) form several layers and do not show signs of degeneration. The figure is of a sporophyte with nearly mature spores. Bar = 10 μm.

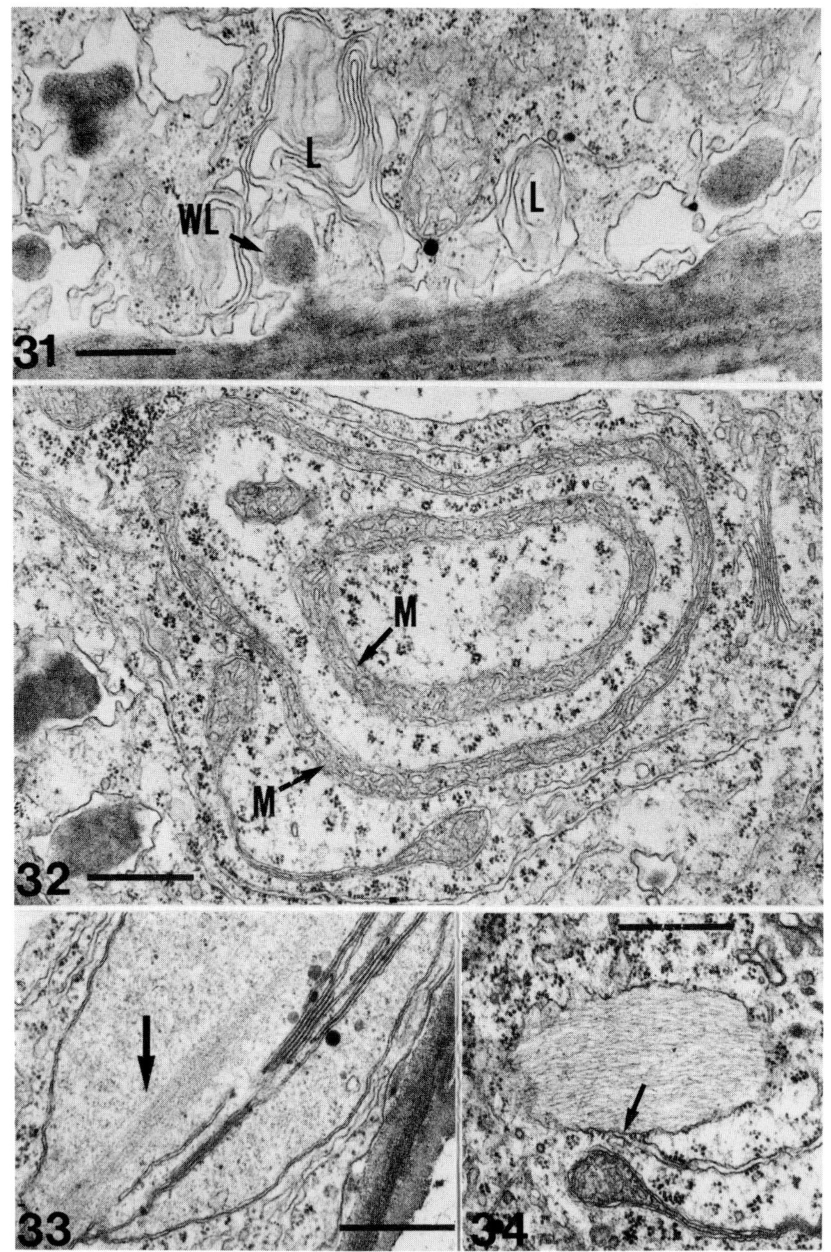

Figures 31–34. *Reboulia hemisphaerica.* **31,** Multivesicular structures (L) are frequently found associated with the wall labyrinth (WL) of sporophyte transfer cells, bar = 0.5 μm. **32–34.** Details of gametophyte transfer cells. **32,** Giant, ring-shaped mitochondria (M) enclosing cytoplasmic areas, bar = 0.5 μm; **33,** Plastid with a bundle (arrow) of thin fibrils parallel to its long axis, bar = 0.5 μm; **34,** Fibrillar deposit delimited by a membrane that appears to be continuous (arrow) with rough endoplasmic reticulum, bar = 0.5 μm.

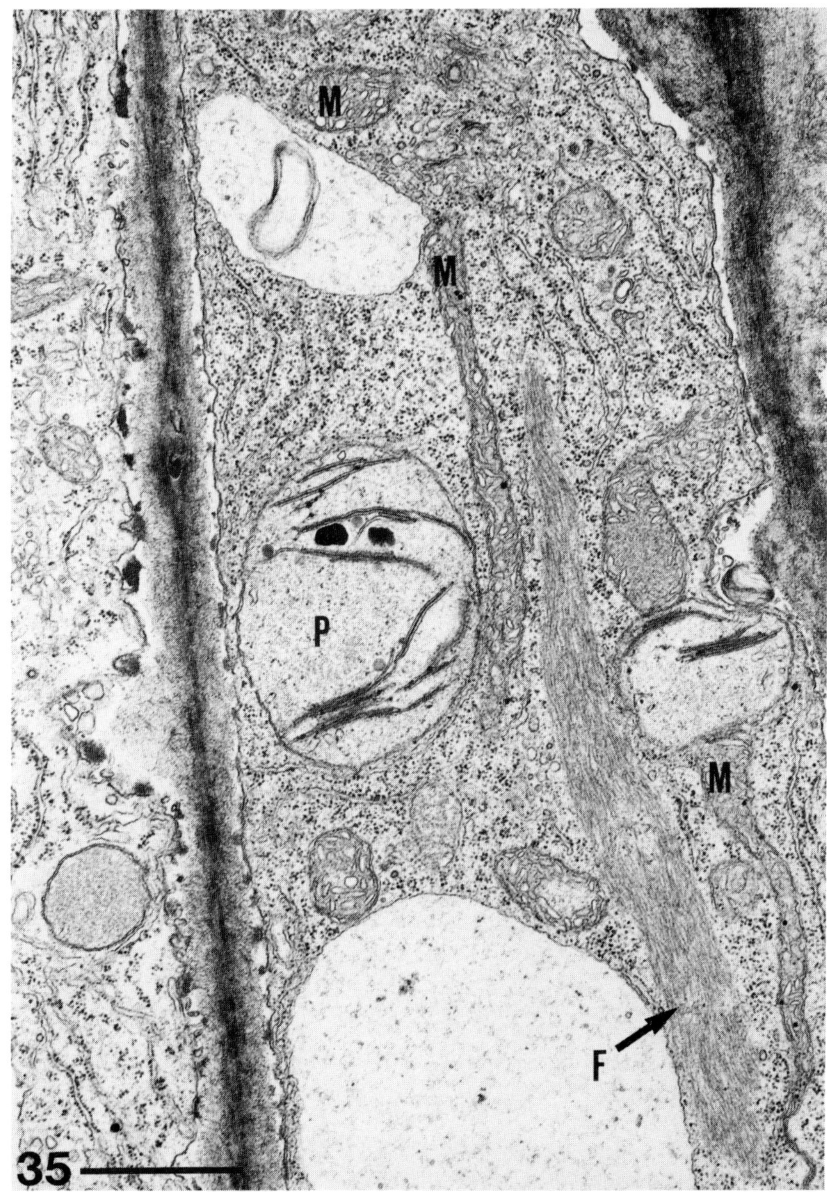

Figure 35. *Reboulia hemisphaerica*. Detail of gametophyte transfer cell showing a prominent fibrillar inclusion (F) parallel to the long axis of the cell and lacking a limiting membrane. Mitochondrial profiles (M) of different size and shape are visible. Plastid (P). Bar = 1 μm.

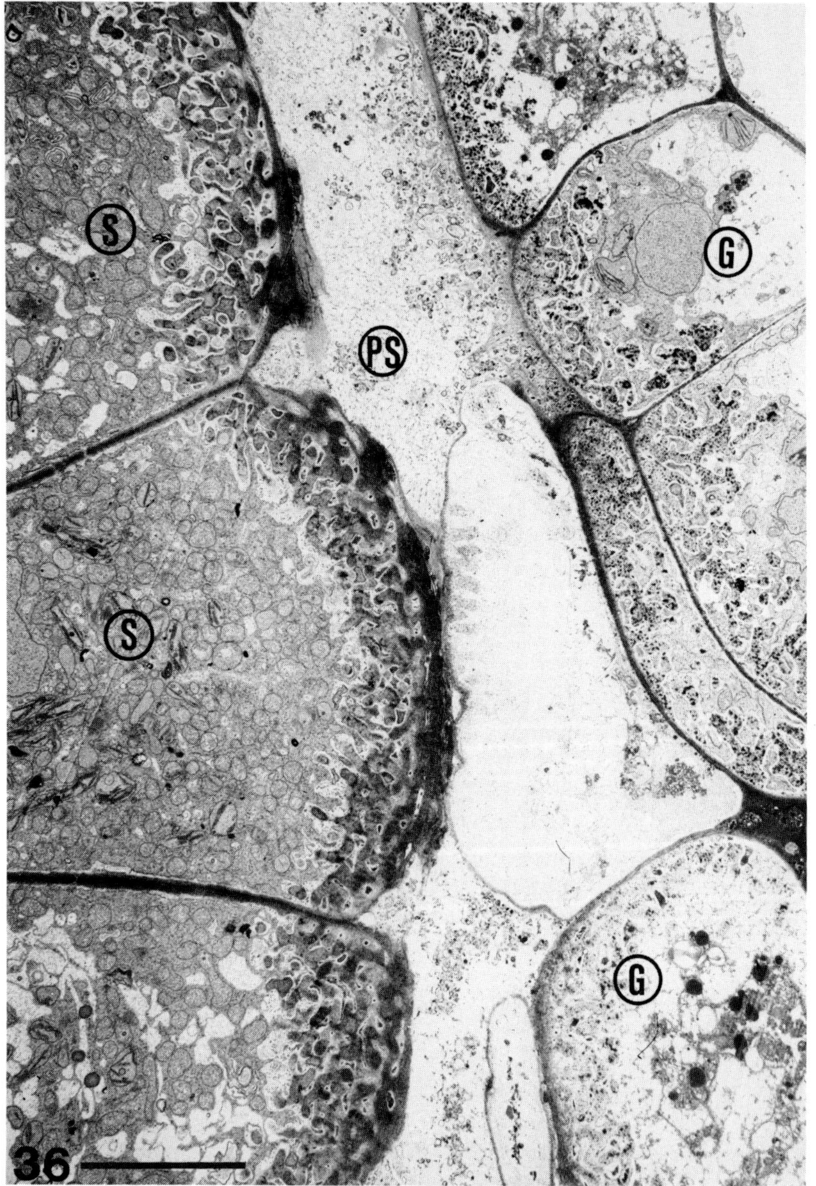

Figure 36. *Fossombronia echinata*. A prominent wall labyrinth is present in both sporophyte (S) and gametophyte (G) placental cells. The latter show cytoplasmic degeneration. The placental space (PS) contains heterogeneous material, probably arising from collapsed gametophyte cells. The figure is of a sporophyte with mature spores. Bar = 10 μm.

close, and with light microscopy it is often difficult to distinguish sporophyte cells from gametophyte cells in the placental region (Fig. 24). Gametophyte placental cells arise from the gametophyte parenchyma cells adjoining the egg cavity that are stimulated to divide when embryogenesis begins (Renzaglia 1978). The proliferation of gametophyte tissue surrounding the embryo also produces an investing involucre from which the sporophyte (except in *Notothylas*) emerges as soon as the first-formed spore mother cells complete meiosis (Crandall-Stotler 1984).

Gunning and Pate (1969) reported that in *Phaeoceros laevis* placental transfer cells were formed in the gametophyte only. This finding was later confirmed by electron microscopy (Gambardella et al. 1981). Sporophyte haustorial cells and gametophyte transfer cells are closely intermingled in the placental region and are separated by placental spaces that frequently enlarge to form prominent lacunae (Fig. 37). These lacunae contain mucilage, vesicles, and dense bodies with a crystalline substructure (Fig. 40, 41; Gambardella et al. 1981). High respiratory enzyme and acid phosphatase activities were found associated with the placental tissue in *P. laevis* (Thomas et al. 1978). Wall ingrowths in gametophyte cells usually lie along cell walls facing placental spaces. They consist of a dense fibrillar core that is continuous with the primary wall, a loose sheath of fibrils arranged concentrically, and a wide light area outlined by the plasmalemma (Gambardella et al. 1981). Haustorial cells lack wall ingrowths. Their cell walls have a compact fibrillar texture and are covered externally by an additional electron-dense layer of uniform thickness (Fig. 38). Mature gametophyte placental cells contain chromoplast-like plastids rich in lipid droplets, while peculiar amyloplast-like plastids occur in haustorial cells (Gambardella et al. 1981).

The crystals present in the placental spaces in *Phaeoceros laevis* react positively to stains specific for protein rather than carbohydrate (Marsh & Doyle 1985). Ultrastructural changes in these crystals, eventually resulting in their complete disintegration, were observed during capsule elongation and spore formation. This was interpreted as an indication that crystals are degraded to provide protein or component aminoacids to the sporophyte for these energy intensive processes (Marsh & Doyle 1985). However, the same authors considered that the observed changes might reflect degradation without utilization. Further investigation is needed.

A prominent placental region develops around the foot well before the capsule emerges from the involucre (Fig. 23). Marsh and Doyle (1985) reported that differentiation of haustorial cells, transfer cells, and crystal-containing placental spaces occurs during early sporophyte development. An interesting point that deserves to be investigated is the origin of placental lacunae and crystals. Crystals like those in the lacunae were also found in vacuoles of gametophyte placental cells in *Phaeoceros laevis* (Fig. 39; Gambardella et al. 1981), but Marsh and Doyle (1985) failed to find intracellular crystals in *P. laevis* var. *caroliniana*. Of great interest would be extending electron microscope investigations to those genera of anthocerotes (e.g., *Anthoceros* L.) in which the gametophyte and sporophyte appear sharply outlined in the placental region (Renzaglia 1978), suggesting a morphological and developmental pattern distinct

(Renzaglia 1978), suggesting a morphological and developmental pattern distinct from that of *P. laevis* and similar taxa.

Gametophyte-Sporophyte Nutritional Relationships

The presence of chlorophyll in even the earliest stages of sporophyte development has been cited in support of the idea that the sporophyte of bryophytes is photosynthetic (Bold 1938, 1940). It is now established that chloroplasts with a highly developed thylakoid system are present in sporophyte tissues of bryophytes. Moreover, in mosses and anthocerotes the sporophyte develops chlorenchyma with air spaces that communicate with the outside by means of functional stomata (Crandall-Stotler 1984, Schuster 1984b).

Haberlandt (1886) reported that excised sporophytes of *Funaria hygrometrica* kept for three weeks in inorganic medium increased in dry weight by 150 % and produced a normal crop of viable spores. Much later, Paolillo and Bazzaz (1968) demostrated, using an infrared gas analyser, that sporophytes of *F. hygrometrica* with expanded capsules have net photosynthetic assimilation of CO_2 over a wide range of temperatures and light intensities, but that in sporophytes of *Polytrichum juniperinum* Hedw. photosynthesis does no more than balance respiration. Utilization of seta reserves during the expansion of the capsule was demonstrated in *P. juniperinum* (Paolillo & Bazzaz 1968). The experiments of Grubb (1970) confirmed that the sporophyte is closely dependent on the gametophyte for photosynthate in *P. formosum*. Close dependence of the sporophyte on the gametophyte was also apparent in *Atrichum undulatum* (Hedw.) P. Beauv. and *Plagiomnium cuspidatum* (Rastorfer 1962), as well as in *Dicranum scoparium* and *Tortella tortuosa* (Hedw.) Limpr. (Atanasiu 1975). According to Krupa (1969), the sporophyte of *F. hygrometrica* requires import of photosynthate from the gametophyte to balance the dry weight gain and respiratory losses of carbon during the phase of elongation.

Proctor (1977) studied $^{14}CO_2$ uptake and movement of labelled assimilate in several mosses. He confirmed that photosynthesis in the sporophyte contributes part of the assimilate necessary for its growth and produced experimental evidence of assimilate transport from the gametophyte to the sporohyte. Sporophyte photosynthesis and assimilate transport from the gametophyte are complementary processes, and both are

Figures 37–38. *Phaeoceros laevis.* Same stage as Fig. 24. **37,** Detail of the placental region showing gametophyte transfer cells (G) and sporophyte haustorial cells (S), the placental space consists of large lacunae (PS) connected by narrow channels, note the compression of the gametophyte parenchyma at the bottom (arrow), seemingly caused by the end of a haustorial cell, bar = 10 μm; **38,** Detail of gametophyte transfer cell (G) with wall ingrowths (WI) and sporophyte haustorial cell (S) with wall coating (arrow), the intervening placental space (PS) contains fibrils, membrane fragments, and dense bodies (Cr) with crystalline substructure, bar = 0.5 μm.

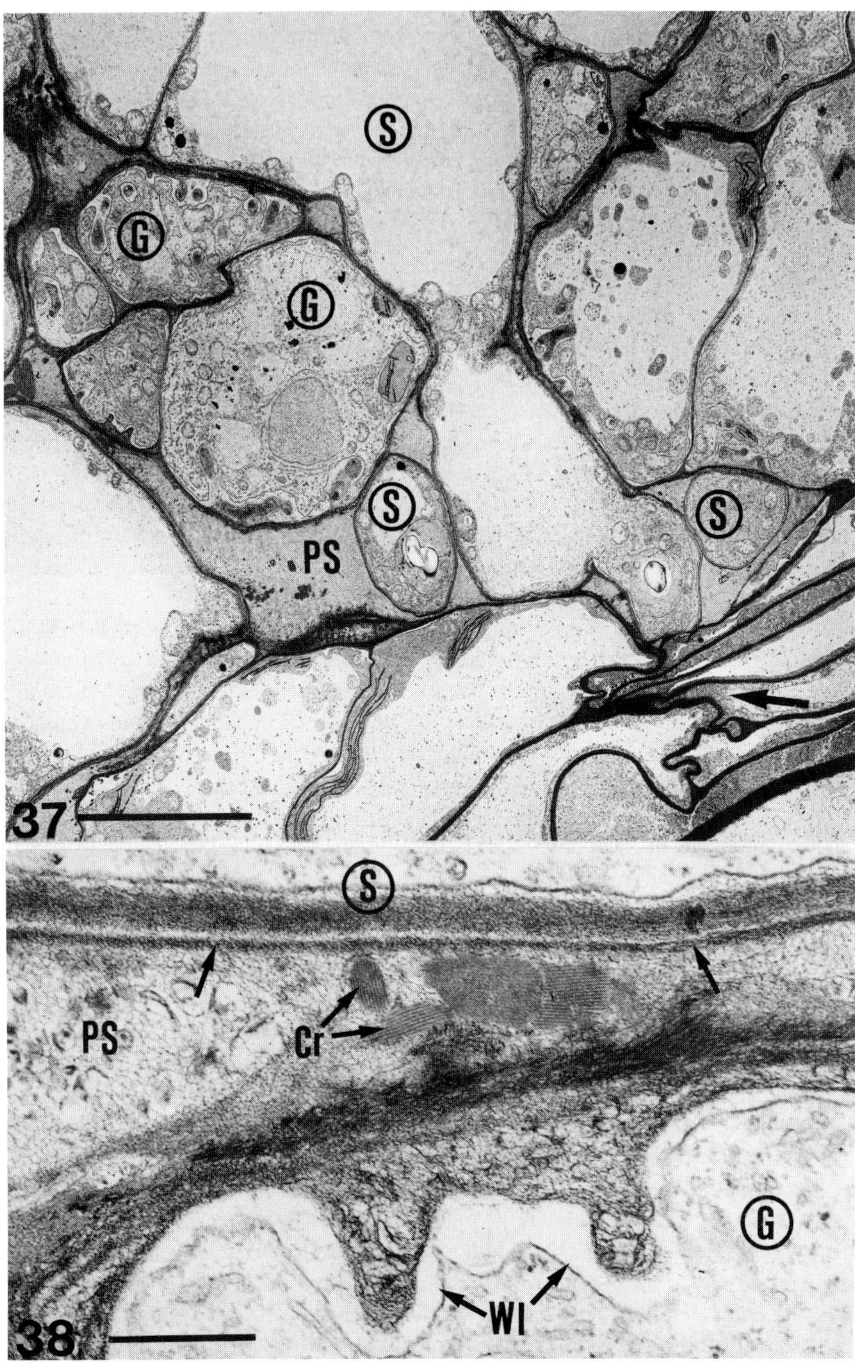

active during the last stages of capsule expansion and through the subsequent phase of differentiation but then decline with the senescence of the green tissue of the capsule (Proctor 1977). These conclusions were substantially confirmed by Browning and Gunning (1979c) who measured ^{14}C content in sporophytes of different age in *Funaria hygrometrica* after exposure of whole plants to ^{14}CO$_2$.

In *Funaria hygrometrica* phosphate is translocated from the gametophyte to the sporophyte capsule at a rate that decreases with the sporophyte aging (Chevallier et al. 1977).

In the anthocerote *Phaeoceros laevis*, the photosynthetic rate of the sporophyte is almost twice that of the gametophyte (per unit fresh weight), but nevertheless the sporophyte depends on the gametophyte for sustained growth (Thomas et al. 1978). Net transfer of ^{14}C-glucose from the gametophyte to the sporophyte occurs through the placental region in *P. laevis*, contradicting the prediction of Gunning and Pate (1969) that the gametophyte would be the absorptive partner. However, a small fraction of ^{14}C-labelled glucose and auxin applied in agar blocks to the tips of decapitated sporophytes was found to pass from sporophytes to the gametophytes (Thomas et al. 1978). This finding is in accordance with the assumption of Stewart and Rodgers (1977) that nutrient flow from illuminated sporophytes can sustain nitrogen fixation by symbiotic cyanobacteria in gametophytes kept in the dark.

Like the sporophytes of mosses and anthocerotes, those of liverworts are clearly capable of photosynthesis, but they depend on the gametophyte for sustained growth (Thomas et al. 1979). Young sporophytes in *Lophocolea heterophylla* (Schrad.) Dum. showed higher photosynthetic activity than older sporophytes. Moreover, substantial transport of ^{14}C-glucose from the gametophyte to the sporophyte, and much lesser transport in the opposite direction, were demonstrated (Thomas et al. 1979).

The active role of the sporophyte foot in the uptake of organic and mineral nutrients was demonstrated by experiments of Browning and Gunning (1979b) and Caussin and co-workers (1979, 1983). Excised sporophytes of *Funaria hygrometrica* were found to absorb radioactively labelled sugars. The rates of absorption were strongly reduced by the removal of the foot, as well as by metabolic inhibitors and darkness. Autoradiography of the foot, fixed immediately after short exposure to ^{3}H-glucose, showed a high density of silver grains over the epidermal layer of transfer cells (Browning & Gunning 1979b). Removal of the foot strongly reduced the absorption of ^{14}C-labelled amino acids and ^{86}Rb by the isolated sporophyte of *Polytrichum formosum* (Caussin et al. 1979, 1983).

Probable mechanisms that facilitate translocation across the wall-membrane apparatus of transfer cells have been discussed by Gunning and Pate (1974). These authors applied the standing gradient osmotic flow hypothesis developed by animal physiologists to both absorption and secretion by transfer cells. A modification of this mechanism is presented in Fig. 42. This hypothesis applies to the most common situation found in mosses, i.e., one layer of transfer cells on each side of the placenta. As Gunning and Pate (1974) emphasized, such a mechanism would be particularly efficient if wall ingrowths are long and thin, although this is not often the case in

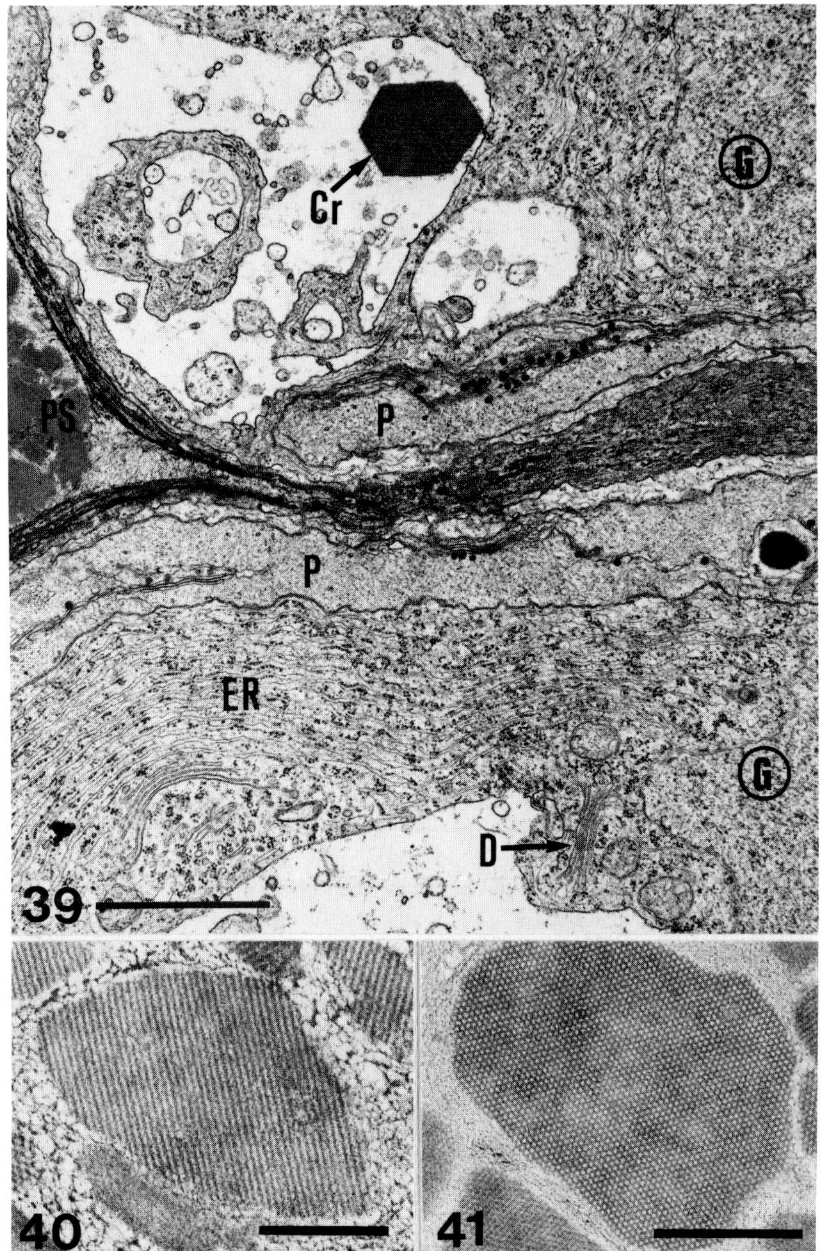

Figures 39–41. *Phaeoceros laevis.* Same stage as Fig. 24. **39,** Gametophyte placental cells (G) contain intravacuolar crystals (Cr) like those found in placental lacunae (PS), abundant rough endoplasmic reticulum (ER) and dictyosomes (D), plastids (P), bar = 10 μm; **40,** Placental crystals in longitudinal section, bar = 0.4 μm; Placental crystals in transverse section, bar = 0.4 μm.

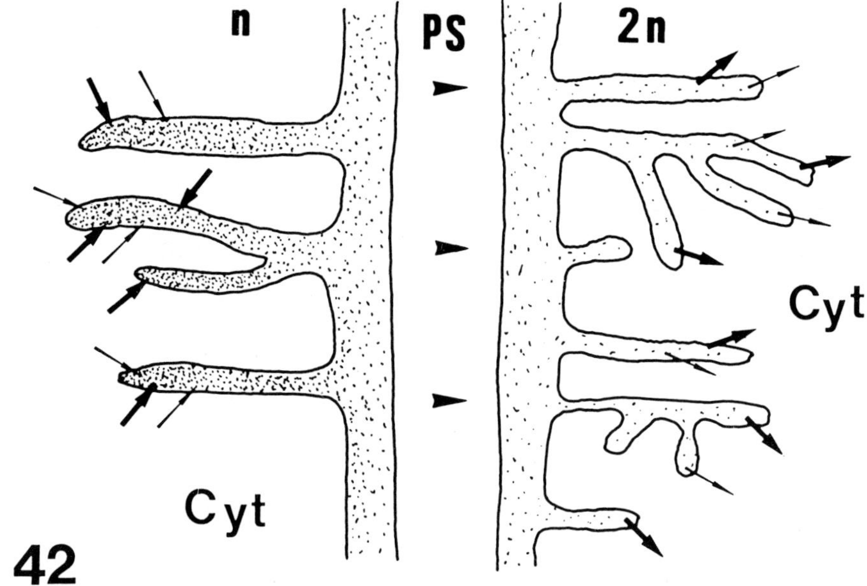

Figure 42. Diagram illustrating the possible pattern of nutrient transport in the sporophyte-gametophyte junction based upon the standing gradient osmotic hypothesis. On the gametophyte side of the placenta (n), the plasmalemma of the transfer cells actively pumps solutes from the cytoplasm (Cyt) into the lumen of the wall ingrowths. Conversely, the plasma membrane in sporophyte transfer cells (2n) pumps solutes from the lumen of wall ingrowths into the cytoplasm. At the steady state, active solute secretion by gametophyte transfer cells and active absorption by sporophyte transfer cells (possibly coupled with proton export/import) maintain a standing concentration gradient across which the solute diffuses. The same mechanism maintains a standing concentration gradient, with the lowest water potential at the extremities of wall ingrowths in the gametophyte transfer cells and the highest at the extremities of wall ingrowths in the sporophyte transfer cells. This causes water loss by the gametophyte cells and water uptake by sporophyte cells, resulting in a mass flow of water from gametophyte to sporophyte through the placental space. The density of the stippling in the figure is proportional to the concentration of solutes. For the sake of clarity, stippling has been omitted in the placental space (PS). Heavy arrows represent solute pumps; fine arrows represent the direction of water movement; arrowheads represent water and solute movement in the placental space.

bryophyte placental cells. Nevertheless, wall labyrinths of bryophytes contain an extensive system of thin cytoplasmic channels lined by the plasmalemma (see, for instance, Maier & Maier 1972). This structure could create the conditions necessary for the standing osmotic gradient mechanism to be operative. Transport patterns predicted by this hypothesis (see Fig. 42) are corroborated by the detection of ATP-ase activity in the wall-membrane apparatus of transfer cells (Maier & Maier 1972). Further support comes from the results of Bopp and Weniger (1971), showing that the absorption of water by isolated sporophytes of *Funaria* is almost entirely suppressed by increasing the external osmotic pressure. Moreover, a fluid was observed to collect in the vaginula cavity after removal of the sporophyte. This was probably due to the unbalanced secretory activity of gametophyte transfer cells. Finally, there is evidence that suggests the uptake of neutral amino acids by sporophyte transfer cells involves proton-cotransport pumps located in the plasmalemma (Caussin et al. 1979, Caussin & Bonnemain 1980). A similar system was suggested to be involved in sugar uptake (Browning & Gunning 1979b).

Failure of a vitamin supplement (added to the culture medium to sustain vigorous gametophyte growth) to reach the zygotes in sufficient quantity may be responsible for self-sterility in vitamin auxotrophic strains of the moss *Physcomitrella patens* (Hedw.) B.S.G. (Courtice et al. 1978). This indicates that the flow of substances translocated from the gametophyte to the sporophyte may be subject to qualitative as well as quantitative constraints. Furthermore, young embryos of *Dicranum scoparium* transplanted to gametophyte plants of *Atrichum undulatum*, from which the embryo had been excised, grew regularly and formed mature sporophytes (Arnaudow 1925). This suggests that the nutritional requirements of the sporophyte are not highly specific, probably involving a limited range of simple organic compounds. On the other hand, it is likely that the sporophyte, at least in early stages of development, absorbs water and mineral nutrients, if available, from the outside. Perichaetial leaves, the calyptra, the pseudoperianth, and other gametophytic protective devices that surround the sporophyte create capillary interstices in which water can circulate freely and reach the sporophyte. Absorption of external water by the sporophyte is documented in *Pellia epiphylla* (L.) Corda (Clee 1939). External water absorption was also found in sporophytes of *Phaeoceros laevis* (Thomas et al. 1978).

The development of the vaginula and other gametophyte structures associated with the sporophyte probably reflects emission of growth regulators by the sporophyte. The possibility of reverse translocation of organic compounds from the sporophyte to the gametophyte (Thomas et al. 1978, 1979) supports this hypothesis. As discussed previously, hormone-like substances produced by the sporophyte may be involved in the induction of cellular lysis in the gametophyte tissue during the growth of the foot. The rapid histological differentiation observed at the gametophyte apex concomitantly with sporophyte development (Roth 1969) also suggests hormone-mediated interactions between the two generations. Participation of hormones in growth regulation in mosses and liverworts is well documented (Bopp 1981, 1983, Maravolo 1980, Thomas 1980, Thomas et al. 1983), but there is currently no information about hormonal regulation of sporophyte-gametophyte relationships.

Conclusions

To keep nutrient translocation at the necessary rate, a close functional interaction is needed between the placental tissue and the adjoining sporophyte and gametophyte tissues. Continual absorption of water and solutes by sporophyte transfer cells must be balanced by concomitant transport of water and nutrients to the internal tissues of the foot. Conversely, gametophyte cells adjoining the foot must continuously absorb water and solutes from neighboring gametophyte cells. Plasmodesmata are probably a preferential pathway for cell-to-cell flow of solutes in these adjoining tissues, especially in those cases in which the apoplastic compartment is poorly developed. A study of the frequency and distribution of plasmodesmata, therefore, could result in information useful to the elucidation of nutrient translocation patterns. Distribution of plasmodesmata in the foot of *Funaria hygrometrica* (Wiencke & Schulz 1975) is indicative of radial translocation of solutes from the epidermal cells to the leptoids and longitudinal translocation through the leptoids toward the capsule. The absence of leptoids in the sporophytes of most mosses (Hébant 1977, 1979) is rather puzzling if we consider that in all species thus far investigated the sporophyte is largely dependent on the gametophyte for photoassimilate.

It has been suggested, based partly on the distribution of starch in foot cortical cells of *Timmiella barbuloides*, that hydroids may have a major role in translocation of nutrients toward the capsule (Ligrone et al. 1982a). If transpiration from the capsule is high enough, a sympathetic flow of water would effectively draw solutes through the hydroids of the foot and seta with little or no expenditure of energy. That hydroids function in nutrient transport is supported by their association with transfer cells in the sporophyte foot of polytrichaceous mosses. This system shows an interesting analogy with vascular bundles and associated transfer cells in legume root nodules (Pate et al. 1969).

The enlargement of the central strand of hydroids, observed in the lower part of the foot, independent of the presence of a collapsed tip, causes the transfer cells to be in contact with hydroids and may reflect functional interactions between these cells. Autoradiographic studies of photosynthate transport in the sporophyte in *Funaria hygrometrica* (Browning & Gunning 1979b) did not show silver grains over hydroids. One cannot exclude, however, that this might have been caused by a high rate of transport out of the hydroids or by loss of radioactive label from hydroids due to absent cytoplasmic contents. Leptoids in the seta of *F. hygrometrica* were found to degenerate at a pre-meiotic stage concomitant with capsule expansion (Schulz & Wiencke 1976), a stage at which the photosynthetic activity of the sporophyte has been found to exceed respiratory activity (Paollilo & Bazzaz 1968). Wiencke and Schulz (1978) argued that, at this stage, the sporophyte of *Funaria* was independent of the gametophyte for photosynthate. However, this is in conflict with the results of Proctor(1977) and Browning and Gunning (1979c), which demonstrate substantial nutrient translocation from the gametophyte to the sporophyte at comparable stages of sporophyte development. Heavy nutritional demand by the sporophyte may account for high rates of both photosynthesis and nutrient uptake. The kinetics of nutrient translocation along the seta (Browning & Gunning 1979c) is consistent with the hypothesis that nutrients are translocated via hydroids when leptoids degenerate, and a fully expanded, transpiring capsule is present.

The apoplastic compartment, comprising cell walls and intercellular spaces, is an ideally suited sink, or source, for transport systems involving transfer cells. The prominent systems of intercellular spaces found in the foot of polytrichaceous mosses and in *Buxbaumia piperi* likely enhance the diffusion of water and solutes. In the absence of more detailed information it is not easy to know whether the transfer cells abutting these spaces in polytrichaceous mosses actually do absorb or secrete substances. The presence of intercellular spaces outlined by cells with the characteristics of transfer cells in the foot of *Conocephalum conicum* is a remarkable convergence with polytrichaceous mosses. The dead empty cells found in the young vaginula of *Isopterygium pulchellum* may act as a capillary system, enhancing water uptake and storage.

Degeneration of the foot tip creates apoplastic continuity between the gametophyte and sporophyte hadrom, allowing water and solutes to pass directly from the former to the latter. The absence of a collapsed tip in the foot of pleurocarps and some acrocarps is probably a derived condition allowing for a more efficient control of the translocated substances.

The presence of several layers of gametophyte transfer cells in the placenta of liverworts probably does not yield a proportional enhancement of nutrient translocation to the foot. The innermost transfer cell layers degenerate during development (Gambardella & de Lucia Sposito 1981–82). Some cases are known (e.g., the foot in polytrichaceous mosses, the vaginula in *Buxbaumia piperi*, and perhaps in *Reboulia hemisphaerica*) of multiple layers of transfer cells that show no signs of cytoplasmic degeneration up to an advanced stage of development. These cells probably function with enhanced efficiency in series through a "push and pull" mechanism.

The cytoplasmic degeneration of transfer cells is preceded by the obliteration of the wall labyrinth, which occurs as a consequence of the production of "tertiary" wall material. In *Funaria hygrometrica* this mechanism was suggested to reduce the movement of water into the sporophyte at a stage when partial desiccation of the seta and capsule is necessary for spore dispersal (Browning & Gunning 1979a). While this is probably true for mosses, it is not for long-lived hornwort sporophytes, in which transfer cell function and spore discharge by capsule desiccation are concomitant events. Rapid elongation of the seta in some hepatic sporophytes is also not affected by premature capsule dehiscence (Schuster 1966).

Transfer cells in bryophytes are not restricted to the sporophyte-gametophyte junction. Wall ingrowths have also been found in parenchymatous cells associated with the central strand of hydroids in the gametophyte stem of *Funaria hygrometrica* (Hébant 1970), in deuters, socci, and passage cells of the leaf nerve in *Polytrichum commune* (Scheirer 1983), in jacket cells of the antheridia in *Phaeoceros laevis* (Duckett 1973), and in gametophyte callus cells of *Physcomitrium pyriforme* (Hedw.) Brid.

(Menon & Bell 1981) and *P. coorgense* Broth. (Lal & Narang 1985). This agrees with the conclusion of Gunning and Pate (1974) that the genetic basis of transfer cell formation is an ubiquitous feature of archegoniate plants, expressed in response to the physiological demand for effective short-distance transport of solutes. The recent discovery of wall ingrowths in gametophyte cells adjoining the zygote of *Coleochaete orbicularis* Pringsheim (Graham & Wilcox 1983) suggests that this feature was shared by the hypothetical green algal ancestor(s) of land plants.

There do not seem to be meaningful differences between mosses and hepatics in the ultrastructural morphology of the placenta tissue, except, perhaps, the multilayered arrangement of the gametophyte transfer cells in hepatics. Conversely, the structural organization of the placenta in *Phaeoceros* is quite different from that of mosses and hepatics and may provide an additional criterion for separating anthocerotes from other bryophytes.

Acknowledgments

The ultrastructural observations were performed at the Centro di Studio di Microscopia Elettronica (University of Naples, Faculty of Sciences). This work was supported by grants from Ministero della Pubblica Istruzione of Italy.

Literature Cited

Arnaudow, N. 1925. Über Transplantieren von Moosembryonen. Flora 118/9: 17–26.

Atanasiu, L. 1975. Photosynthesis and respiration in sporophytes and gametophytes of two species of mosses. Revue Roumaine de Biologie 20: 43–48.

Blaikley, N. M. 1933. The structure of the foot in certain mosses and in *Anthoceros laevis*. Transactions of the Royal Society of Edinburgh 57: 669–709.

Bergtrom, G., Schaller, M. & Eickmeier, W. G. 1982. Ultrastructural and biochemical bases of resurrection in the drought-tolerant vascular plant, *Selaginella lepidophylla*. Journal of Ultrastructural Research 78: 269–282.

Berrie, G. K. & Webster, P. M. 1982. Ultrastructure of plastids and mitochondria in gemmae of *Marchantia polymorpha*. Annals of Botany 50: 199–206.

Bold, H. C. 1938. The nutrition of the sporophyte in the Hepaticae. American Journal of Botany 25: 551–557.

—— 1940. The nutrition of the sporophyte in the Musci. American Journal of Botany 27: 318–322.

Bopp, M. 1981. Entwicklungsphysiologie der Moose. Advances in Bryology 1: 1–67.

—— 1983. Developmental physiology of bryophytes, pp. 276–324. In: Schuster, R. M. (Ed.), New Manual of Bryology, Vol. I. Nichinan, Japan. Hattori Botanical Laboratory.

Bopp, M. & Weniger, H. P. 1971. Wassertransport vom Gametophyten zum Sporophyten bei Laubmoosen. Zeitschrift für Pflanzenphysiologie 64: 190–198.

Brighigna, L., Cecchi Fiordi, A. & Palandri, M. R. 1980. The possible role of a fibrous body in the generative cell of the pollen grain of *Tillandsia caput-medusae* Morr. (Bromeliaceae). American Journal of Botany 67: 1493–1494.

Browning, A. J. & Gunning, B. E. S. 1977. An ultrastructural and cytochemical study of the wall-membrane apparatus of transfer cells using freeze-substitution. Protoplasma 93: 7–26.

—— 1979a. Structure and function of transfer cells in the sporophyte haustorium of *Funaria hygrometrica* Hedw. I. The development and ultrastructure of the haustorium. Journal of Experimental Botany 30: 1233–1246.

—— 1979b. Structure and function of transfer cells in the sporophyte haustorium of *Funaria hygrometrica* Hedw. II. Kinetics of labelled sugars and localization of absorbed products by freeze-substitution and autoradiography. Journal of Experimental Botany 30: 1247–1264.

—— 1979c. Structure and function of transfer cells in the sporophyte haustorium of *Funaria hygrometrica* Hedw. III. Translocation of assimilate into the attached sporophyte and along the seta of attached and excised sporophytes. Journal of Experimental Botany 30: 1265–1273.

Caussin, C. & Bonnemain, J. L. 1980. Proton-amino acid cotransport in transfer cells of bryophytes. Plant Physiology 65 S: 60.

Caussin, C., Despeghel, J. P., Faucher, M., Leger, A. & Bonnemain, J. L. 1979. Étude du mechanisme des échanges entre le gamétophyte et le sporophyte chez les Bryophytes. Comptes Rendus des Séances de l'Academie des Sciences Paris, Série D, 289: 1329–1334.

Caussin, C., Fleurat-Lessard, P. & Bonnemain, J. L. 1983. Absorption of some amino acids by sporophytes isolated from *Polytrichum formosum* and ultrastructural characteristics of the haustorium transfer cells. Annals of Botany 51: 167–173.

Chahuan, E. & Lal, M. 1980. Histochemical and ultrastructural investigations on the foot of the moss *Physcomitrium cyathicarpum* Mitt.–Evidence for the presence of Ca^{2+}-activated ATPase. Journal of Cytology and Genetics 15: 191–195.

—— 1982. Localization of some hydrolases and succinate dehydrogenase in the sporophyte-gametophyte junction in *Physcomitrium cyathicarpum* Mitt. Annals of Botany 50: 763–769.

—— 1984. Ultrastructural investigations on the haustorial foot in the moss *Physcomitrium cyathicarpum*. Cytobios 41: 85–93.

Chevallier, D., Nurit, F. & Pesey, H. 1977. Orthophosphate absorption by the sporophyte of *Funaria hygrometrica* during maturation. Annals of Botany 41: 527–531.

Clee, D. A. 1939. The morphology and anatomy of *Pellia epiphylla* considered in relation to the mechanisms of absorption and conduction of water. Annals of Botany 9: 105–111.

Corley, M. F. V., Crundwell, A. C., Düll, R., Hill, M. O. & Smith, A. J. E. 1981. Mosses of Europe and the Azores; an annotated list of species, with synonyms from the recent literature. Journal of Bryology 11: 609–689.

Courtice, G. R. M., Ashton, N. W. & Cove, D. J. 1978. Evidence for the restricted passage of metabolites into the sporophyte of the moss *Physcomitrella patens* (Hedw.) Br. Eur. Journal of Bryology 10: 191–198.

Crandall-Stotler, B. 1984. Musci, Hepatics and Anthocerotes - an essay on analogues, pp. 1093–1129. In: Schuster, R. M. (Ed.), New Manual of Bryology, Vol. 2. Nichinan, Japan. Hattori Botanical Laboratory.

Duckett, J. G. 1973. Wall ingrowths in the jacket cells of the antheridia of *Anthoceros laevis* L. Journal of Bryology 7: 405–412.

Eymé, J. & Suire, C. 1967. Au sujet de l'infrastructure des cellules de la région placentaire de *Mnium cuspidatum* Hedw. (Mousse bryale acrocarpe). Comptes Rendus des Séances de l'Academie des Sciences Paris, Série D, 265: 1788–1791.

Frey, W. 1981. Morphologie und Anatomie der Laubmoose. Advances in Bryology 1: 399–477.

Gambardella, R., Ligrone, R. & Castaldo, R. 1981. Ultrastructure of the sporophyte foot in *Phaeoceros*. Cryptogamie, Bryologie et Lichénologie 2: 23–45.

Gambardella, R. & de Lucia Sposito, M. L. 1981–82. Ultrastructure of the placental region in a liverwort *Mannia androgyna*. Delpinoa, n.s., 23–24: 177–184.

—— 1983. Placenta ultrastructure in *Plagiochasma rupestre* (Forst.) Steph. (Marchantiales, Hepaticae). Giornale Botanico Italiano 117: 166–167.

Graham, L. E. & Wilcox, L. W. 1983. The occurrence and phylogenetic significance of putative placental transfer cells in the green alga *Coleochaete*. American Journal of Botany 70: 113–120.

Grolle, R. 1983. Hepatics of Europe including the Azores: an annotated list of species, with synonyms from the recent literature. Journal of Bryology 12: 403–459.

Grubb, P. J. 1970. (In report of the autumn meeting of the British Bryological Society, 1969). Transactions of the British Bryological Society 6: 216–218.

Gunning, B. E. S. & Pate, J. S. 1969."Transfer cells" – Plant cells with wall ingrowths, specialized in relation to short distance transport of solutes – Their occurrence, structure, and development. Protoplasma 68: 107–133.

—— 1974. Transfer cells, pp. 441–480. In: Robards, A. W. (Ed.), Dynamic aspects of plant ultrastructure. London & New York. McGraw Hill.

Gunning, B. E. S., Pate, J. S. & Briarty, L. G. 1968. Specialized "transfer cells" in minor veins of leaves and their possible significance in phloem translocation. Journal of Cell Biology 37: C7–C12.

Haberlandt, G. 1886. Beiträge zur Anatomie und Physiologie der Laubmoose. Jahrbuch der Wissenschaften Botanik 17: 359–498.

Hallam, N. D. & Luff, S. E. 1980. Fine structural changes in the leaves of the desiccation-tolerant plant *Talbotia elegans* during extreme water stress. Botanical Gazette 141: 180–187.

Hébant, C. 1970. A new look at the conducting tissues of mosses (Bryopsida): Their structure, distribution, and significance. Phytomorphology 20: 390–410.

— 1973. Acid phosphomonoesterase activities (ß-glycerophosphatase and naphtol AS-MX phosphatase) in conducting tissues of bryophytes. Protoplasma 77: 231–241.

— 1975. Organization of the conducting tissue-system in the sporophytes of *Dawsonia* and *Dendroligotrichum* (Polytrichales, Musci). Journal of Hattori Botanical Laboratory 39: 235–254.

— 1977. The conducting tissues of bryophytes. Bryophytorum Bibliotheca 10: 1–157.

— 1979. Conducting tissues in bryophyte systematics, pp. 365–383. In: Clarke, G. C. S. & Duckett, J. G. (Eds.), Bryophyte Systematics. London & New York. Academic Press.

Hébant, C. & Suire, C. 1974. Mise en évidence d'activités enzymatiques au niveau de la zone de transfert gamétophyte-sporophyte chez quelques bryophytes. Revue Bryologique et Lichénologique 40: 171–181.

Kelley, C. 1969. Wall projections in the sporophyte-gametophyte junction of *Sphaerocarpos*. Journal of Cell Biology 41: 910–914.

Krupa, J. 1969. Photosynthetic activity and productivity of the sporophyte of *Funaria hygrometrica* during ontogenesis. Acta Societatis Botanicorum Poloniae 38: 207–215.

Lal, M. & Chauhan, E. 1981. Transfer cells in the sporophyte-gametophyte junction of *Physcomitrium cyathicarpum* Mitt. Protoplasma 107: 79–83.

Lal, M. & Narang, A. 1985. Ultrastructural and histochemical studies of transfer cells in the callus and apogamous sporophytes of *Physcomitrium coorgense* Broth. New Phytologist 100: 225–231.

Ligrone, R., Gambardella, R. & de Lucia Sposito, M. L. 1982a. Ultrastructure of the sporophyte foot-gametophyte vaginula complex in *Timmiella barbuloides* (Brid.) Moenk. Planta 154: 414–425.

Ligrone, R., Gambardella, R., Castaldo, R., Giordano, S. & de Lucia Sposito, M. L. 1982b. Gametophyte and sporophyte ultrastructure in *Buxbaumia piperi* Best (Buxbaumiales, Musci). Journal of Hattori Botanical Laboratory 52: 465–499.

Lorch, W. 1925a. Über die Saugzellen im Fusse und in der Vaginula bei den Laubmoosen. Berichte der Deutschen Botanischen Gesellschaft 43: 120–127.

— 1925b. Über die Haustorialschläuche am Fusse der Laubmoose. Berichte der Deutschen Botanischen Gesellschaft 43: 262–270.

Maier, K. 1967. Wandlabyrinthe im Sporophyten von *Polytrichum*. Planta 77: 108–126.

Maier, K. & Maier, U. 1972. Localization of beta-glycerophosphatase and Mg^{++}-activated adenosine triphosphatase in a moss haustorium, and the relation of these enzymes to the cell wall labyrinth. Protoplasma 75: 91–112.

Maravolo, N. C. 1980. Control of development in hepatics. Bulletin of the Torrey Botanical Club 107: 308–324.

Marsh, B. H. & Doyle, W. T. 1985. Intercellular protein crystals from the gametophyte-sporophyte junction of the hornwort *Phaeoceros laevis* Prosk. Protoplasma 129: 223–226.

Menon, M. K. & Bell, P. R. 1981. Ultrastructural and cytochemical aspects of induced apogamy following abscisic acid pre-treatment of secondary moss protonemata. Planta 151: 427–433.

Oliver, M. J. & Bewley, J. D. 1984. Desiccation and ultrastructure in bryophytes. Advances in Bryology 2: 91-111.

Paolillo, D. J., Jr. & Bazzaz, F. A. 1968. Photosynthesis in sporophytes of *Polytrichum* and *Funaria*. The Bryologist 71: 335–343.

Pate, J. S. & Gunning, B. E. S. 1972. Transfer cells. Annual Review of Plant Physiology 23: 173–196.

Pate, J. S., Gunning, B. E. S. & Briarty, L. G. 1969. Ultrastructure and functioning of the transport system of the leguminous root nodule. Planta 85: 11–34.

Proctor, M. C. F. 1977. Evidence on the carbon nutrition of moss sporophytes from $^{14}CO_2$ uptake and subsequent movement of labelled assimilate. Journal of Bryology 9: 375–386.

Rastorfer, J. R. 1962. Photosynthesis and respiration in moss sporophytes and gametophytes. Phyton 19: 169–177.

Renzaglia, K. S. 1978. A comparative morphology and developmental anatomy of the Anthocerotophyta. Journal of Hattori Botanical Laboratory 44: 31–90.

Roth, D. 1969. Embryo und Embryotheca bei den Laubmoosen. Eine histogenetische und morphologische Untersuchung. Bibliotheca Botanica 129: 1–49.

Scheirer, D. C. 1983. Leaf parenchyma with transfer cell-like characteristics in the moss, *Polytrichum commune* Hedw. American Journal of Botany 70: 987–992.

Schulz, D. & Wienke, C. 1976. Sporophytenentwicklung von *Funaria hygrometrica* Sibth. II. Differenzierung des Wasser- und Stoffleitungssystems in der Seta. Flora 165: 47–60.

Schuster, R. M. 1966. The Hepaticae and Anthocerotae of North America East of the Hundredth Meridian. New York & London. Columbia University Press.

— 1984a. Evolution, phylogeny and classification of Hepaticae, pp. 892–1070. In: Schuster, R. M. (Ed.), New Manual of Bryology, Vol. 2. Nichinan, Japan. Hattori Botanical Laboratory.

— 1984b. Morphology, phylogeny and classification of the Anthocerotae, pp. 1071–1092. In: Schuster, R. M. (Ed.), New Manual of Bryology, Vol. 2. Nichinan, Japan. Hattori Botanical Laboratory.

Stevenson, D. W. 1977. Histochemical and ultrastructural observations on the nacreous walls of the sieve elements of *Atrichum undulatum*. Annals of Botany 41: 849–853.

Stewart, W. D. P. & Rodgers, G. A. 1977. The cyanophyte-hepatic symbiosis. II. Nitrogen fixation and the interchange of nitrogen and carbon. New Phytologist 78: 459–471.

Thomas, R. J. 1980. Cell elongation in hepatics: the seta system. Bulletin of the Torrey Botanical Club 107: 339–345.

Thomas, R. J., Harrison, M. A., Taylor, J. & Kaufman, P. B. 1983 Endogenous auxin and ethylene in *Pellia* (Bryophyta). Plant Physiology 73: 395–397.

Thomas, R. J., Stanton, D. S. & Grusak, M. A. 1979. Radioactive tracer study of sporophyte nutrition in hepatics. American Journal of Botany 66: 398–403.

Thomas, R. J., Stanton, D. S., Longendorfer, D. H. & Farr, M. E. 1978. Physiological evaluation of the nutritional autonomy of a hornwort sporophyte. Botanical Gazette 139: 306–311.

Vaizey, J. R. 1888. On the anatomy and development of the sporogonium of the mosses. The Journal of the Linnean Society 24: 262–285.

Vitt, D. H. 1981. Adaptative modes of the moss sporophyte. The Bryologist 84: 166–186.

— 1984. Classification of the Bryopsida, pp. 696–759. In: Schuster, R. M. (Ed.), New Manual of Bryology, Vol. 2. Nichinan, Japan. The Hattori Botanical Laboratory.

Wiencke, C. & Schulz, D. 1975. Sporophytenentwicklung von *Funaria hygrometrica* Sibth. I. Strukturelle Grundlagen der Wasser- und Nährstoffaufnahme im Haustorium. Protoplasma 86: 107–117.

— — 1978. The development of transfer cells in the haustorium of the *Funaria hygrometrica* sporophyte. Bryophytorum Bibliotheca 13: 147–167.

Index[1]

A page number followed by a dash indicates that the entry occurs at many places in the article beginning on the cited page.

Illustrations pertaining to an index entry are indicated by italicized page numbers.

[1] Prepared by N. G. Miller and C. L. Tribley